BIOANALYSIS OF DRUGS AND METABOLITES, Especially Anti-Inflammatory and Cardiovascular

METHODOLOGICAL SURVEYS IN BIOCHEMISTRY AND ANALYSIS

Series Editor: Eric Reid

Guildford Academic Associates
72 The Chase
Guildford GU2 5UL, United Kingdom

The series is divided into Subseries A: Analysis, and B: Biochemistry
Enquiries concerning Volumes 1–11 should be sent to the above address.

Volumes 1–10 edited by Eric Reid

Volume 1 (B): Separations with Zonal Rotors

Volume 2 (B): Preparative Techniques

Volume 3 (B): Advances with Zonal Rotors

Volume 4 (B): Subcellular Studies

Volume 5 (A): Assay of Drugs and Other Trace Compounds in Biological Fluids

Volume 6 (B): Membranous Elements and Movement of Molecules

Volume 7 (A): Blood Drugs and Other Analytical Challenges

Volume 8 (B): Cell Populations

Volume 9 (B): Plant Organelles

Volume 10 (A): Trace-Organic Sample Handling

Volume 11 (B): Cancer-Cell Organelles
Edited by Eric Reid, G. M. W. Cook, and D. J. Morré

Volume 12 (A): Drug Metabolite Isolation and Determination
Edited by Eric Reid and J. P. Leppard
(includes a cumulative compound-type index)

Volume 13 (B): Investigation of Membrane-Located Receptors
Edited by Eric Reid, G. M. W. Cook, and D. J. Morré

Volume 14 (A): Drug Determination in Therapeutic and Forensic Contexts
Edited by Eric Reid and Ian D. Wilson

Volume 15 (B): Investigation and Exploitation of Antibody Combining Sites
Edited by Eric Reid, G. M. W. Cook, and D. J. Morré

Volume 16 (A): Bioactive Analytes, Including CNS Drugs, Peptides, and Enantiomers
Edited by Eric Reid, Bryan Scales, and Ian D. Wilson

Volume 17 (B): Cells, Membranes, and Disease, Including Renal
Edited by Eric Reid, G. M. W. Cook, and J. P. Luzio

Volume 18 (A): Bioanalysis of Drugs and Metabolites, Especially Anti-Inflammatory and Cardiovascular
Edited by Eric Reid, J. D. Robinson, and Ian D. Wilson

A Continuation Order Plan is available for this series. A continuation order will bring delivery of each new volume immediately upon publication. Volumes are billed only upon actual shipment. For further information please contact the publisher.

BIOANALYSIS OF DRUGS AND METABOLITES, Especially Anti-Inflammatory and Cardiovascular

Edited by

Eric Reid
Guildford Academic Associates
Guildford, United Kingdom

J. D. Robinson
Hoechst UK
Milton Keynes, United Kingdom

and

Ian D. Wilson
ICI Pharmaceuticals Division
Macclesfield, United Kingdom

Springer Science+Business Media, LLC

Library of Congress Cataloging in Publication Data

International Bioanalytical Forum (7th: 1987: Guildford, Surrey)
Bioanalysis of drugs and metabolites, especially anti-inflammaory and cardiovascular.
(Methodological surveys in biochemistry and analysis; v. 18 (A))
"Proceedings of the Seventh International Bioanalytical Forum . . . held September 8–11, 1987, in Guildford, United Kingdom"–T.p. verso.
Includes bibliographies and indexes.
1. Anti-inflammatory agents–Analysis–Congresses. 2. Cardiovascular agents–Analysis–Congresses. 3. Arachidic acid–Derivatives–Analysis–Congresses. 4. High performance liquid chromatography–Congresses. 5. Blood–Analysis–Congresses. 6. Chemistry, Clinical–Technique–Congresses. I. Reid, Eric, date. II. Robinson, J. D. III. Wilson, Ian D. IV. Title. V. Series: Methodological surveys in biochemistry and analysis; v. 18. [DNLM: 1. Anti-Inflammatory Agents–analysis–congresses. 2. Cardiovascular Agents–analysis–congresses. 3. Chromatography, Gas–methods–congresses. 4. Chromatography, High Pressure Liquid–methods–congresses. 5. Drugs–analysis–congresses. 6. Eicosanoic Acids–metabolism–congresses. W1 ME9612NT v.18 / QV 25 I61 1987b]
RB56.5.A56I68 1987 615′.77 88-22412
ISBN 978-1-4757-9426-7 **ISBN 978-1-4757-9424-3 (eBook)**
DOI 10.1007/978-1-4757-9424-3

Based on proceedings of the Seventh International Bioanalytical Forum entitled Bioanalysis of Drugs and Metabolites, Especially Anti-Inflammatory and Cardiovascular, held September 8–11, 1987, in Guildford, United Kingdom

Originally published by Plenum Press, New York in 1988
Softcover reprint of the hardcover 1st edition 1988

Senior Editor's Preface

Whereas the 'bioactive analytes' coverage of Vol. 16 included endogenous peptides, the present book features eicosanoids, besides drugs. Awareness of the therapeutic potential of such agents was evidenced by the keen participation of company staff in the Bioanalytical Forum (September 1987) which gave rise to this book.

The Forum series, whilst still focused on blood-level determinations, has shown trends which mostly match changes in the analytical scene. Reflecting the tendency of newer drugs to be even more potent than the CNS-active drugs that featured in Vol. 16, analytical concern is increasingly with ng/ml or even pg/ml levels. Yet reliable measurement is still needed at the μg/ml level too, notably in re-investigating old-established drugs for pharmacokinetic properties indicated in the text. Both Vol. 16 and the present book manifest the increasing attention to drug chirality.

Metabolite investigation is increasingly aided by NMR (see 'Notes' in Sect. #F). For bioanalysis in general, chromatographic separations have played a cardinal role since the 1970's. Capillary GC has at last burgeoned in the drug-assay area, somewhat reducing the dominance gained by HPLC (for which fluorescence detection is gaining ground). Analysts increasingly seek instrumentation that may be of daunting cost even if the aim is subtlety rather automation. For sample preparation, solvent extraction still serves skilled practitioners as a powerful tool, although largely superseded by 'cartridge' extraction notwithstanding evident pitfalls. In respect of publishing assay methodology, our policy that method-descriptions must include the 'rationale' no longer needs strong advocacy; moreover, it is no longer rare for analysts in pharmaceutical companies to publish their methods. Now an analyst faced with setting up a method may be daunted by the swollen literature. For the analyte in question, the customary **Analyte Index** in this book (or earlier volumes) may list the actual compound, or chemically kindred compounds with analytical pertinence. The Table on p. 218 augments the coverage for two therapeutic classes that feature in this book, viz. cardiovascular and anti-inflammatory drugs. (Eicosanoids may likewise be anti-inflammatory, although some may be pro-inflammatory.)

There has been a trend, noted with relief by the Editors, towards better compilation of publication texts. Yet some texts have undergone extensive editing, acceptable to the authors. Erratic alternation of the terms 'serum' and 'plasma', and of the synonymous terms 'precision' and 'reproducibility', caused some trouble in the editing, as did non-appreciation that % values for bought-in solutions (notably ammonia) may be on a weight basis, not made evident by the manufacturer. Notwithstanding the shortcomings or lateness of some texts, authors are thanked for compiling them amidst other pressures. Elsevier and the American Chemical Society are also thanked, for Figures now reproduced with source acknowledgement.

This Editor has generally respected authors' phrasing, whilst shuddering when the term 'incubate' is encountered in a 0° context. He remains a 'diehard' in certain respects, notably in favouring 'M' rather than 'mol/l', and a wt./ml basis for drug concentrations in test samples; he regards 'mmol/l' as a fatuous fashion. Concerning infelicitous **abbreviations**, a distinction is made between electron-capture (detector context; 'ECD') and electrochemical ('EC', never 'ECD'); the hallowed GC term 'FID' means free induction decay to NMR practitioners, who may pardon the term 'Fid' as introduced editorially. The convention for '°C' throughout the book is '°'.

Undefined but well-known abbreviations include GC, HPLC and TLC. MS (mass spectrometry), NPD (nitrogen-phosphorus detector), t_r (retention time) and RIA (radioimmunoassay) are usually defined in the article concerned, as are the HPLC modes NP (normal-/straight-phase) and RP (reversed-phase; C-18 and ODS are synonymous), and i.s. denoting internal standard, about which E. Reid made a plea in Vol. 7 (1978; #A-6).- "To ascertain what a particular author means by the term 'internal standard', one often has to scrutinize the small print in the paper. When taken through the full procedure, it allows for any idiosyncrasy (wobble) in the particular sample." Happily the latter usage now prevails.

Support for the Forum, much appreciated, came from U.K. pharmaceutical companies: Beechams, Glaxo, ICI and Smith Kline & French.

Guildford Academic Associates ERIC REID
72 The Chase, Guildford
Surrey, GU2 5UL, U.K. *12 April 1988*

Contents

The 'NOTES & COMMENTS' ('NC' items) at the end of each Section include comments (listed on each 'NC' title page) made at the Forum on which the book is based, along with some supplementary material.

Groupings of 'Notes' on particular themes:
Column switching (cf. #E-1): #NC(E)-2 & -3; also p. 308
Solid-phase sample processing: #NC(E)-8 to -11; also p. 308
NMR approaches: #NC(F)-2 to -6

List of Authors

Primary author	*Co-authors, with relevant name to be consulted in left column*
M.F. Barkworth - p. 129 iphar Inst, Höhenkirchen-Siegertsbrunn, W. Germany	W. Adams - Bye L.A. Allan - Vose (i) B.A. Bailey - Hill (iii)
T.J.A. Blake - pp. 353-359 Smith Kline & French Research, Welwyn, Herts.	I.G. Beattie - Blake K. Belsner - Höller H. Bippi - Frölich M.B. Bottorff - Wainer
U.A.Th. Brinkman - pp. 321-338 Free Univ., Amsterdam	O.A.M. Brockhoff - de Bree (ii) R.D. Brownsill - Vose (i), (ii) & (iii)
A.P. Bruins - pp. 339-351 Rijksuniversiteit, Groningen	F. Bucheli - Wyss
A. Bye - pp. 97-105 Upjohn Research Labs., Crawley	
J. Caldwell - pp. 257-261 (& see Hutt) St. Mary's Hosp. Med. Sch., London W2	
F. Carey - pp. 43-50 ICI Pharmls., Alderley Park	
S.H. Curry - pp. 213-214 Coll. of Pharmacy, Univ. of Florida, Gainesville, FL	R.A. Clare - Doig J.F. Darbyshire - Caldwell
B.E. Davies - pp. 179-183 Beecham Pharmls. Medicinal Res. Cent., Harlow, Essex	G.P. Davidson - Caldwell L. Dehelean - Hill (iii) B.J. de Jong - de Bree (iv)
H. de Bree (& see Ruijten) - pp. (i) 209-212; (ii) 251-255; (iii) 273-276; (iv) 301-303 Duphar BV, Weesp, The Netherlands	G.J. de Jong - Brinkman A. DeLacroix - Lecaillon R.A. de Zeeuw - Feitsma
P.H. Degen - pp. (i) 107-114 (ii) 193-200 Ciba-Geigy, Basle	D. Donnell - Woodward B.F.H. Drenth - Feitsma J.P. Dubois - Lecaillon C.J. Dyde - Barkworth
M.V. Doig - pp. 53-55 Wellcome Res. Labs., Beckenham	
J.G. Dorsey - pp. (i) 235-244; (ii) 305-306 Univ. of Florida, Gainesville, FL	J.C. Edwards - Summers C. Elcombe - Nicholson (1) L. Embrechts - Woestenborghs (ii)

Primary author

J. Fauler (& see Frölich) - pp. 35-39
Medical Sch., Hannover, W. Germany

K.G. Feitsma - pp. 265-268
Rijksuniversiteit, Groningen

J.C. Frölich - pp. 15-21
Medical Sch., Hannover, W. Germany

H.M. Hill - pp. (i) 269; (ii) 277-278; (iii) 299-300
Bio-Research Labs., Senneville, Quebec

M. Höller - pp. 23-27
Janssen Res. Foundation, Neuss, W. Germany

A.J. Hutt (& see Caldwell) - pp. 115-125
Brighton Polytechnic, Brighton

W.N. Jenner - pp. 77-86
Glaxo Group Res., Ware, Herts.

J.B. Lecaillon - pp. 225-233
Ciba-Geigy Pharm. Res. Centre, Rueil-Malmaison, France

R.D. McDowall - pp. (i) 201-206: (ii) 393-396
Smith Kline & French Research, Welwyn, Herts.

J.C. McElnay - pp. 87-95
Queen's Univ., Belfast

P.S.B. Minty - pp. 159-161
Charing Cross & Westminster Med. Sch., London W6

G.P. Mould - pp. 185-191
St. Luke's Hosp., Guildford

J.K. Nicholson - pp. (i) 375-381; (ii) 383-388
Birkbeck Coll., London WC1

K.D. Rainsford - pp. 67-74
Strangeways Res. Lab., Cambridge

A. Rakhit - pp. 135-142
Ciba-Geigy Pharmls. Divn., Summit, NJ

Co-authors, with relevant name to be consulted in left column

K.A. Fernandes - McDowall (i)
J. Firth - Vose (i)
R.A. Forder -Carey
U. Förstermann - Frölich
A. Gescher - Threadgill
C. Gooijer - Brinkman
A. Gray - Vose (ii) & (iii)
A.J.E. Green - Mould
C.P. Groen - de Bree (iv)

A.J. Hawkins - Vose (i)
D. Haworth - Carey
J. Heykants - Woestenborghs (i) & (ii)
J. Holian - Summers
J.J.M. Holthuis - Vendrig
P.R. Hurst - Whelpton

J.A. Jelly - McDowall (i)
V. Kaever - Fauler
D. Kaiser - Bye
D. J. Kirkham - Jenner
H-R. Kwon - Curry
R.L. Lalonde - Wainer
N. Lammers - de Bree (iv)
W.J. Leavens - McDowall (i)
R.M. Lee - McDowall (i)
D. Lessard - Hill (ii)
L. Letarte - Hill (ii)
P.A. Lewis - Woodward
S. Lewis - J.D. Robinson
I. Luhmann - Höller
C.M. Macdonald - Nicholson (ii)
J. Maddock - Woodward
J.K. Martin - Summers
K-H. Marx - Frölich & Fauler
K. Mashiter - Summers
D.K.F. Meijer - Feitsma
A. Mekking - Vendrig
G.S. Murkitt - McDowall (i)
B. Osikowska-Evers - Höller
R.P.B. Passas - Stevenson
J.C. Pearce - McDowall (i) & (ii)
R. Pickford - Wilson (iii)

Primary author	*Co-authors, with relevant name to be consulted in left column*
K-D. Rämsch - pp. 163-164 Bayer Inst. f. Klin. Pharmacologie, Wuppertal, W. Germany	K.D. Rehm - Barkworth
C. Robinson - pp. 3-13 Univ. of Southampton	B. Rosenkranz - Frölich R.J. Ruane - Wilson (i)
J.D. Robinson - pp. 143-148 Hoechst UK, Walton Milton Keynes	S.M. Sanins - Nicholson (i) D. Scherling - Rämsch K. Selinger - Hill (i)
H.M. Ruijten [& see de Bree (iv)] - pp. 397-399 Duphar, Weesp, The Netherlands	K. Sierat - de Bree (i) I.K. Smith - Wilson (iii) C. Souppart - Lecaillon
D. Stevenson - pp. 279-281 Univ. of Surrey, Guildford	J.A. Steiner - Vose (i)
M.R. Summers - pp. 401-402 Amersham Internatl., Amersham, Bucks.	T. Teeuwsen - Vendrig J.A. Timbrell - Nicholson (i)
M.D. Threadgill - pp. 389-392 Univ. of Aston, Birmingham	B.E. Timmerman - Ruijten P. Timmerman - Woestenborghs (i)
D.E.M.M. Vendrig - pp. 283-288 Rijksuniversiteit, Utrecht	G.P. Tomkinson - Wilson (i) J. Troke - Nicholson (ii)
C.W. Vose - pp. (i) 29-35; (ii) 363-365; (iii) 367-369 Hoechst UK, Walton Milton Keynes*	F.A. Tucker - Minty M.P. van Berkel - de Bree (i)
I.W. Wainer - pp. 169-177 St. Jude Children's Res. Hosp., Memphis, TN	D.J.K. van der Stel - de Bree (ii) A. van Peer - Woestenborghs (i)
R. Whelpton - pp. 289-294 London Hosp. Med. Coll., London El	
I.D. Wilson - pp. (i) 295-298; (ii) 313-320 (iii) 371-374; (iv) 383-388. ICI Pharmls., Alderley Park	
R. Woestenborghs - pp. (i) 149-155; (ii) 215-216 Janssen Res. Foundn., Beerse, Belgium	
A.J. Woodward - pp. 245-250 Simbec Res., Merthyr Tydfil, Wales	K.E. Wade - Nicholson (ii)
P.M. Woollard - pp. 57-58 Wellcome Res. Labs., Beckenham†	C.M. Walls - Vose (iii) C. Weber - Höller
R. Wyss - pp. 271-272 Hoffmann-La Roche, Basle	S.M. Winter - Caldwell

---

Former 'affiliations': *G.D. Searle, High Wycombe; †Inst. of Dermatology

Section #A

EICOSANOIDS AND RELATED ANALYTES

#A-1

THE USE OF HPLC IN STUDIES OF THE STEREOSELECTIVE METABOLISM OF PROSTAGLANDIN D_2

Clive Robinson

Department of Clinical Pharmacology
University of Southampton
Southampton General Hospital
Southampton SO9 4XY, U.K.

PGD_2 may be an important mediator of inflammatory reactions. Because of its short biological half-life and its tendency to isomerize or dehydrate, its measurement in complex biological fluids entails problems. To circumvent these, attention has been turned to metabolites of PGD_2 that may have longer half-lives and be less susceptible to dehydration. Metabolism in primates including man was known to proceed via the formation of PGF-ring compounds, and later it was verified for urine that the ring hydroxyls have the 9α,11β-orientation. We have reached similar conclusions for plasma, in volunteers who received ^3H-PGD_2 by i.v. infusion or by inhalation. Enzymic 11-ketoreduction gives 9α,11β-$PGF_{2\alpha}$, which is further metabolized by PGdh. Consideration is given to the identification of these metabolites, their sites of formation, to analytical implications, and especially to the actual methods used, notably HPLC, GC after derivatization, and MS.*

The past 10 years have seen a gradual increase of interest in the pharmacology and biochemistry of PGD_2, despite an early, incorrect report that it was biologically inert. Immunological activation of human pulmonary [1], colonic [2] or cutaneous [3] mast cells results in massive synthesis and release of PGD_2, which has prompted interest in its putative role as a mediator of inflammation. So as to implicate PGD_2 as a pathophysiological mediator it is necessary to demonstrate its formation and release. Similarly to other arachidonic

---

* *Abbreviations.*- PG, prostaglandin; 15-PGdh, 15-hydroxyprostaglandin dehydrogenase (EC 1.1.1.41); ^{3}H-, tritiated; MoMeTMS, *O*-methyloxime, methyl ester, trimethylsilyl ether; MS, mass spectrometry; SIM, selected ion monitoring; LSC, liquid scintillation counting; (relative) retention time, (rel.) t_r *[Editor's preference; conventionally* t_R*]*.

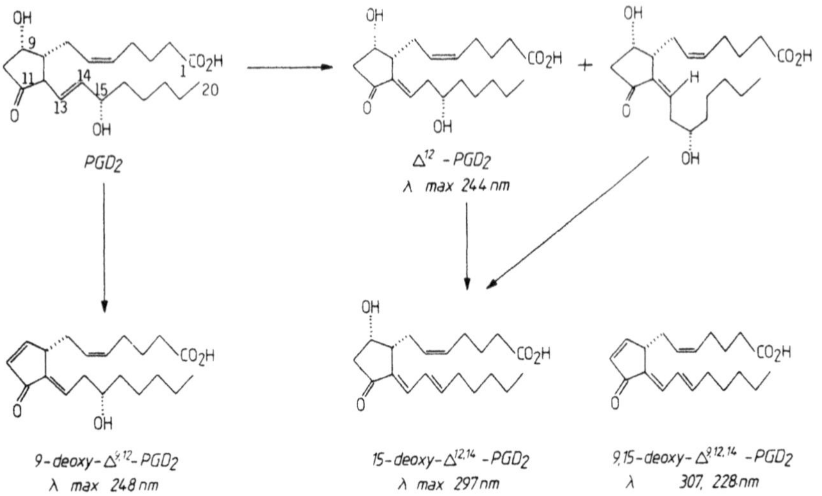

Fig. 1. The chemical decomposition of PGD_2 by facile isomerization and dehydration. These products can be formed likewise from PGD_2 in the presence of serum albumin.

acid metabolites, PGD_2 measurement in complex biological fluids is difficult because of its short plasma half-life and its low plasma concentrations (1-5 pg/ml). In addition, PGD_2 is susceptible to facile isomerization of the C-13 double bond to C-12; also at basic pH the molecule undergoes dehydration reactions (Fig. 1), which frustrated attempts to raise antisera to PGD_2 without protection of the C-11 function [4]. Despite these difficulties a few groups have described MS assays for PGD_2 in biological fluids.

A proven approach to prostanoid measurement has been to select metabolites of the parent compound [5]. These usually have longer plasma half-lives and are often present in higher concentration. Till recently rather little was known about PGD_2 metabolism. Work now outlined has provided information useful for study of its pathophysiological role.

THE METABOLISM OF PGD_2[⊗]

Studies with purified preparations of 15-PGdh have shown that PGD_2 is a poor substrate for this enzyme [6]. Alternative routes of metabolism have therefore been sought. One early report indicated that sheep blood contained an enzyme capable of reducing PGD_2 to $PGF_{2\alpha}$ [7]. The belief that PGD_2 metabolism involves the reduction of the C-11 keto function gained support from a study on the urinary metabolites of PGD_2 in a cynomolgus monkey [8]; although the stereochemistry of the hydroxyl groups was not investigated, evidently a significant

⊗ The author's text has been abridged.- *Ed.*

route of metabolism was reduction of the 3-hydroxycyclopentanone ring to a cyclopentane-1,3-diol, i.e. a PGF ring system. This study did not exclude the possibility that such metabolism occurred after the formation of other PGD ring metabolites.

Increased plasma concentrations of 13,14-dihydro-15-keto-$PGF_{2\alpha}$ had been reported in 1974 by Gréen et al. [9] in allergen-provoked asthmatic patients. From other literature [1-3, 10] pointing to a role of mast cells in the early asthmatic reaction, seemingly there was PGD_2 release from lung and then 11-ketoreduction, as later observations [11, 12] apparently bore out. However, it was only an untested presumption [9, 11] that the metabolite's ring hydroxyls have co-planar α stereochemistry.

Amongst 25 identified urinary metabolites found in a volunteer after i.v. infusion of ^3H-PGD_2, 23 had PGF ring structures, generally (in 13 out of 15 tested for butylboronate derivative formation) with 9α,11β hydroxyl geometry [12; cf.13]. These results are pertinent to an observed rise in the plasma metabolite concentration (measured by a very selective RIA) in humans after inhalation of $PGF_{2\alpha}$ but not of PGD_2 [14]: possibly inhaled differs in metabolism from i.v.-infused PGD_2, or else (taking account of the specificity of the RIA antibody) 11-ketoreductase had acted but led to 15-keto-9α,11β-PGF_2 and then 13,14-dihydro-15-keto-9α,11β-PGF_2 [14]. To identify which epimer is in fact formed, and to highlight suitable analytes for studying PGD_2 turnover *in vivo*, plasma has now been analyzed after giving volunteers ^3H-PGD_2 by inhalation or i.v. infusion.

EXPERIMENTAL

PG administration.- For PGD_2 infusion, over 20 min into an antecubital vein, 8 male subjects (mean age 29 ±4 yrs) were given 128 ng PGD_2/Kg/min containing ^3H-PGD_2, 86 ±4.7 µCi/min. Blood samples taken from the contralateral antecubital vein both during and after infusion were collected into K-EDTA. Venous samples were taken frequently from the subjects (6 males; 28 ±1 yrs) who inhaled ~30 µCi PGD_2 from a breath-actuated Inspiron nebulizer, containing 4 mCi ^3H-PGD_2.

Sample preparation.- From plasma centrifuged at 4°, 0.1 ml was taken for determination of total ^3H content by LSC. The remaining plasma was acidified to pH 3 with 5 M HCl and extracted into 5 ml ethyl acetate using C-18 SepPak cartridges which had been pre-conditioned with 10 ml each of methanol and distilled water.

***In vitro* studies.**- Fresh human lung or animal organs were homogenized in pH 7.4 K phosphate buffer containing 1 mM EDTA and 1 mM L-cysteine and then centrifuged at 400 **g** (10 min, 4°). The supernatant was further centrifuged at 100,000 **g** (1 h, 4°) to yield a cytosol preparation, used to study prostanoid metabolism with added cofactors (NAD and a NADH-generating system).

Instrumental analyses.- HPLC of PG's was performed using a Spectra-Physics SP 8700 ternary solvent delivery system with either a SP 8440 or a Kratos SF783 UV detector. The mobile phase usually comprised 0.017 M phosphoric acid/acetonitrile of far-UV grade (67.2:32.8 by vol.), flowing at 1 ml/min through a C-18 column (12.5 × 0.46 cm; Nucleosil 5) with its exit coupled to a μBondapak C-18 column (length 25 cm). Samples were reconstituted in mobile phase containing a mixture of unlabelled PG standards. Column effluent was collected in fractions (0.5 ml) using a Gilson model 202 collector; to each was added 4 ml of Optiphase scintillant and the ^{3}H content was determined by LSC.

GC-MS was performed using a Varian-MAT 44S spectrometer interfaced to a Varian 3700 gas chromatograph. The MoMeTMS derivatives were prepared by standard methods and the sample redissolved in heptane prior to injection. Injections were performed in the split mode (ratio 30:1) onto a 10 m × 0.32 mm i.d. DB-5 capillary column with a helium flow of 1 ml/min. GC was performed isothermally at 240° with the inlet and separator at 240°. The ion source was kept at 200° and MS performed with an electron energy of 55 eV.

RESULTS AND DISCUSSION

In view of the small quantities of material involved in the *in vivo* studies the analytical strategy adopted was to perform initial analysis by HPLC using a method which resolves well the major PG's and their known metabolites. The next step was to identify individual reactions using subcellular preparations from human and animal tissues. In view of the mounting evidence that metabolism of PGD_2 involved reduction of the C-11 keto function, it had to be ensured that the chosen system afforded separation between 9α,11α- and 9α,11β-PGF_2 epimers. As shown in Table 1, good resolution (R_s) was obtained between most prostanoids, the exceptions (R_s as stated) being the following

Table 1. Separation of PG's by HPLC: t_r's (absolute) and rel. t_r's.

Compound (& no. of expts.)	t_r, min	rel. t_r
6-keto $PGF_{1\alpha}$ (28)	16.00 ±0.07	0.317 ±0.001
9β,11α-PGF_2 (15)	24.66 ±0.37	0.510 ±0.001
9α,11β-PGF_2 (124)	28.19 ±0.20	0.560 ±0.001
9α,11α-PGF_2 (124)	35.49 ±0.48	0.690 ±0.001
PGE_2 (111)	42.69 ±0.35	0.850 ±0.001
15-keto-$PGF_{2\alpha}$ (111)	44.61 ±0.34	0.890 ±0.001
PGD_2 (124)	50.78 ±0.41	(1.000)
15-keto-PGE_2 (110)	55.59 ±0.42	1.110 ±0.001
13,14-dihydro-15-keto -9α,11β-PGF_2 (14)	57.77 ±0.78	1.120 ±0.002
ditto -$PGF_{2\alpha}$ (101)	63.83 ±0.61	1.270 ±0.001
ditto -PGE_2 (114)	71.14 ±0.55	1.420 ±0.002
ditto -PGD_2 (3)	92.14 ±0.71	1.860 ±0.002

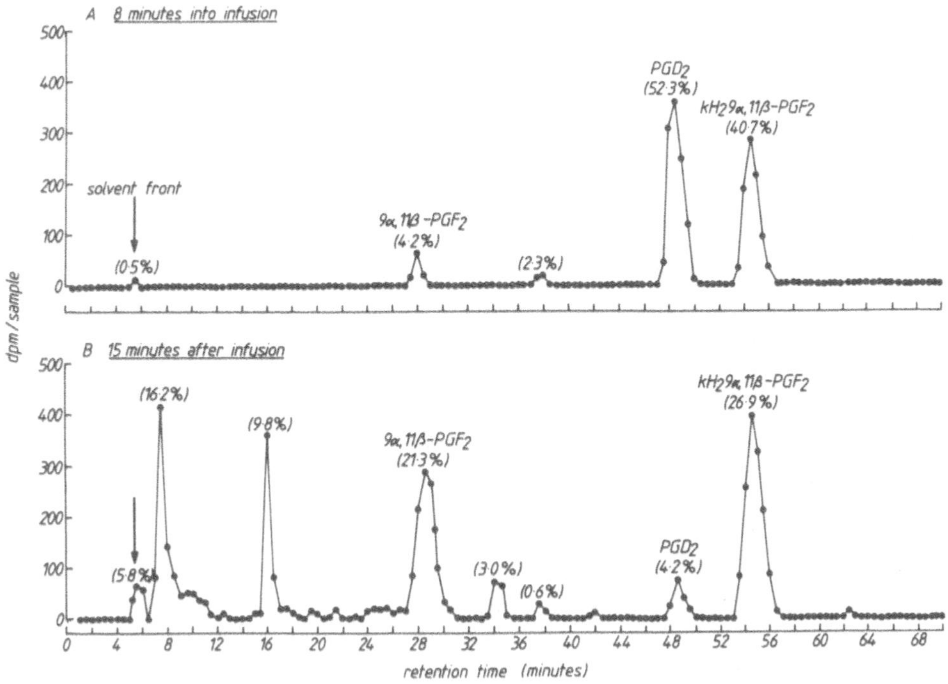

Fig. 2. Separation (as in the text) of PGD$_2$ and its plasma metabolites formed during and after i.v. infusion of 86 µCi of ^{3}H-PGD$_2$.

pairs: PGE$_2$/15-keto-PGF$_{2\alpha}$ (0.8; qualitative separation) and 15-keto-PGE$_2$/13,14-dihydro-15-keto-9α,11β-PGF$_2$ (0.5). However, it was considered unlikely that PGE$_2$ and its metabolites would be formed from PGD$_2$.

A feature of the RP-HPLC of poly-^{3}H compounds is their shorter t_r's when compared to their unlabelled counterparts. This is most obvious in long high-resolution separations and arises because the shorter carbon-^{3}H bonds interact less efficiently with the stationary phase. So as to quantify this HPLC effect, mixtures of ^{3}H-PG's and their unlabelled analogues were co-injected. For the compounds with $^3H_{6-8}$ the t_r's plotted against those for the analogues yielded a straight-line relationship (r = 0.9993) whose slope (1.0339) was used to predict the t_r of ^{3}H compounds in the study samples, as is feasible if there are 6-8 ^{3}H's per molecule.

Compounds formed after administration to man

Samples from two subjects taken during and for 3 h after infusion were analyzed by HPLC. Fig. 2 shows a pair of representative chromatograms from one subject. Blood samples taken up to 90 min after PGD_2 inhalation by two subjects were similarly processed. For the successive peaks, signified by rel. t_r values, the individual time-averaged amounts of associated radioactivity [as %, ±S.E.M. (with no. of expts.)] were as follows, where rel. t_r (*vs.* PGD_2) is 0.56 for 9α,11β-PGF_2 and 1.12 for 13,14-dihydro-15-keto-9α,11β-PGF_2.-

Infusion:

0.11 ±0.02: 6.9 ±1.0 (26); 0.15 ±0.01: 19.7 ±3.8 (28); 0.21 ±0.02: 8.5 ±1.5 (9); 0.32 ±0.01: 2.8 ±0.4 (25); 0.56 ±0.01: 11.4 ±1.1 (16); 0.68 ±0.02: 3.7 ±0.9 (19); 0.77 ±0.01: 1.6 ±0.3 (19); 0.88 ±0.01: 1.7 ±0.6 (14); **1.00**, by definition: 13.8 ±2.8 (16); 1.12 ±0.01: 32.8 ±4.6 (16); 1.28 ±0.01: 5.1 ±2.1 (8); 1.45 ±0.00: 6.6 ±1.8 (3).

Inhalation:

0.11 ±0.01: 10.4 ±1.9 (20); 0.16 ±0.01: 11.5 ±1.8 (15); 0.23 ±0.01: 7.2 ±2.8 (7); 0.34 ±0.01: 6.7 ±1.5 (15); 0.50 ±0.03: 3.7 ±0.6 (7); 0.56 ±0.01: 3.9 ±0.9 (12); **1.00**, by definition: 9.2 ±2.7 (11); 1.12 ±0.01: 19.2 ±4.5 (12).

In both studies one of the fast-running metabolites, which was more polar than PGD_2, had a t_r identical to 9α,11β-PGF_2 prepared by stereo-controlled synthesis from the Corey lactone intermediate. Application of the isotope correction factor to the absolute t_r of unlabelled 9α,11β-PGF_2 successfully predicted the t_r of its ^3H analogue. This identity was confirmed by observed co-elution of the sample peak with authentic [5,6,8,9,12,14,15-(n)-^{3}H]PGF_2 prepared by sodium borohydride reduction of ^3H-PGD_2. Except for a small quantity formed only with i.v. infusion, the foregoing radioactivity results give scant evidence for the formation of PGF_2 and its metabolites.

One of the metabolites of PGD_2 was less polar than the parent compound, with t_r (PGD_2 =1.00) of 1.12, identical to that of chemically synthesized 13,14-dihydro-15-keto-9α,11β-PGF_2. The formation of this metabolite is consistent with the metabolism of 9α,11β-PGF_2 by the sequential action of 15-PGdh and 15-keto-PG Δ^{13}reductase (Δ^{13}R). However, there remains the possibility that PGD_2 is first metabolized to a 13,14-dihydro-15-keto compound prior to formation of the corresponding 9α,11β-PGF ring compound. Pooled samples from one participant in the infusion study were subsequently analyzed for the presence of 13,14-dihydro-15-keto-PGD_2 as an index of possible 15-PGdh/Δ^{13}R-initiated metabolism. This compound constituted 4.3% of the recovered ^3H, compared to 7.8% as PGD_2, 16.9% as 9α,11β-PGF_2, 35.0% as 13,14-di-15-keto-9α,11β-PGF_2 and 2.7% as 15-keto-$PGF_{2\alpha}$.

Metabolism of PGD_2

Purified NAD^+-dependent 15-PGdh from various sources hardly attacks PGD_2. Reported activity of platelet [15] or ($NADP^+$-dependent) swine brain [16] preparations may be non-specific for PG and physiologically irrelevant. More is known about 11-ketoreductase, active towards PGD_2: the rabbit liver cytosolic enzyme, M_r 66,000, requires NADPH [17].

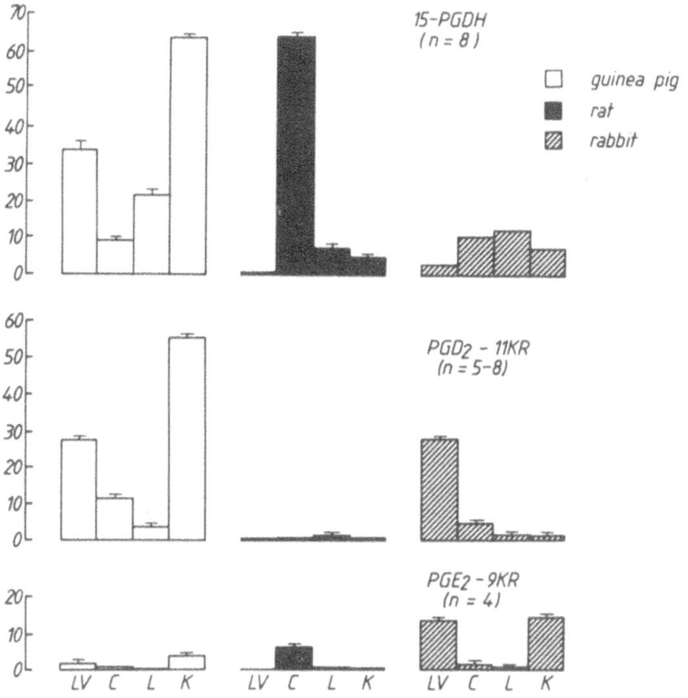

Fig. 3. PGD_2 11-ketoreductase activity in different species, and comparison with the distributions of 15-PGdh and PGE_2 9-ketoreductase. ***LV***, liver; ***C***, caecum; ***L***, lung; ***K***, kidney. Bars show S.E.M.'s.

Activity is present too in human liver and lung. The bovine lung enzyme, M_r 30,500, also catalyzes the stereoselective reduction of PGH_2 to $PGF_{2\alpha}$ [18].

In this work-area the literature on the stereochemistry of the reaction product(s) was conflicting. We undertook experiments *in vitro* to address several questions relating to PGD_2 metabolism, viz. (i) the relative distribution of 15-PGdh, 11-ketoreductase and PGE_2 9-ketoreductase; (ii) to investigate whether PGD_2 11-ketoreductase and PGE_2 9-ketoreductase are alternative activities of a single protein; and (iii) to study the further metabolism of $9\alpha,11\beta$-PGF_2 itself. [Cf. suggestion on p. 59 that the latter be termed $PGF_{2\beta}$.-*Ed.*]

Fig. 3 shows results for different organs and species. Clearly the enzyme activies metabolizing the four substrates tested vary strikingly, both between-organ and between-species. Evidently 11-ketoreductase does not have a universal distribution, a predictable finding in view of the tissue selectivity of PGD_2 generation, and furthermore there is no link between 11- and 9-ketoreductase activities.

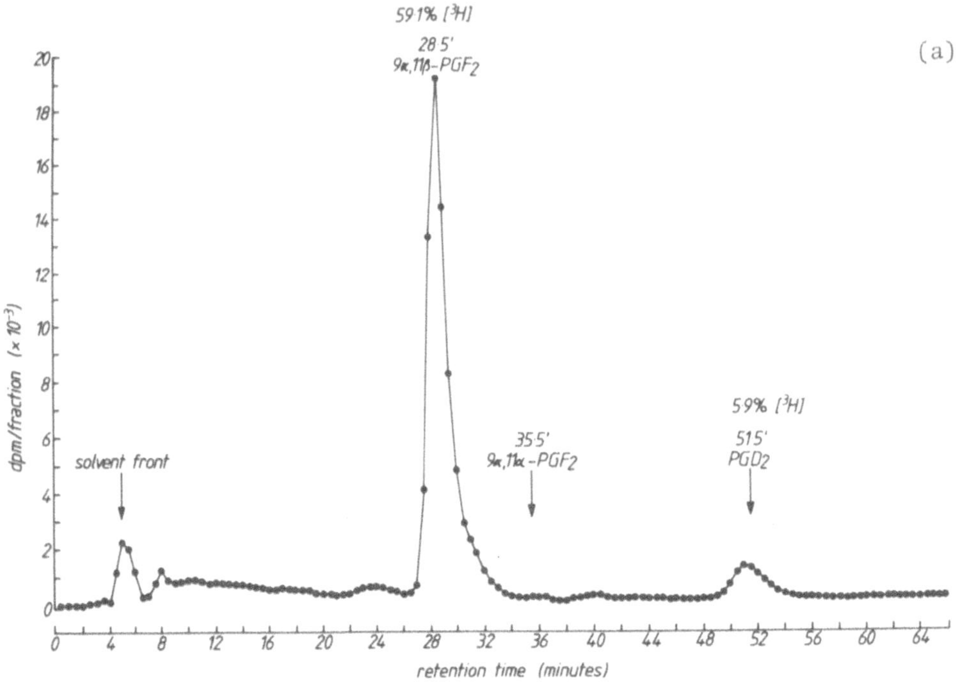

Fig. 4. *Above.*- NADPH-dependent 11-ketoreduction of PGD_2 in guinea pig kidney cytosol: the product formed co-elutes in HPLC with chemically synthesized ^{3}H-labelled 9α,11β-PGF_2. No evidence for $PGF_{2\alpha}$ formation could be obtained. *Right.*- EI-MS of the product. The GC t_r and fragmentation patterns are identical with those for the synthetic material.

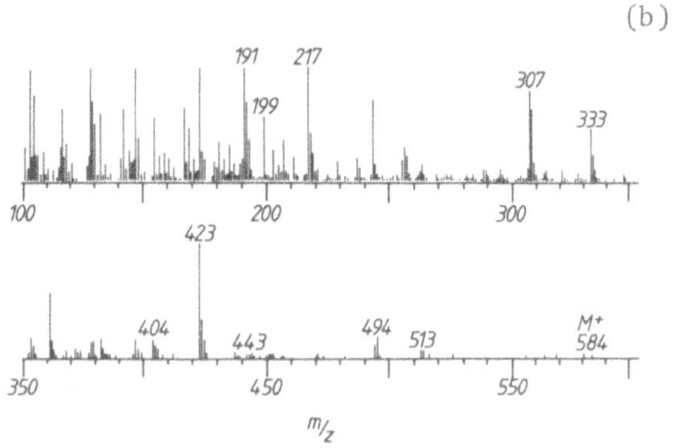

In the case of the 11-ketoreductase experiments, illustrated for guinea pig kidney in Fig. 4, HPLC of the product(s) from all organs rich in the enzyme revealed only 9α,11β-PGF_2 This was confirmed by GC-MS (EI) analysis of the product as its MoMeTMS derivative. The key fragment ions were: m/z 584 [M+]; 513 - loss of $.C_5H_{11}$; 494 - loss of $(CH_3)_3SiOH$; 443 - loss of $.CH_2CH{:}CH(CH_2)_3COOCH_3$; 423 - loss of $(CH_3)_3SiOH$ and $.C_5H_{11}$; 404 - loss of 2 × $(CH_3)_3SiOH$; 333 - loss of 2 × $(CH_3)_3SiOH$ and $.C_5H_{11}$; 217; 191 and 199 - $CH{:}CHCH[OSi(CH_3)_3](CH_2)_4CH_3$. In all cases the GC t_r's, fragmentation patterns and relative abundance of the fragment ions were identical with those of authentic $9\alpha,11\beta$-PGF_2. Taken together with the HPLC data these experiments clearly predicate the β-configuration of the C-11 hydroxyl function.

When incubated in the presence of 4 mM NAD^+, $9\alpha,11\beta$-PGF_2 itself was transformed in cytosol preparations rich in 15-PGdh as assayed using PGF_2. To investigate the identities of the metabolites formed, the cytosol reaction mixtures from guinea pig liver and rat caecum were subjected to HPLC. In caecum 34.4% of the recovered radioactivity eluted with $9\alpha,11\beta$-PGF_2 (rel. t_r taken as 1.00), whereas 44.5% eluted as a peak with rel. t_r 1.33 (metabolite I) and 7.4% (metabolite II) co-eluted with 13,14-dihydro-15-keto-$9\alpha,11\beta$-PGF_2 (at rel. t_r 1.93). Similar results were obtained with guinea pig liver: 54.9% of the sample was $9\alpha,11\beta$-PGF_2, 24.7% appeared as metabolite I and 4.7% as metabolite II.

Activity in human lung.- Under similar incubation conditions favouring 15-PGdh activity, $9\alpha,11\beta$-PGF_2 was an acceptable substrate for metabolism in human lung cytosol. HPLC analysis indicated metabolites with similar t_r's to those seen in the animal organ experiments. Structural studies with GC-MS indicated the presence of a major and a minor metabolite, besides untransformed $9\alpha,11\beta$-PGF_2. The MoMeTMS derivative of the major metabolite gave the following MS fragment ions: m/z 539 [M+]; 508 (M - 31, loss of $.OCH_3$); 449 (M - 90, loss of $(CH_3)_3SiOH$; 418 (M - [90 + 31]); 468 (M - 71, loss of $.C_5H_{11}$); 328 (M - [90 + 90 + 31]), i.e. two hydroxyls and one keto function in the molecule, consistent with the structure 15-keto-$9\alpha,11\beta$-PGF_2. In view of the small amount of metabolite formed, MS verification of the presence of 13,14-dihydro-15-keto-$9\alpha,11\beta$-PGF_2 could only be obtained in the SIM mode using fragment ions at m/z 510, 420 and 330 in the MoMeTMS derivative and by comparison with authentic material.

CONCLUDING COMMENTS

Evidently, as summarized in Fig. 5, in man and various animal organs PGD_2 initially gives rise to $9\alpha,11\beta$-PGF_2. This itself appears to be further transformed in man, since there is a major plasma metabolite that co-elutes with 13,14-hydro-15-keto-$9\alpha,11\beta$-PGF_2 (as likewise found, with a 15-keto intermediate, in lung cytosol incubates with NAD^+ present). This metabolite may prove to be a useful analyte for measuring PGD_2 turnover *in vivo*. The observed metabolic pattern and its strict cofactor dependency suggests that the further reactions are mediated by 15-PGdh/Δ^{13}R, as supported by recent evidence [19] that $9\alpha,11\beta$-PGF_2 breakdown is inhibited by sulphasalazine or two azobenzene analogues which are known inhibitors of 15-PGdh. From recent *in vitro* experiments (unpublished) on the formation of the $9\alpha,11\beta$-PGF_2 metabolites there seems to be little contribution from 11-ketoreduction of the corresponding PGD_2 ring compounds.

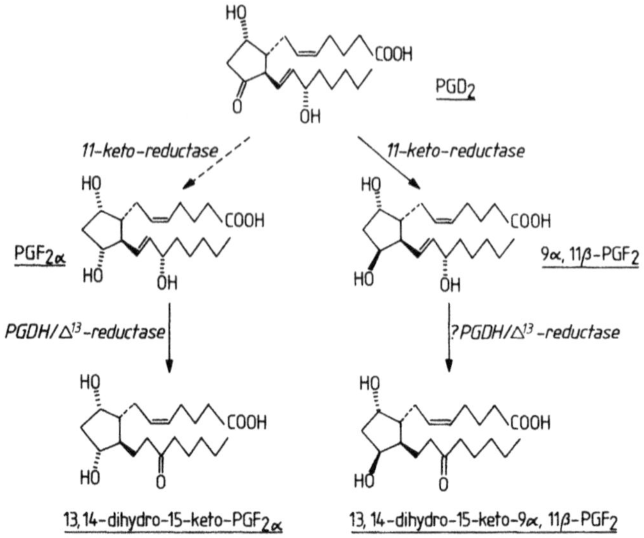

Fig. 5. Summary of initial events in the metabolism of PGD_2. The degradation of PGD_2 by an 11-ketoreductase enzyme produces 9α,11β-PGF_2 which is biologically active. Inactivation of this product may proceed via the concerted actions of 15-PGdh and Δ^{13}R (15-keto-prostaglandin Δ^{13}-reductase). A small amount of $PGF_{2\alpha}$ may be formed from PGD_2, ultimately producing the corresponding metabolites.

These studies have provided other analytical messages. Although some previous studies have employed GC-MS techniques, the failure to observe the stereochemistry of the PGF ring hydroxyl groups undoubtedly led to the misassignment of metabolite identity. Common derivatization techniques for prostanoids do not provide stereochemical information, and as the GC-MS characteristics of 9,11-PGF epimers are very similar the assignment of metabolite identity based solely on these criteria is open to error. As Roberts and colleagues [12, 13] have discussed, it may be necessary to re-evaluate all previous studies reporting the formation of '$PGF_{2\alpha}$'. In contrast, RIA and RP-HPLC or inclusion-complexing HPLC (C. Robinson, unpublished) all easily discriminate between the 9α,11β-PGF_2 epimers and their respective metabolites. Under ideal circumstances analytical strategies should employ a variety of analytical techniques to avoid such problems.

Acknowledgements

I am indebted to the Research Fund of the University of Southampton for financial support and to Dr Robin Hoult, Dept. of Pharmacology, King's College London, for collaboration in the *in vitro* studies. Ms. Dianne Wilson provided excellent secretarial assistance.

References

1. Lewis, R.A., Soter, N.A., Diamond, P.T., Austen K.F., Oates, J.A. & Roberts, L.J. (1982) *J. Immunol. 129*, 1627-1631.
2. Fox, C.C., Dvorak, A.M., Peters, S.P., Kagey-Sobotka, A. & Lichtenstein, L.M. (1985) *J. Immunol. 135*, 483-491.
3. Robinson, C., Benyon, C., Holgate, S.T. & Church, M.K. (1987) *Br. J. Pharmacol. 92*, 635-638.
4. Kelly, R.W., Dean, S., Cameron, M.J. & Seamark, R.F. (1986) *Prostaglandins Leuk. Med. 24*, 1-14.
5. Granström, E. & Kindahl, H. (1978) in *Advances in Prostaglandin and Thromboxane Research, Vol. 5* (Frölich, J.C., ed.), Raven Press, New York, pp. 119-210.
6. Sun, F.F., Armour, S.B., Bockstanz, V.R. & McGuire, J.C. (1976) in *Advances in Prostaglandin Research, Vol. 1* (Samuelsson, B. & Paoletti, R., eds.), Raven Press, New York, pp. 163-169.
7. Hensby, C.N. (1974) *Prostaglandins 8*, 369-375.
8. Ellis, C.K., Smigel, M.D., Oates, J.A., Oelz, O. & Sweetman, B.J. (1979) *J. Biol. Chem. 254*, 4152-4163.
9. Gréen, K., Hedqvist, P. & Svanborg, N. (1974) *Lancet ii*, 1419-1421.
10. Holgate, S.T., Twentyman, O.P., Rafferty, P., Beasley, R., Hutson, P.A., Robinson, C. & Church, M.K. (1987) *Int. Arch. Allergy Appl. Immunol. 82*, 498-506.
11. Barrow, S.E., Heavey, D.J., Ennis, M., Chappell, C.G., Blair, I.A. & Dollery, C.T. (1984) *Prostaglandins 28*, 743-754.
12. Liston, T.E. & Roberts, L.J. (1985) *J. Biol. Chem. 260*, 13172-13180.
13. Roberts, L.J. & Sweetman, B.J. (1985) *Prostaglandins 30*, 383-401.
14. Hardy, C.C., Holgate, S.T. & Robinson, C. (1986) *Br. J. Pharmacol. 87*, 563-568.
15. Watanabe, T., Shimizu, T., Narumiya, S. & Hayaishi, M. (1982) *Arch. Biochem. Biophys. 216*, 372-379.
16. Watanabe, K., Shimizu, T., Sadahiko, I., Watasuka, H., Hayaishi, M. & Hayaishi, O. (1980) *J. Biol. Chem. 255*, 1779-1782.
17. Wong, P.Y-K. (1981). *Biochim. Biophys. Acta 659*, 169-178.
18. Watanabe, K., Iguchi, Y., Iguchi, S., Arai, Y., Hayaishi, O. & Roberts, L.J. (1986) *Proc. Nat. Acad. Sci. 83*, 1583-1587.
19. Bacon, K.B., Hoult, J.R.S., Osborne, D.J. & Robinson, C. (1987) *Br. J. Pharmacol. 91*, 322P.

#A-2

ANALYSIS OF EICOSANOIDS BY GC-MS/MS

J.C. Frölich, K-H. Marx, H. Bippi, J. Fauler, B. Rosenkranz and U. Förstermann

Department of Clinical Pharmacology
Hannover Medical School, Konstanty-Gutschow-Str. 8
3000 Hannover 61, W. Germany (FRG)

Analysis of eicosanoids in biological matrices is difficult. Their levels are very low, and they are rapidly metabolized to numerous structurally similar substances often found in much higher concentration. Hence past assays of insufficient specificity have given erroneous values. GC-MS, which is more specific than RIA, is effective for structure elucidation and assay of eicosanoids and metabolites, yet may be insufficiently selective at low eicosanoid concentrations. PGE_2 *and* $PGF_{2\alpha}$ *of renal origin are present in human urine, as now shown also for* PGD_2 *by GC-MS/MS of the methoxime-TMS* ether of its methyl ester.*

There has been increasing interest over the last few years in identifying the biological roles of eicosanoids in health and disease. This interest stems from the evidence that these substances are involved in very many physiological responses such as the regulation of blood pressure, Na excretion and water balance. Furthermore, eicosanoids have been implicated in a number of diseases ranging from hypertension to asthma and rheumatism. Numerous drugs are under development to substitute for or antagonize eicosanoids. Altogether there has been a notable increase in demand for analysis of these compounds, in clinical investigation as well as in drug development.

There are three major causes of difficulty in assaying eicosanoids. (1) The receptor affinity is high; hence very small amounts of eicosanoids will exert biological effects. Our early studies showed that PGE_2, similarly to peptide hormones, has a binding constant of 10^{-9} M for its receptor [1]. The conclusion that very low concentrations

* *Abbreviations.*- NCI, negative-ion chemical ionization; PG, prostaglandin; TMS, trimethylsilyl; i.s., internal standard; ^2H, deuterium.

suffice for the biological functions is borne out by all later studies where total body synthesis of eicosanoids has been evaluated; thus a normal human studied by B. Samuelsson's team produced no more than ~60 μg/day of PGE_2.

(2) Eicosanoid metabolism is very rapid (most products having half-lives of ~2 min) and leads to numerous closely related substances. Hence (i) there is virtually no steady-state level of the biologically active compounds in the circulation, and (ii) many metabolites interfere with assay methods that cannot readily distinguish them and, moreover, most metabolites are present in much higher concentrations than the parent PG's.

(3) Artifact generation may disturb the assessment of eicosanoid levels.

PGA_2 — FACT OR FICTION?

GC-MS provides unique opportunities to investigate these problems. The appearance of PGA_2 in dog and human kidney has been described [2]. It has also been reported that PGA_2 is resistant to metabolism in the lung [3], sharply contrasting with both PGE_2 and $PGF_{2\alpha}$ which are metabolized almost completely in a single passage through the lung. Because of this metabolic stability, PGA_2 might be a true circulating hormone. Measurements of PGA_2 by RIA were soon forthcoming and showed levels in plasma of 1.4 ng/ml in normal volunteers [4], these amounts being quite relevant for blood pressure regulation.

We were puzzled that reported levels of PGA_2 in rabbit kidney varied 3-fold [5, 6]. PGE_2 is easily dehydrated under acidic or basic conditions to form PGA_2, and since acidic conditions are commonly employed for the extraction of PG's we considered the possibility that the discrepancy in the reported values was partly because of variable chemical dehydration.

Initially it was attempted to minimize chemical dehydration by employing rapid extraction and purification methods that minimized the time of contact with acid. Analyses were carried out by GC-MS using 3,3,4,4-tetra-^2H-PGE_2 and -PGA_2 as i.s.; PGE_2 was derivatized to the methyl ester methoxime, and PGA_2 to the methyl ester TMS ether. Freshly obtained rabbit adrenal medulla was homogenized, and one portion extracted immediately after adding the i.s. and another portion was incubated and then extracted. On incubation the amounts of PGE_2 increased, showing active PGE_2 synthesis. The levels of PGA_2 found were very low, <3% of PGE_2, and did not increase in three experiments [7].

These experiments showed that PGA_2 levels stated previously may have been too high. However, our approach did not serve to answer the question whether these small amounts were originating from biosynthesis or from chemical hydration, and a definitive approach was needed: uniquely this lay in GC-MS. Again we split the homogenate into two portions, one that was immediately worked up and the other that was incubated. The difference from the previous experiment was the use of two internal standards, one for PGE_2, namely tetra-^2H-PGE_2, and one for PGA_2, namely PGA_1. Any tetra-^2H-PGA_2 appearing in the samples would originate from tetra-^2H-PGE_2. The % conversion of tetra-^2H-PGE_2 to tetra-^2H-PGA_2 would, of course, also apply to the conversion of PGE_2 to PGA_2 and serve as an index of chemical dehydration during work-up.

On incubation, PGE_2 was found to increase significantly from 4.4 to 14.9 ng/g of tissue. PGA_2 appeared to increase from 35 to 150 ng/g; but in the 6 experiments the increase was quite variable and did not become statistically significant. When the PGA_2 levels were corrected with the help of each i.s. for chemical dehydration during *in vitro* work-up, PGA_2 levels in both the incubated and the non-incubated samples were indistinguishable from zero. Dog and human kidney medulla gave similar results [8].

These studies excluded the renal medulla as a source of any circulating PGA_2. However, there might be other sources. We explored this using a GC-MS assay which enabled PGA_2 to be measured at a level of 35 pg/ml with S.D. $<10\%$. Samples (23) of plasma from 12 normal volunteers on random sodium intake showed PGA_2 levels indistinguishable from zero. As sodium depletion has been reported to increase PGA_2 levels in plasma to 2.1 ng/ml [9], we looked for PGA_2 in plasma from 6 normal volunteers after severe sodium depletion, and again found none [10]. Hence PGA_2 is not synthesized by the kidney and is not a circulating hormone. GC-MS provided a singularly useful approach to this problem.

INVESTIGATIONS ON PGI_2 SYNTHESIS IN MAN

Recently PGI_2, another vasodilator prostanoid which also inhibits platelet aggregation and dissolves aggregates, was discovered by John Vane's group [11]. PGI_2 itself is unstable and is readily converted in aqueous media to 6-keto-$PGF_{1\alpha}$. This was therefore measured in human blood plasma by various groups of investigators. There was disconcerting disagreement amongst the reported values, which ranged from 180 to 0.5 pg/ml [12]. We too were interested in studying the rate of PGI_2 synthesis in man, but took a different approach. Firstly we investigated PGI_2 metabolism in man. We infused ^3H-labelled PGI_2 at the rate of 1 ng/kg/min for 10 h. This resulted in a virtual steady state of ^3H levels in plasma, surprisingly taking so long to attain. This finding has recently been confirmed by others who infused PGI_2 for 20 h without reaching a steady state completely [13], suggesting that one or more metabolites are present with a very long half-life.

We have identified by GC-MS the following metabolites in urine: 6-keto-$PGF_{1\alpha}$, 2,3-dinor-6-keto-$PGF_{1\alpha}$, 6,15-diketo-13,14-dihydro-$PGF_{1\alpha}$ and dinor-6,15-diketo-13,14-dihydro-20-carboxyl-$PGF_{1\alpha}$. Urinary

Fig. 1. Six normal volunteers were given 20 ml water/kg body wt. Urinary excretion rates of 2,3-dinor-6-keto-$PGF_{1\alpha}$ and 6-keto-$PGF_{1\alpha}$ were determined (1) before, (2) for 1 h following the water load at the height of the urine flow, and (3) for 3 h afterwards. The data shown are means ±S.D.'s; * denotes $p < 0.05$ by paired t-test.

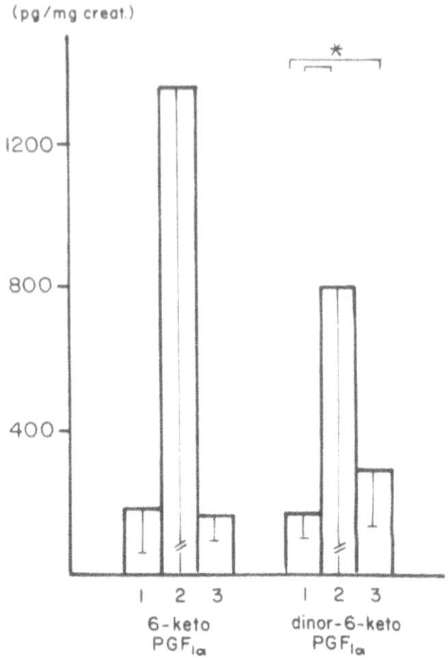

2,3-dinor-6-keto-$PGF_{1\alpha}$ comprises ~10% of all PGI_2 infused, and emerged as the most prominent metabolite of PGI_2 and 6-keto-PGF in man. We subsequently developed a method for its analysis by GC-MS based on the biosynthesis of its ^2H analogue obtained from ^2H-labelled 6-keto-$PGF_{1\alpha}$ by bacterial β-oxidation. Normal excretion values in males and females were found to be about equal at 350 ng/day.

Assessing PGI_2 biosynthesis rates⊗.- From the excretion of the metabolite the maximal delivery of PGI_2 into the blood stream is calculated to be only 2 ng/min. Viewing endogenous PGI_2 synthesis as a constant infusion into the blood stream, the equation for the steady-state plasma level attained by a constantly infused drug is pertinent. With 2 ng/min infused, then assuming a distribution only into plasma (4.0 l), and a 30 min half-life [15], the plasma level of 6-keto-$PGF_{1\alpha}$ would be maximally 22 pg/ml. But if the half-life is only 10 min [16] and if cell types besides endothelium can generate PGI_2 and so increase the distribution volume, the value becomes much lower. Such levels, then, are not reliable indices of PGI_2 biosynthesis, the assessment of which is evidently artifact-prone and has inherent uncertainty unconnected with faults in analysis.

MEASUREMENT OF A CIRCULATING PGI_2 METABOLITE

One might still hope that 2,3-6-keto-$PGF_{1\alpha}$ might serve as a reliable index of PGI_2 synthesis; but even this parameter has its shortcomings. Its excretion escalated with a rise in urine flow (Fig. 1), as a typical kidney response, dependent on urine flow,

⊗ The authors' account has been shortened.- *Ed.* (Ref. [14] is pertinent.)

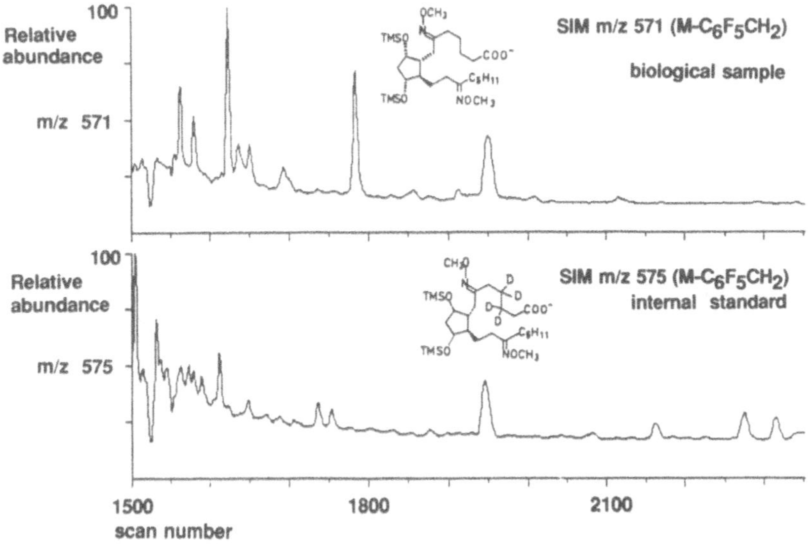

Fig. 2. GC-MS with multiple ion detection tracings for 3,3,4,4-tetradeutero-6,15-diketo-Δ13-dihydro-$PGF_{1\alpha}$-pentafluorobenzyl ester-bis-methoxime-bis-TMS ether; monitoring at m/z 574 and, for the endogenous metabolite, at m/z 571 (NCI mode). Note the peak interferences.

when handling weak organic acids as exemplified by all PG's and their metabolites.

In view of this and probably other influences on the excretion, an alternative parameter for PGI_2 synthesis would appear desirable. We therefore investigated whether 6,15-diketo-Δ13-dihydro-$PGF_{1\alpha}$ was detectable in plasma. This metabolite cannot be formed *in vitro* and thus lead to artifacts.

Our first attempt involved the pentafluoro-benzyl ester-bis-methoxime-bis-TMS ether from which, by NCI, the carboxylate fragment of m/z 571 was the most prominent. Plasma was extracted after addition of 3,3,4,4-tetra-^{2}H-6,15-diketo-Δ13-dihydro-$PGF_{1\alpha}$ and purification by HPCL. Its derivative was analyzed by NCI, monitoring m/z 571, after GC on a 15 m capillary column with ~14 min retention time. However, this procedure resulted in a quite unsatisfactory tracing with many interfering peaks (Fig. 2).

GC-MS/MS provides a new approach in trace analysis [cf. other articles in this book -*Ed.*]. While over the last few years MS analytical sensitivities have been enhanced so that prostanoids can now be measured in the low-pg range, this has not been accompanied by significant improvements in selectivity. In general it is possible to increase selectivity of magnetic focusing instruments by increasing resolution,

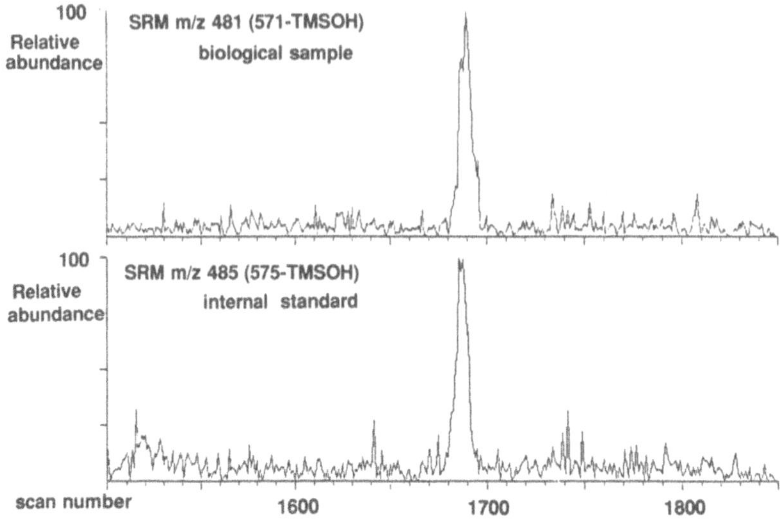

Fig. 3. GC-MS/MS: selected ion monitoring for daughter ions m/z 481 and 485 representing 6,15-diketo-Δ13-dihydro-$PGF_{1\alpha}$ and the tetra-^2H i.s. respectively.

but loss of sensitivity is a severe penalty. Thus, our own experience with biological samples indicates that increasing resolution from 1,000 to 5,000 results in a large loss of sensitivity but does not regularly result in suppression of undesirable peaks and so allow quantification of a sample that was hopeless at 1,000. Furthermore, this approach is precluded with the widely used quadrupole instruments.

GC-MS/MS provides a new dimension in MS analysis with very significant gains in specificity, achieved by subjecting a fragment to a second fragmentation step to obtain a daughter spectrum. Monitoring a fragment in the latter is termed selected reaction monitoring. In the case of our derivative the most prominent fragment in the daughter spectrum was m/z 481. Selected reaction monitoring of this fragment and of m/z 485 gave acceptable tracings with a very evident gain in selectivity (Fig. 3). This chromatogram proves that 6,15-diketo-Δ13-dihydro-$PGF_{1\alpha}$ is an additional metabolite of prostacyclin in the human circulation [17].

In summary, the GC-MS and GC-MS/MS analytical approaches have made decisive contributions to our understanding of the eicosanoids. The sensitivities of the current generation of equipment allows investigations that were formerly restricted to RIA measurements. GC-MS/MS analysis opens a new dimension in selectivity which will be particularly useful in investigations of the eicosanoids in man.

References

1. Smigel, M. & Frölich, J.C. (1974) *Prostaglandins 6*, 537-539.
2. Lee, J.B., Covino, B.J., Takman, B.H. & Smith, E.R. (1965) *Circ. Res. 17*, 57-60.
3. McGiff, J.C., Terragno, N.A., Strand, J.C., Lee, J.C., Lonigro, A.J. & Ng, K.K.F. (1969) *Nature 223*, 742-745.
4. Zusman, R.M., Caldwell, B.V., Speroff, L. & Behrman, H.R. (1972) *Prostaglandins 2*, 41-53.
5. Lee, J.B., Crowshaw, K., Takman, B.H., Attrep, K.A. & Gougoutas, J.Z. (1967) *Biochem. J. 105*, 1251-1260.
6. Daniels, E.G., Hinman, J.W., Heath, B.E. & Muirhead, E.E. (1967) *Nature 215*, 1298-1299.
7. Frölich, J.C., Williams, W.M., Sweetman, B.J., Smigel, M., Carr, K., Hollifield, J.W., Fleischer, S., Nies, A.S., Frisk-Holmberg, M. & Oates, J.A. (1975) in *Prog. in Prostaglandin and Thromboxane Res.* (Samuelsson, B. & Paoletti, R., eds.), Raven Press, New York, pp. 65-80.
8. Frölich, J.C., Sweetman, B.J., Carr, K. & Oates, J.A. (1975) *Life Sci. 17*, 1105-1112.
9. Zusman, R.M., Spector, D., Caldwell, B.V., Speroff, L., Schneider, G. & Mulrow, P.J. (1973) *J. Clin. Invest. 52*, 1093-1098.
10. Frölich, J.C., Sweetman, B.J., Carr, K., Hollifield, J.W. & Oates, J.A. (1975) *Prostaglandins 10*, 185-195.
11. Harrod, M.J.E. & Sherrod, P.S. (1981) *Obstet. & Gynec. 57*, 673-676.
12. Frölich, J.C. & Rosenkranz, B. (1984) *Prostaglandins 27*, 354-355.
13. Brash, A.R., Jackson, E.K., Lawson, J.A., Brauch, R.A., Oates, J.A., & Fitzgerald, G.A. (1983) in *Adv. in Prostaglandin, Thromboxane and Leukotriene Res. 11* [Samuelsson, B., et al., eds.], 119-122.
14. Rosenkranz, B., Fischer, C., Weimer, K.E. & Frölich, J.C. (1980) *J. Biol. Chem. 255*, 10194-10198.
15. Ylikorkala, O. & Viinikka, L. (1981) *Prost. Med. 6*, 427-436.
16. Taylor, B.M., Shebnski, R.J. & Sun, F.F. (1983) *J. Pharm. Exp. Ther. 224*, 692-698.
17. Rosenkranz, B., Fischer, C. & Frölich, J.C. (1981) *Clin. Pharmacol. Ther. 29*, 420-424.

#A-3

QUANTITATIVE DETERMINATION OF SEVERAL PROSTAGLANDINS AND THROMBOXANE B_2 IN THE ISOLATED PERFUSED RAT BRAIN

Claudia Weber, Klaus Belsner, Barbara Osikowska-Evers, Inge Luhmann and Michael Höller*

Department of Experimental Medicine
Janssen Research Foundation
Raiffeisenstr. 8, D-4040 Neuss 21, W. Germany

A GC-MS assay was developed to investigate the time course of the tissue concentrations of PGE_2, $PGF_{2\alpha}$, PGD_2, 6-keto-$PGF_{1\alpha}$ as well as TXB_2 in the isolated rat brain, perfused with a synthetic fluorocarbon perfusion medium. After perfusion, the brains were frozen in isopentane at -56°, freeze-dried at -40° and pulverized (liq. N_2). The powder (~100 mg) was spiked with 4 deuterated analytes and extracted with methanol/water. The supernatants were applied to a BondElut RP-18 cartridge. The eluate was dried down, the residue dissolved in ether/p.e.• and the solution applied to a silica gel column. The eluate was dried down and the residue subjected to methoximation, PFB esterification and silylation. The derivatized COP's were separated and quantitated by NCI GC-MS.

The absolute detection limit was 0.5 pg per column injection of standards for each COP. For determination in brain tissue the limit was 100 pg per sample. Calibraton curves in the range 0.2-50 ng/50 mg dry tissue were linear for each COP. The method is suitable for determining various PG's and TXB_2 in the normoxic brain.

In ischaemia of the brain, activation of phospholipases leads to liberation of arachidonic acid with subsequent formation of COP's [1, 2]. It has been suggested that COP's play an important role in the development of ischaemic brain damage [see 3]. Here a GC-MS method for determining PG's and TXB_2 in rat brain is described.

---

* Address correspondence to Prof.Dr.med. M. Höller.

• *Abbreviations (some introduced by Ed.).*- 'ether' = diethyl; p.e., petroleum ether (light petroleum); COP, cyclooxygenase product; PFB, pentafluorobenzyl; NCI, negative-ion chemical ionization.

TISSUE PREPARATION, EXTRACTION AND ANALYTE DERIVATIZATION

Isolated perfused rat brains were used as biological models [4]. After 30 min of perfusion, the isolated heads from which the lower jaw and most of the skin had been removed beforehand were frozen in isopentane at -56°. The tissue was freeze-dried and the brains were removed under absolutely dry N_2 at 0°. The brains were pulverized at liquid N_2 temperature and weighed. To 45-55 mg of tissue powder were added 2.5 ng each of $[^2H_4]PGE_2$, $[^2H_4]PGF_{2\alpha}$, $[^2H_4]$6-keto-$PGF_{1\alpha}$, and $[^2H_4]TXB_2$. In separate experiments, recovery was checked with 900 Bq $[^{14}C]PGE_2$ added to tissue: from a stock solution of internal standards (0.1 ng of each ^2H-COP/ml of acetonitrile), 25 µl was added to 50 mg dry tissue powder and mixed by homogenization.

The tissue powder was extracted twice with 3 ml methanol/water (9:1 by vol.) according to Mayer et al. [5]. The suspension was centrifuged at 5°; the pellet was discarded and the supernatants were combined and diluted with 27.5 ml of dil. formic acid at pH 3.5. The solution was applied to a BondElut RP-18 cartridge, pre-washed with 20 ml methanol, 20 ml water and 5 ml methanol/pH 3.5 dil. formic acid (15:85 by vol.). The loaded column was washed with 5 ml methanol/dil. formic acid, and the COP's eluted with 5 ml ethanol. The eluate was evaporated to dryness under N_2 and the residue dissolved in 0.5 ml ether/p.e. (25:75).

The solution was applied to a silica gel column pre-prepared thus: 0.5 g Silica CC-4 (Mallinckrodt, St. Louis) was put into a pasteur pipette, washed with 2 ml methanol, dried overnight at 80° and washed with 2 ml of ether/p.e. (25:75). The loaded column was washed with 3 ml ether/p.e. (75:25), and the COP's eluted with 3 ml ethyl acetate/methanol (9:1).

The eluate was evaporated to dryness under N_2 at 40°, and the residue derivatized, for GC, to methoxime PFB ester tris(trimethylsilyl) ethers according to [6]. The derivatized samples were stored at 4° until analyzed.

GC-MS AND QUANTITATION

A Varian GC 3700 equipped with a cooled injection system (Gerstel, Mühlheim, FRG) and an Ultra 2 fused silica capillary column (12 m, 0.2 mm i.d., 0.3 µm film thickness; Hewlett Packard) were used. A retention gap (1 m length, 0.2 mm i.d.) was coupled to the analytical column. The injection (0.5 µl) was made in the splitless mode; the port, starting at 60°, was programmed at 10°/sec up to 250°, with 100 sec hold time. The GC oven was at 140° initially with a heating rate of 35°/min up to 290°. The GC was coupled to a Finnigan MAT 8230 MS with the column directly introduced into the ion source. The derivatized COP's were determined by NCI GC-MS with ammonia as the reagent gas and with multiple ion detection.

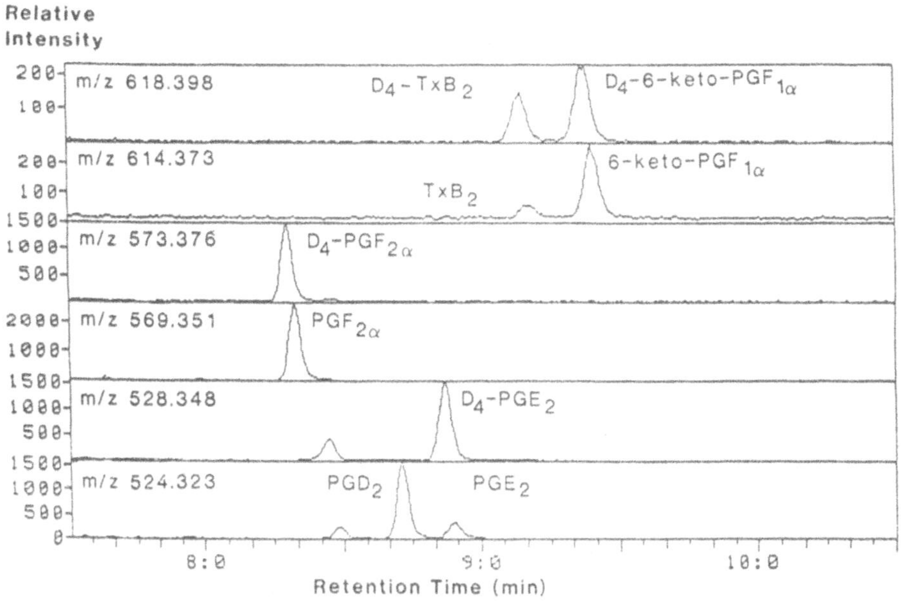

Fig. 1. GC-MS chromatogram, with multiple ion detection, of endogenous PG's and TBX_2 in the perfused rat brain, with ^2H internal standards (see text).

Calibration curves were calculated from each peak-area ratio (unlabelled/labelled COP) obtained from spiked brain. The curves from brain tissue analyses were linear for each of the 5 COP's. The lowest correlation coefficient was obtained for TXB_2 ($r = 0.997$). The absolute detection limit was 0.5 pg per injection of each pure COP and 100 pg per tissue sample. The blank values subtracted from the curves came from separate determination of endogenous COP's, as illustrated for a normoxic brain (after 30 min of perfusion) in Fig. 1.

For $[^{14}C]PGE_2$ the mean recovery (±S.D.; n = 6) was 82 ±3.2%. The lowest C.V. (highest reproducibility; n = 6) was obtained for 6-keto-$PGF_{1\alpha}$ (1.6%), and the highest for 6-keto-PGD_2 (5.7%; no ^2H standard available).

VALUES FOR COP's IN PERFUSED BRAIN

Isolated brain perfused for 30 min gave the following values, as ng/g dry wt. (mean ±S.E.M; n = 4): PGD_2, 13.5 ±10.7; PGE_2, 3.1 ±0.4; $PGF_{2\alpha}$, 32.1 ±8.6; 6-keto-$PGF_{1\alpha}$, 10.9 ±3.9; TXB_2, 5.2 ±2.0. As no thrombocytes were present in the perfusion medium, the observed TXB_2 levels represent TX synthesis in brain structures.

When comparing these results with data from the literature, various aspects must be considered: e.g. regional distribution of COP's in the brain, influence of species and sex, method of tissue sampling and preparation, basis for expressing the COP content (wet wt., dry wt., mg protein) and analytical method. In the present experiments the heads of the rats were immersed at -56°, most of the skin and the lower jaw having been removed during perfusion. It may therefore be suggested that the freezing process was very rapid and that the ischaemic period above 0° was very short.

Narumiya et al. [7] inactivated the enzymes of PG metabolism in rat brain by microwave irradiation and determined various PG's by RIA. PGD_2 and PGE_2 levels were similar, while $PGF_{2\alpha}$ was ~8 times lower. Hiroshima et al. [8] prepared the tissue by decapitation and freezing the rat brains in liquid N_2; PGD_2 was determined by enzyme immunoassay. Its basal levels were ~7 times lower than the present ones; however, the levels in ischaemic brains were also very low compared with those obtained in the ischaemic isolated perfused brain (C. Weber, F. Tegtmeier, D. Scheller, K. Belsner & M. Höller, in preparation).

Wolfe et al. [9] determined $PGF_{2\alpha}$ and PGE_2 in rat brain by GC-MS, and obtained values in the same range as the present ones. Brain levels of TXB_2 and 6-keto-PGF_1 in the rat were determined by RIA after microwave fixation or decapitation [10]. The present results are ~30-50% higher for both compounds than in their microwave fixation group and are considerably lower than in the decapitation group.

CONCLUDING COMMENTS

With the method presented here, PGD_2, PGE_2, $PGF_{2\alpha}$, 6-keto-PGF_1 and TXB_2 can easily be determined in non-ischaemic brain. For an experienced analyst the work-up and GC-MS analysis of 10 samples and a calibration curve with 4 concentrations takes 2 days. Use of isolated perfused brain enables the tissue to be frozen rapidly and easily so that true basal levels of COP's in whole brain tissue can be determined.

Acknowledgements

The authors are grateful to Dr. C.O. Meese, of the Dr. Margarethe-Fischer-Bosch-Institut, Stuttgart, for synthesizing deuterated TXB_2 and 6-keto-$PGF_{1\alpha}$.

References

1. Siesjö, B.K. (1981) *J. Cerebr. Blood Flow Metab. 1*, 155-185.
2. Wolfe, L.S. (1982) *J. Neurochem. 38*, 1-14.
3. Schrör, K. (1986) in *Pharmacology of Cerebral Ischaemia* (Krieglstein, J., ed.), Elsevier, Amsterdam, pp. 199-209.
4. Höller, M., Breuer, H. & Fleischhauer, K. (1983) *J. Pharmacol. Meths. 9*, 170-182.

5. Mayer, B., Moser, R., Leis, H.J. & Gleisbach, H. (1986) *J. Chromatog.* *378*, 430-436.
6. Fischer, C. & Meese, C.O. (1985) *Biomed. Mass Spectrom.* *12*, 399-404.
7. Narumiya, S., Ogoeochi, T., Nakao, K. & Hayaishi, O. (1982) *Life Sci.* *31*, 2093-2103.
8. Hiroshima, O., Hayashi, H., Ito, S. & Hayaishi, O. (1986) *Prostaglandins* *32*, 63-80.
9. Wolfe, L.S., Pappius, H.M. & Marion, J. (1976) in *Advances in Prostaglandin and Thromboxane Research*, Vol. 1 (Samuelsson, B. & Paoletti, R., eds.), Raven Press, New York, pp. 345-355.
10. Petroni, A., Socini, A., Blasevich, M., Borghi, A. & Galli, C. (1985) *Prostaglandins* *29*, 579-587.

#A-4

ANALYTICAL METHODS FOR [^{3}H]-ENISOPROST, AN ANTI-SECRETORY PGE_1 ANALOGUE, AND ITS METABOLITES

L.A. Allan, A.J. Hawkins, [1]J. Firth, [2]R.D. Brownsill, [3]J.A. Steiner and [4]C.W. Vose

G.D. Searle and Co. Ltd.
High Wycombe, Bucks. HP12 4HL, U.K.

This article describes analytical methods developed to study the disposition of [^{3}H]enisoprost, a PGE_1 analogue, in man. Solid-phase extraction and HPLRC were used. These methods were scaled up to allow isolation and purification of urinary metabolites. These were identified by GC-MS of their ME-MO-TMS and PFB-MO-TMS derivatives. The latter provided mol. wt. data from the intense [M-PFB]$^-$ ions formed by NCI with ammonia, and the former provided information on sites of metabolism from the characteristic EI fragmentation pathways. The addition of a cholinesterase inhibitor, 2 M pyridostigmine bromide, served to stabilize enisoprost in whole blood immediately after collection; in its absence enisoprost was rapidly degraded to its carboxylic acid metabolite (SC-36067).*

The naturally occurring PG's have been shown to be rapidly and extensively metabolized [1-5]. The development of synthetic analogues of PGE_1 has yielded compounds which show potent gastric anti-secretory activity. Some of these compounds were synthesized with the intention of blocking certain PG metabolic pathways. One such analogue, enisoprost (Fig. 1), had 16-hydroxy-16-methyl substituents in place of the 15-hydroxy group of natural PGE_1's. This substitution blocked the metabolic pathway to the corresponding 13,14-dihydro-15-keto compound, normally found for the natural PG's.

---

[1]now at Finnegan-MAT Ltd., Hemel Hempstead; [2]now at Servier R & D, Fulmer, Slough; [3]now at Roche Products Ltd., Welwyn Garden City; [4]now at Hoechst U.K. Ltd., Milton Keynes (& handles correspondence).

* *Abbreviations, besides* GC.— HPLRC, high performance liquid radiochromatography; MS, mass spectrometry (EI, electron-impact mode; NCI, negative-ion chemical ionization); ME - methyl ester, & MO - *O*-methyloxime [rendered as MoMe in art. #A-1]; PFB, pentafluorobenzyl; TMS, trimethylsilyl; PG, prostaglandin.

Fig. 1. Relevant formulae.

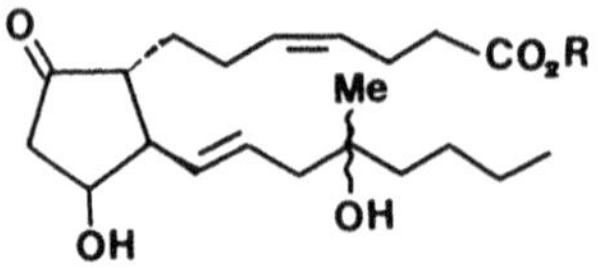

ENISOPROST, R = Me
SC-36067, R = H

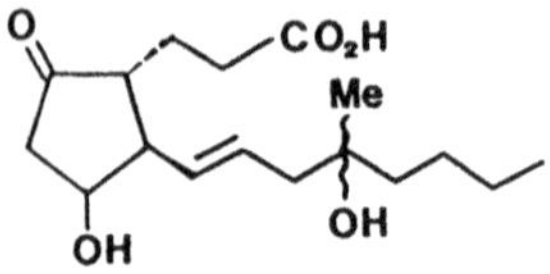

SC-41411

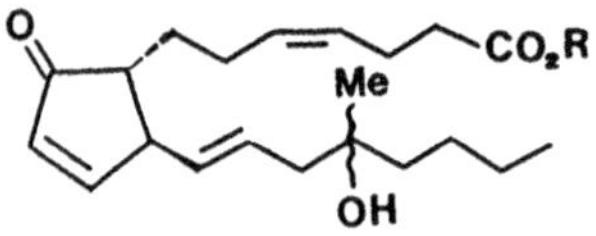

ENISOPROST **A** form, R = Me
SC-36067, **A** form, R = H

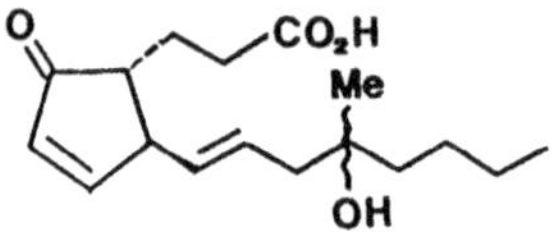

SC-41411, **A** form

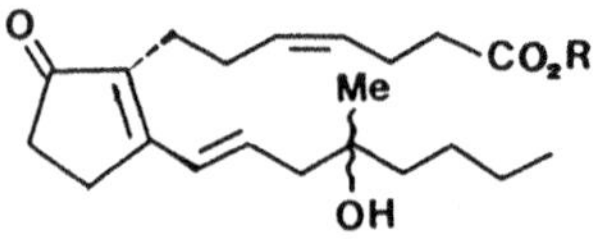

ENISOPROST **B** form, R = Me
SC-36067, **B** form, R = H

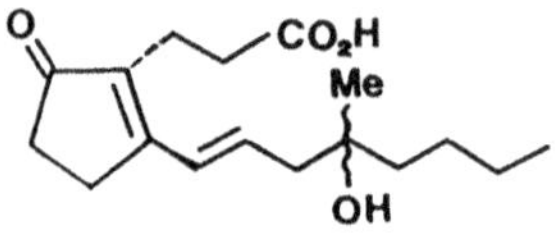

SC-41411, **B** form

Suitable analytical methods were required to investigate the pharmacokinetics and metabolism of enisoprost and its primary metabolite (SC-36067, Fig. 1). The compound posed a number of analytical challenges:

- enisoprost was metabolically unstable in plasma;
- enisoprost and its primary metabolite SC-36067 were expected to be rapidly and extensively metabolized in the organism, by analogy with naturally occurring PG's;
- no sensitive, specific assay was available for either compound;
- each subject would receive only one 450 μg (= 6 μg/kg) dose of the drug, resulting in low concentrations of drug and metabolites;
- an 11β-^{3}H-labelled form of enisoprost would be used, which could result in the formation of tritiated water *in vivo*.

This article describes aspects of the analytical methods developed to investigate the disposition of enisoprost in man.

EXTRACTION, ANALYSIS AND STABILITIES

Reference compounds (Fig. 1).- [^{3}H]Enisoprost, [^{3}H]SC-36067 and [^{3}H]SC-41411 were supplied by G.D. Searle & Co., Skokie, IL, U.S.A. The A and B forms of these compounds were prepared by heating each (0.1 mg) in 0.1 ml, respectively, of 1M HCl or 0.1 M methanolic KOH in 0.9 ml tetrahydrofuran.

Extraction.- Solvent extraction of acidified biological fluids has been the commonly used method for many PG's (Scheme 1, *overleaf*), but is slow, needs quite large solvent volumes, results in emulsion formation and, for enisoprost and its potential metabolites, gave only ~70% recoveries. In contrast, solid-phase extraction on C-18 BondElut columns (Scheme 1) gave >95% recovery of these compounds into a small volume of organic solvent (ethanol), and some 10-20 samples could be processed in parallel using the VacElut system. The aqueous eluates could be combined for each sample and analyzed for tritiated water by azeotropic distillation with toluene in a Dean & Stark apparatus.

Analytical method.- The method had to separate enisoprost from its metabolites and allow estimation of the enisoprost and SC-36067 concentrations in plasma, and of major metabolites in urine. HPLRC was selected for metabolite profiling. The HPLRC properties of [^{3}H]-enisoprost and its potential metabolites (Fig. 1) were investigated on a C-18 Novapak (5 μm) radial compression column in a Z-module (Waters Insts., Harrow, U.K.). The final choice of conditions (cf. Fig. 2) came from trial of a series of gradient elution conditions

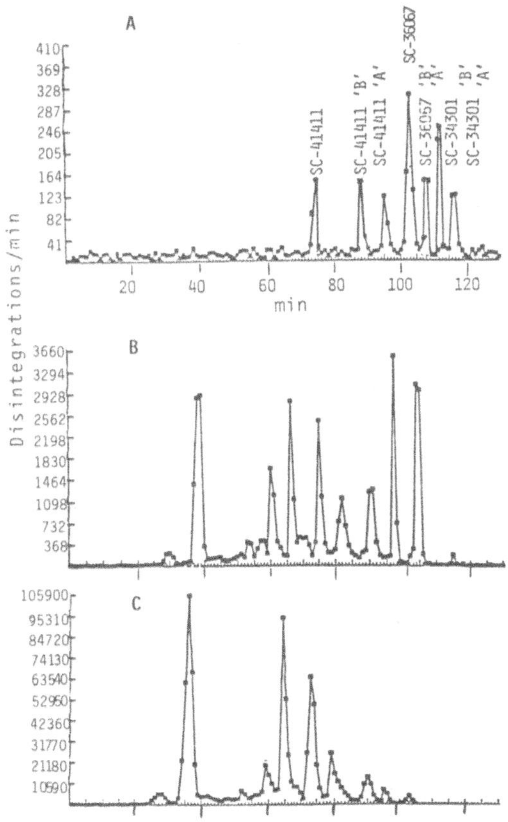

Fig. 2. [^{3}H]Enisoprost HPLRC profiles for **(A)** reference compounds, and for **(B)** plasma and **(C)** urine from a volunteer who had taken the compound orally.

See text for description of column (10 × i.d. 0.8 cm). Mobile phase: methanol/ acetonitrile/acetic acid (49.5:49.5:1 by vol.), 15% → 99% in water/acetic acid (99:1) over 65 min; total flow-rate 1 ml/min. Fractions (0.5 ml) were collected and analyzed for ^{3}H by liquid scintillation counting.
From ref. [6], by permission.

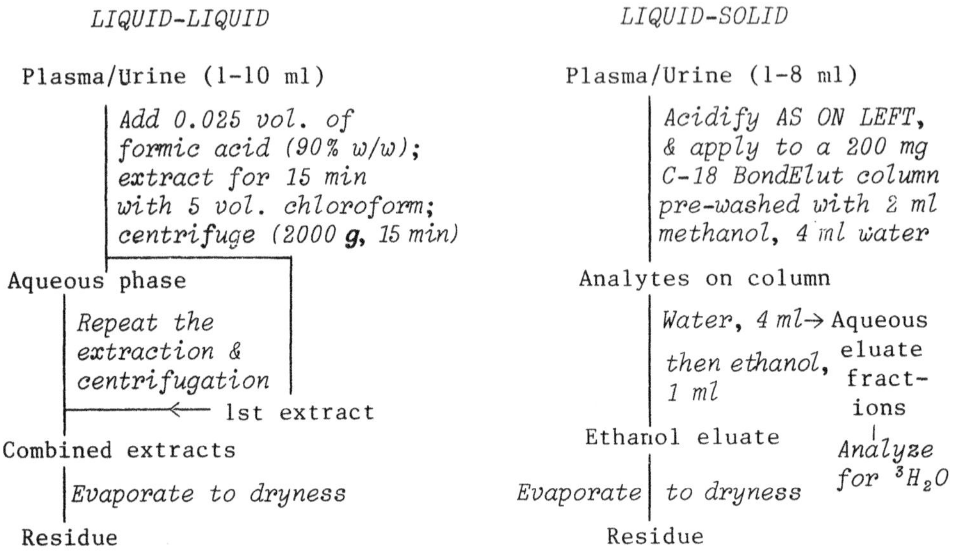

Scheme 1. Alternative extraction procedures for enisoprost and its metabolites.

Table 1. Mean recoveries (±S.D., and n) through the analytical procedure for [^{3}H]enisoprost, [^{3}H]SC-36067, [^{3}H]SC-41411 and ^3H_2O. The plasma containing [^{3}H]enisoprost was kept 3 h at 4° with 0.1 M pyridostigmine bromide present [cf. #NC(E)-8 by R. Whelpton - *Ed.*]

Compound	Plasma	Urine
[^{3}H]enisoprost	91 ±0.6 (6)	91 ±2 (6)
[^{3}H]SC-36067	95 ±1.3 (6)	96 ±3 (6)
[^{3}H]SC-41411	95 (2)	97 (2)
Tritiated water	105 ±0.6 (6)	104 ±1.5 (5)
^3H_2O from [^{3}H]enisoprost	0.4 ±0.2 (5)	0.2 ±0.1 (5)
^3H_2O from [^{3}H]SC-36067	0.8 ±0.4 (5)	0.8 ±0.2 (5)

using water/acetic acid (99:1 by vol.) as the polar component. Good resolution of enisoprost and its potential metabolites was obtained when methanol/acetonitrile/acetic acid (49.5:49.5:1 by vol.) was used as the non-polar component, as shown in Fig. 2 for plasma and urine from a volunteer following a 450 µg (500 µCi) dose of ^{3}H-drug.

Stability and recovery of the reference compounds in urine were evaluated using the solid-phase extraction and HPLRC methods, prior to the metabolism study. Table 1 shows that the potential metabolites were stable in and efficiently recovered from human plasma and urine through the analytical procedures. Enisoprost was stable in urine but was rapidly degraded in plasma at room temperature (~20°) due to hydrolysis to SC-36067 by plasma esterases. This degradation was inhibited by collection of blood into ice-cooled heparinized tubes containing 2 M pyridostigmine bromide (0.1 ml).

METABOLITE ISOLATION AND IDENTIFICATION

The solid-phase extraction and HPLRC procedures were modified to allow, for the major urinary metabolites of enisoprost, **(1) isolation** (cf. **C** in Fig. 2), and **(2) purification** by a second HPLRC procedure; then **(3) identification** was performed by GC-MS after derivatization. The steps were as follows.

- **(1a)** A 500 mg C-18 BondElut column was pre-washed with 5 ml methanol and 10 ml water.
These 500 mg columns allowed large urine volumes to be efficiently extracted (supplier: Analytichem Inc.).
- **(1b)** The column was loaded with 20 ml of urine acidified as in Scheme 1, and washed through with 10 ml water.
- **(1c)** Analytes were eluted with water/methanol/acetonitrile/acetic acid (39.5:29.7:29.7:1 by vol.).
The eluting solvent was the most polar, in the analytical gradient HPLRC method, that would elute all 5 major urinary metabolites, so minimizing the amount of endogenous contaminants co-eluted.

- **(2a)** For further purification, metabolite fractions from (1c) were loaded onto a column as in Fig. 2.
- **(2b)** Elution with a linear gradient was performed over 70 min, collecting 1 min fractions, with increasing concentrations, 1% to 50% v/v, of acetonitrile/acetic acid (99:1 by vol.) in water/acetonitrile/acetic acid (89:10:1). For each metabolite peak the combined fractions were dried down and redissolved in the derivatizing agent.

- **(3a)** Each isolated component was converted to its ME-MO-TMS [7] or PFB-MO-TMS [8] derivative, with final dissolution in dodecane (25 µl) from which 3 µl was taken for on-column injection (OCI-3 injector, SGE Ltd.).
The fused-silica column was 25 m × 0.025 mm, with a 25 µm film of DB-1.
- **(3b)** GC, with coupling to MS, was performed using helium at 7 psi, 0.3 m/sec linear velocity; column at 235° for 2 min, then raised to 300° at 3.5°/min and kept 10 min at 300°.
- **(3c)** MS was performed with alternative ionization modes: EI, 220°, 50 eV, or ammonia NCI, 180°, 120 eV, 26.6 Pa.

GC-MS results obtained from the two types of derivative are complementary. The ME-MO-TMS derivatives of PG's do not yield molecular ions: the ion of highest mass is generally the $[M-CH_3]^+$ ion. This is of low intensity and sometimes absent from the EI mass spectra. However, these derivatives show characteristic fragmentations which allow the assignment of sites of metabolism (Fig. 3). In contrast, the PFB-MO-TMS derivatives yield intense $[M-PFB]^-$ ions under NCI

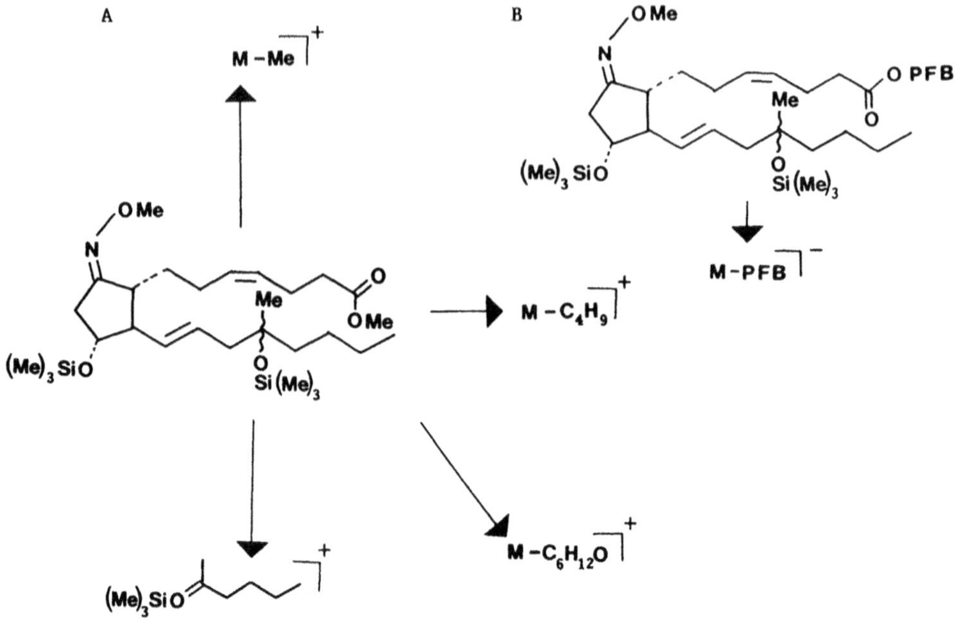

Fig. 3. Key fragmentation pathways for (**A**) ME-MO-TMS derivatives, and (**B**) PFB-MO-TMS derivatives. *From ref. [6], by permission.*

conditions with ammonia as the reagent gas. Essentially little or no fragmentation is seen in the mass spectra of these derivatives. Waddell et al. [8] used the intense $[M-PFB]^-$ ion of the PFB-MO-TMS derivatives for quantitation of prostanoids in biological fluids. We found that this derivative provided complementary mol. wt. information for the enisoprost metabolites which was not available from the ME-MO-TMS derivatives.

Thus, the PFB-MO-TMS derivatives provide mol. wt. information indicating the extent of oxidation or reduction for the metabolites. The extensive fragmentation of the ME-MO-TMS derivatives provides the information to assign the sites of metabolism.

CONCLUSIONS

The combination of solid-phase extraction, RP-HPLRC and GC-MS has provided an efficient, sensitive method to investigate the pharmacokinetics and metabolism of a synthetic PGE_1 analogue. The methods described have been used effectively in a human metabolism study for [^{3}H]enisoprost (see [6] for detailed results). The methods appear to be an improvement on those previously described in the literature, allowing rapid pharmacokinetic and metabolic data to be collected following a single oral dose (450 μg) of the ^{3}H-labelled PG. The methods should, with suitable modifications, be applicable to other synthetic PG's.

References

1. Jones, R.L. (1974) *Biochem. Soc. Trans.* *2*, 1192-1196.
2. Oates, J.A., Sweetman, B.J., Green, K. & Samuelsson, B. (1976) *Anal. Biochem.* *74*, 546-559.
3. Hamburg, M. & Samuelsson, B. (1971) *J. Biol. Chem.* *246*, 6713-6721.
4. Lands, W.E.M. (1979) *Ann. Rev. Physiol.* *41*, 633-652.
5. Polet, H. & Levine, L. (1975) *J. Biol. Chem.* *250*, 351-357.
6. Allan, L.M., Hawkins, A.J., Vose, C.W., Firth, J., Brownsill, R.D. & Steiner, J.A. (1987) *Xenobiotica* *17*, 1233-1246.
7. Maclouf, J., Rigaud, M., Durand, J. & Chebroux, P. (1976) *Prostaglandins* *11*, 999-1017.
8. Waddell, K.A., Barrow, S.E., Robinson, C., Orchard, M.A., Dollery, C.T. & Blair, I.A. (1984) *Biomed. Mass Spectrom.* *11*, 68-74.

#A-5

IDENTIFICATION OF A NEW METABOLITE OF LEUKOTRIENE B_4 BY RP-HPLC AND GC-MS ANALYSIS

Joachim Fauler, Karl-Heinz Marx and ®Volkhard Kaever

Divisions of Clinical Pharmacology and ®Molecular Pharmacology, Department of Pharmacology & Toxicology Medical School, D-3000 Hannover 61, FRG

RP-HPLC with UV detection is the method of choice for rapid identification and quantification of LT's. The large number of structurally related compounds precludes separating all of them within one system. Therefore RP-HPLC systems have been developed to isolate particular groups of compounds such as peptidoLT's or LTB_4 and its isomers. A RP-HPLC system now described allows the separation of a new metabolite of LTB_4 from LTB_4 and its* trans *isomers.*

LTB_4, namely 5(*S*),12(*R*)-dihydroxy-6,14-*cis*-8,10-*trans*-eicosatetraenoic acid, is among the most potent inflammatory mediators produced by inflammatory cells. It has been shown to be a potent chemoattractant to neutrophils, eosinophils and monocytes. Its biological effects include the induction of neutrophil aggregation, lysosomal enzyme release and superoxide anion production. Furthermore, LTB_4 promotes an increase in vascular permeability in the presence of prostaglandins and/or peptidoLT's, the accumulation of neutrophils and the adherence of granulocytes to endothelial cells.

Neutrophils metabolize LTB_4 by ω-oxidation very rapidly to 20-OH-LTB_4 and 20-COOH-LTB_4 (Fig. 1). While the former metabolite possesses some of the biological activities of LTB_4 the latter is completely inactive. Human inflammatory cells other than granulocytes do not metabolize LTB_4 to a significant extent. Interestingly, cultivated rat mesangium cells challenged with synthetic LTB_4 did not form any ω-oxidation products, but the amount of added LTB_4 decreased with time, which suggested metabolism of LTB_4 by a different pathway.

**Abbreviations*.- LT, leukotriene; RP, reversed phase; MS, mass spectrometry; PFB, pentafluorobenzyl; TMS, trimethylsilyl.
General references are given, without citations in the text.

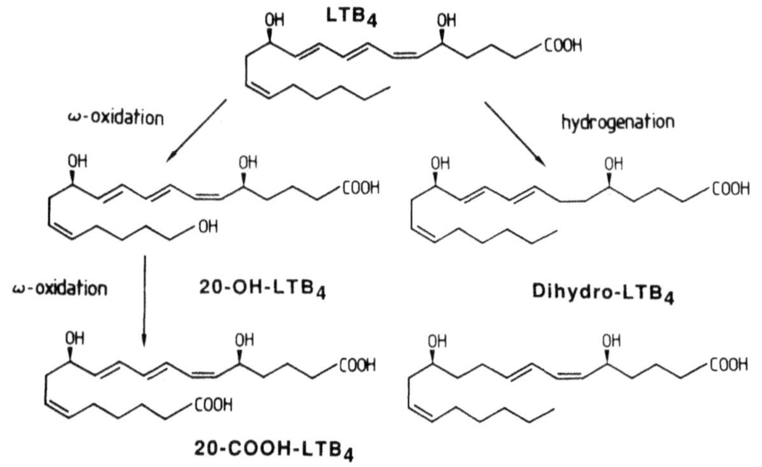

Fig. 1. Metabolism of LTB_4. It can be metabolized by ω-oxidation *(left)* or by a new pathway leading to dihydro-LTB_4.

The present study concerns HPLC approaches for this new metabolite and its characterizaton by GC-MS. In our experience this metabolite was more abundant than the ω-oxidation products in macrophage incubations.

METHODS

Incubation and extraction.- Cultivated mesangial cells (2×10^6 cells; rat) were incubated using a phosphate-buffered solution containing Ca^{2+} (1.5 mM), Mg^{2+} (1.0 mM) and glucose (4.0 mM). Cells were challenged for 4 h with 100 ng synthetic LTB_4.

Solid-phase extraction.- The cell supernatants were acidified to pH 3 with 1 M phosphoric acid, and applied to an ODS-Hypersil Sep-Pak cartridge pre-conditioned with 10 ml of water/methanol (1:1 by vol.). The cartridge was washed with 10 ml water and 10 ml hexane. LTB_4 and its metabolite were eluted with 1.0 ml of acetonitrile/water (4:1) and evaporated under reduced pressure.

RP-HPLC analysis

RP-HPLC was carried out using an LKB solvent delivery and control system and a variable wavelength UV detector (Spectroflow 783, Kratos). Separation was performed on C-18 (ODS) Hypersil 5 µm particles (Shandon Southern Products, Runcorn, U.K.) packed in a stainless steel column 250 × i.d. 4.6 mm, with 1 ml/min flow rate.

Mobile phases were prepared by addition of phosphoric acid to distilled water followed by titration to the desired pH with triethyl-

Fig. 2. RP-HPLC analysis of supernatants from neutrophils (**A**) and from cultivated mesangium cells (**B**) challenged with 100 ng LTB_4 for 1 and 4 h respectively.

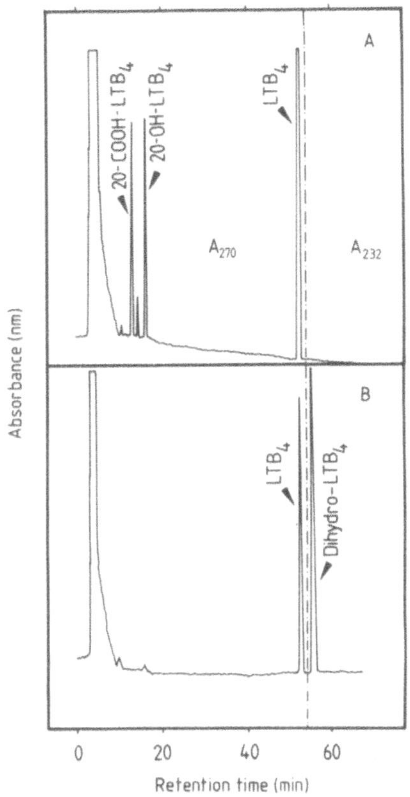

amine; acetonitrile and methanol were added, and the mixtures degassed with He for at least 10 min. The commonly used mobile phase consisting of a methanol/water/acetic acid mixture (68:32:0.02) did not separate the new metabolite from LTB_4. The addition of acetonitrile has been shown to separate LTB_4 from its 6-*trans* isomers in RP-HPLC.

Addition of acetonitrile and simultaneous reduction of the concentration of methanol to give a final 30:30:40:0.02 mixture of acetonitrile/methanol/water/acetic acid resulted in a separation of LTB_4 and its new metabolite by >3 min. The *trans* isomers of LTB_4 do not interfere in this system as they elute 2 min before LTB_4. A further advantage of this system is its ability to separate the ω-products of LTB_4 (Fig. 2). The reproducibility of this RP-HPLC system is very good as retention times of LTB_4 and its new metabolite vary by <20 sec from run to run and by <30 sec from day to day.

Investigation of structure by GC-MS

The new metabolite is distinguishable by being less polar than LTB_4 in RP-HPLC and by having a characteristic UV spectrum with the LTB_4 absorption maximum at 271 nm shifted to 232 nm. For elucidation of the structure of this metabolite, GC-MS was performed.

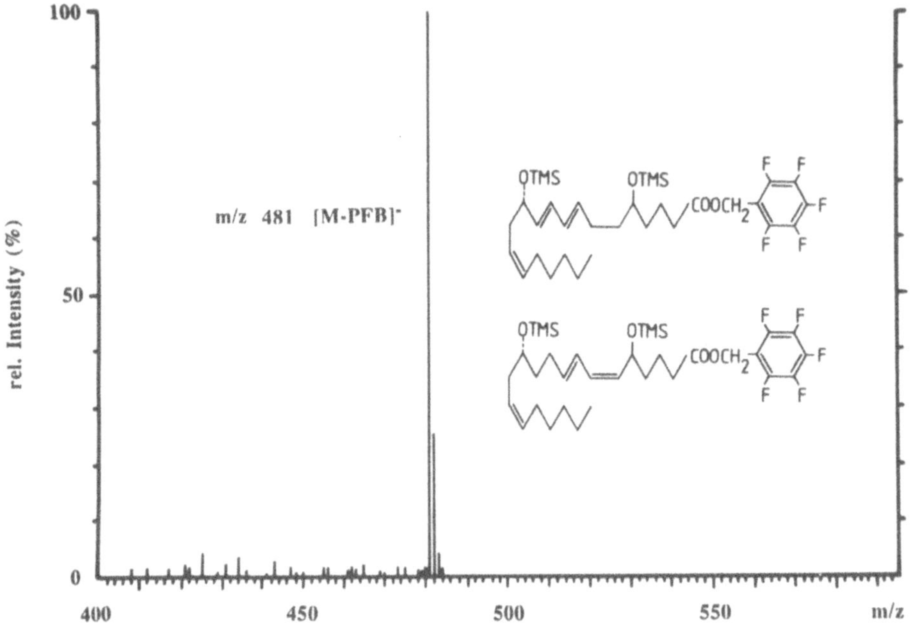

Fig. 3. Negative-ion CI-MS pattern for the metabolite, as PFB ester TMS ether, showing base peak at m/z 481 $(M-PFB)^-$.

The metabolite was converted to the PFB ester TMS ether (see refs. 1 & 2). A Finnigan 9611 GC equipped with a glass-lined splitless injector and a 15 m WCOT fused-silica column was used with He as carrier gas (0.560 hPa pressure at the head of the column). The column was inserted directly into the ion source of a Finnigan MAT TSQ45 triple quadrupole MS. This was operated in the negative-ion CI mode with methane as reagent gas at a pressure of 65 Pa, a source temperature of 110°, an electron current of 0.3 mA, and a multiplier voltage of 1.3 kV.

After injection the GC column was kept at 60° for 2 min, then programmed to 180° at 30°/min, and to 260° at 5°/min. The injection port was kept at 280°, the interface at 260° and the transfer line at 250°.

RESULTS

As PFB esters give high abundances of M-181 (M-PFB) ions, this derivative readily yields information about the mol. wt. of a compound. The corresponding LTB_4-derivative yields a base peak of m/z 479. The new metabolite has a base peak of m/z 481, indicating the metabolite to be a dihydro-LTB_4 (Fig. 3).

CONCLUDING COMMENT

A RP-HPLC system has been described that allows quick and reproducible isolation of dihydro-LTB_4, a new metabolite of LTB_4. The biological relevance of this new metabolite must be further evaluated.

Notable references

1. Kaever, V., Martin, M., Fauler, J., Marx, K-H. & Resch, K. (1987) *Biochim. Biophys. Acta 922*, 337-344.
2. Powell, W.S. (1987) *Biochem. Biophys. Res. Comm. 145*, 991-998.
3. Matthews, W.R., Bundy, G.L., Wynalds, M.A., Guido, D.M., Schneider, S.P. & Fitzpatrick, F.A. (1988) *Anal. Chem. 60*, 349-353.
4. Westcott, J.Y., Stenmark, K.R. & Murphy, R.C. (1986) *Prostaglandins 31*, 227-237.

Editor's note: J. Fauler's Forum abstract, of the nature of a broad sketch that includes RIA of LT's, is reproduced on p. 60.

#A-6

RADIOIMMUNOASSAY OF EICOSANOIDS: ITS APPLICATION IN THE DEVELOPMENT OF ANTI-INFLAMMATORY AND ANALGESIC DRUGS

Frank Carey, Duncan Haworth and Robert A. Forder

Research Department II
ICI Pharmaceuticals plc
Alderley Park, Macclesfield, SK10 4TG, U.K.

Membrane-bound arachidonic acid is a precursor for PG and LT biosynthesis. Members of this family of autocoids have been implicated in pathophysiology including regulation of microvascular blood flow and permeability, leukocyte chemotaxis, hyperalgesia and anaphylaxis. Consequently, in recent years considerable effort has been expended in the search for agents that block the synthesis and antagonize the actions of these autocoids. Most of the biologically stable arachidonic metabolites can now be measured by RIA, and this technique finds widespread use in the search for AI and analgesic drugs. Within a mechanistic approach to new drug development, RIA of eicosanoids can be used at the pre-clinical and clinical stages of drug development and now plays an integral part in hypothesis testing and in the progression of pharmacological curiosities to effective medicines.*

The discovery that aspirin-like drugs inhibit PG biosynthesis paved the way for a mechanistic approach to the discovery of NSAID's. More recently it was discovered that arachidonic acid, the substrate for PG biosynthesis, was also a substrate for LT biosynthesis. Arachidonate metabolites are neither pre-stored nor sequestered in intracellular compartments but are synthesized in response to stimuli which bring about an increase in intracellular free calcium. As such, basal or quiescent levels of PG's and LT's are extremely difficult to quantitate, a problem that has led frequently to confusion in the literature. Accordingly, the experimental paradigm most frequently used in the study of AI and analgesic drugs is the measurement of eicosanoids in response to inflammatory agents, e.g. immune

**Abbreviations*.- (NS)AI(D), (non-steroidal) anti-inflammatory (drug); HETE, hydroxy-eicosatetraenoic acid; LT, leukotriene; PG, prostaglandin; MS, mass spectrometry; RIA, radioimmunoassay; RP, reversed phase.

complex, carrageenin, zymosan, endotoxin and FMLP. Whilst bioassay has played an important role in the discovery of many of the short-lived metabolites of arachidonate, immunoassay (most frequently RIA) has become the method of choice and now fulfils an important role in the discovery of new AI and analgesic drugs. Thus, RIA of eicosanoids can be used at the pre-clinical and clinical stages of drug development to measure pharmacodynamics, investigate mechanism of action and determine selectivity of action with regard to different pathways of arachidonate metabolism. RIA also has a role to play in monitoring the safety and efficacy of this class of drug in the clinic.

DEVELOPMENT OF EICOSANOID RADIOIMMUNOASSAYS

It is beyond the scope of this article to detail the synthesis of immunoreactive conjugates and procedures for raising antibodies (Ab's) to PG's and LT's (see [1] for an overview). Instead, we pass on salient comments based on fully 20 years of combined experience in raising Ab's to eicosanoids. We find that useful Ab's are obtained if covalent incorporation of ligand into the carrier protein is demonstrable, as is particularly important for the more lipophilic arachidonate metabolites (e.g. mono- and di-HETE's) since these will also bind non-covalently to hydrophobic domains in the carrier proteins. Whilst reaction with ***N*,*N***-carbonyldiimidazole or water-soluble carbodiimide has given useful antisera for use with primary PG's, we found this not to be the case with LT's. Instead, we [2] and others [3, 4] have described procedures in which LTB_4 and peptidyl LT's are linked to carrier proteins via a spacing group resulting in antisera with adequate sensitivity and specificity. If a conjugate fails to generate Ab's within 2-3 months of commencement of immunization, in our experience it is unlikely to yield useful antisera. Although a variety of procedures have been described for the separation of Ab-bound and free ligand, simple aqueous buffers in the presence of dextran-coated charcoal are ideal for this purpose.

VALIDATION OF THE RIA PROCEDURE

The oxidative metabolism of arachidonate gives rise to prostanoids and LT's of closely related structures with similar physicochemical properties, and both may be measured by direct assay or following extraction. Whilst the effect of solvents/extraction procedures, linear dilution of immunoreactivity, assay blanks, effect of plasma on ligand binding, etc., are all important aspects of developing a valid RIA procedure, it is essential to characterize the immunoreactive eicosanoid by independent criteria. This may be achieved by extracting the immunoreactive eicosanoid from the biological sample followed by separation/purification by TLC or RP-HPLC, then examining it by RIA or, for structure determination, by GC-MS. Independent corroboration to identify the immunoreactive eicosanoid in question is essential to avoid ambiguous interpretation of data.

These comments do not reflect an analyst's obsession with methodology but have more profound pharmacological consequences. We cite just two examples.

(1) It was a long-held view that selective inhibition of TX synthetase would give rise to elevated plasma levels of prostacyclin - the so-called 'endoperoxide steal or redirection hypothesis'. Indeed, this was shown to be the case by several workers [5-7]; however, only when more rigorous studies were performed was it shown that the apparent increase (× 2.6) in prostacyclin (measured as 6-oxo-$PGF_{1\alpha}$) was due to elevation of $PGF_{2\alpha}$ (× 4.1) and crossreactivity in the RIA [8]. Since prostacyclin and $PGF_{2\alpha}$ may exert opposing actions on vascular patency, these findings have more far-reaching consequences than mere analytical vagaries.

(2) Similarly we demonstrated that stimulation of arachidonate metabolism in whole blood led to elevation of 12-HETE and LTB_4 levels. Both these molecules contain a hydroxyl group at C-12 of the eicosatetraenoic acid backbone. When plasma LTB_4 was measured directly by RIA, particularly in rats and mice, a significant proportion of the immunoreactivity was due to elevated levels of 12-HETE [9]. Since these eicosanoids are products of two different enzymatic pathways, direct RIA of LTB_4 may give misleading information on the specificity of putative 5-lipoxygenase inhibitors [10]. However, prior separation of 12-HETE fractions followed by RIA may be used to distinguish between selective 5-lipoxygenase inhibitors which interact directly with the enzyme and anti-oxidant or free-radical-scavenging types which may be less specific.

ADVANTAGES OF RIA IN THE DRUG-DISCOVERY PROCESS

Provided that due attention is paid to identifying the immunoreactive eicosanoids being measured and the general guidelines outlined in [1] are followed, then RIA of eicosanoids has considerable potential in the search for new AI/analgesic drugs. In cellular assays it obviates the need to pre-label cells with radioactive arachidonic acid and negates concerns over equilibrium labelling of cellular phospholipid pools. The use of RIA enables the investigator to measure mass and therefore determine the relative contribution of 5-lipoxygenase and cyclo-oxygenase metabolites to the response or pathophysiology under study. It follows that the extent of 're-direction' of arachidonate metabolism, from one pathway to another, may also be assessed following selective inhibition of either 5-lipoxygenase or cyclo-oxygenase. The latter, if it occurs, may have important pathological consequences.

Whilst bioassay, GC-MS, HPLC and TLC procedures all have a role in the identification/measurement of eicosanoids, RIA offers specificity, sensitivity, relatively low cost and the potential for high throughput. However, more importantly it has enabled the pharmacodynamics of 5-lipoxygenase and cyclo-oxygenase inhibitors to be measured in laboratory animals [10, 11]. It is hard to foresee how this might otherwise have been achieved on a scale and with the throughput necessary to support a drug-discovery programme.

APPLICATION OF RIA TO DRUG DISCOVERY

In pre-clinical research the application of RIA of eicosanoids falls into two areas: compound selection and hypothesis testing. During the early phases of a drug-hunting programme, compounds are subjected to an increasingly complex hierarchy of tests to provide data on a number of aspects: enzyme inhibition in broken-cell preparations; cell penetration; cellular enzyme inhibition in complex matrices (e.g. blood or inflammatory exudate); and pharmacodynamics by measurement of eicosanoids in an assay of functional *in vivo* models of hyperalgesia and inflammation. In using a mechanistic approach to drug research, RIA may be used at all stages of the screening cascade and plays an important role in determining the predictive nature of *in vivo* experimental models.

Such a system of tests suitable for the testing of dual cyclo-oxygenase/5-lipoxygenase inhibitors with analgesic/AI properties is described below. In the primary *in vitro* test, compounds are evaluated for their inhibitory effects on 5-lipoxygenase and cyclo-oxygenase to provide the investigator with quantitative structure-activity and selectivity data. Compounds which survive this analysis may then be tested in whole-cell assays to determine whether they can penetrate cells and inhibit enzymes of intracellular origin. Macrophages, PMN's, mast cells or platelets, stimulated with a calcium ionophore or other suitable agonists are ideal for this purpose. Such assays are usually performed under conditions of serum- or plasma-protein depletion and so may be followed by whole-blood assays [12] to determine the effect of plasma protein binding prior to *in vivo* studies.

In such a scheme the first *in vivo* test combines measurement of the anti-nociceptive properties of compounds with effects on plasma eicosanoid levels. The details of this test have been described elsewhere [13]. It is important at the outset that the pharmacologist and medicinal chemist retain a flexible attitude as to which compounds progress to the *in vivo* tests. It may transpire, for example, that optimization of activity at the *in vitro* level may inadvertently select for compounds with poor bioavailability (or *vice versa*). This can be checked by initially profiling a range of structural types (and each new type) in both *in vitro* and *in vivo* tests. In the first instance compounds may be administered parenterally and the nociceptive test completed within 1 h of administration so as to avoid overlooking compounds with a short biological half-life. Since measurement of eicosanoids in mouse peritoneal lavage fluid proved

Table 1. Phenylbenzoquinone-induced abdominal constriction: effect of eicosanoid inhibitors on nociceptive response (Noc. resp.) and plasma eicosanoids. Values represent % of control, ± S.E.M. (n = 6). The Comment column indicates for which enzyme the agent is an inhibitor.

Compound (& dose, mg/kg)	Noc. resp.	Plasma TBX_2	Plasma LTB_4	Inhib. comment
Indomethacin (10, s.c.)	94 ±5	97 ±2	10 ±8	cyclo-oxy'ase
Flurbiprofen (10, s.c.)	86 ±8	100 ±10	-54 ±20	cyclo-oxy'ase
BW755C (100, s.c.)	65 ±8	67 ±9	47 ±6	dual
Nafazatrom (100, s.c.)	-15 ±22	-19 ±30	33 ±7	5-lipoxy'ase

difficult, we opted to measure plasma TXB_2 and LTB_4 as indices of cyclo-oxygenase and 5-lipoxygenase inhibition. Thus, data on the anti-nociceptive properties of compounds together with measurement of *ex vivo* cyclo-oxygenase and 5-lipoxygenase inhibition were obtained simultaneously.

It was not our intention at this stage to determine whether LT's were involved in hyperalgesia but rather to establish whether compounds retained inhibitory activity *in vivo*. Table 1 summarizes the properties of 4 standard eicosanoid inhibitors in this test. Clearly, as expected, both indomethacin and flurbiprofen inhibited cyclo-oxygenase activity but were without effect on 5-lipoxygenase activity. In dose-response studies the IC_{50} for thromboxane inhibition was lower than that required to inhibit abdominal constrictions, as would be predicted if cyclo-oxygenase metabolites mediate hyperalgesia. The prototypic inhibitor of both oxygenases, BW775C, inhibited plasma TXB_2 and LTB_4 whereas nafazatrom inhibited only LTB_4 levels. The relative non-specificity of BW775C was revealed by inhibition of 12-HETE also.

This primary screen provides a simple filter to weed out compounds that fail to show the required inhibitory effects on eicosanoid biosynthesis. In addition, the use of mice in place of rats reduces (6- to 10-fold) the demand for compound at this stage of the screening cascade. Refinements of this test include different routes of administration and dosing regimes to determine the biological half-life of the compound.

As part of our screening and hypothesis-testing efforts we have attempted to elucidate the role (if any) that LT's play in the development of experimental hyperalgesia. We have focused out efforts on the yeast-inflamed rat paw model, and our approach has been as follows:

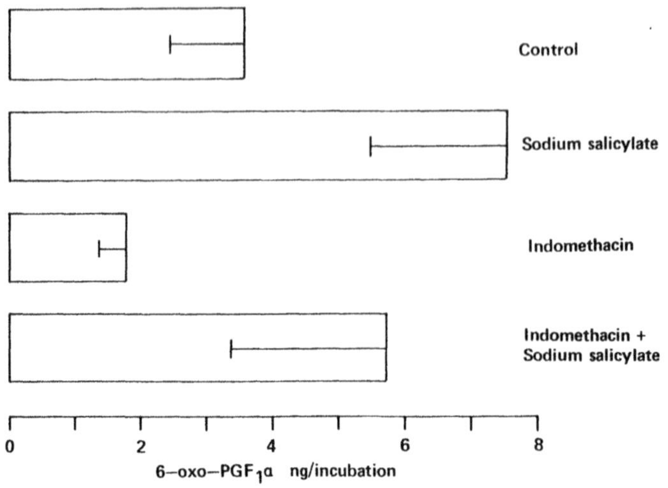

Fig. 1. Effect of sodium salicylate and indomethacin, alone and in combination, on gastric biopsy PG biosynthesis (amplified in text).

- (1) determine whether the direct sub-plantar application of LT's causes hyperalgesia;
- (2) study the effects of prototypic inhibitors of 5-lipoxygenase and dual 5-lipoxygenase/cyclo-oxygenase inhibitors on experimental hyperalgesia and inflammation;
- (3) seek evidence for the involvement of LT's in hyperalgesia and inflammation by measurement of LT's (and prostanoids) in inflammatory exudates;
- (4) determine the pharmacodynamics of 5-lipoxygenase and dual 5-lipoxygenase/cyclo-oxygenase inhibitors by measurement of plasma and inflammatory exudate eicosanoids.

Temporal studies of the levels of cyclo-oxygenase and 5-lipoxygenase products revealed that at the time of onset of hyperalgesia both LTB_4 and PGE_2 were present in the inflammatory exudate. The identity of immunoreactive LTB_4 was further characterized by combined RP-HPLC/RIA; the immunoreactive LTB_4 was shown to co-elute with authentic standard [14]. Furthermore, pharmacodynamic studies using BW755C, a prototypic dual cyclo-oxygenase/5-lipoxygenase inhibitor, revealed that both plasma and inflammatory exudate LTB_4 levels were reduced following administration of this compound. Unlike the NSAID's, which reduce only hyperalgesia in this acute model, BW775C also reduced paw swelling, suggesting a possible role for LT's in oedema formation.

THE ROLE OF RIA IN PREDICTING PRE-CLINICAL TOXICITY

One of the predicted toxicological consequences of receiving a NSAID is faecal blood loss arising from gastro-intestinal erosions. Debate continues as to whether inhibition of mucosal PG biosynthesis *per se* or other properties of these agents underlies their toxicological effects on the gut. Taking gastric mucosal biopsies and measuring eicosanoid biosynthesis following compound administration enables the investigator to assess their propensity to inhibit PG and LT biosynthesis. Such studies allow compounds with low gastric irritant potential to be identified early in a 'drug-hunting' programme. We have demonstrated (see Fig. 1) that prostacyclin is the major cyclo-oxygenase product biosynthesisized by rat gastric biopsies (~5-fold PGE_2 levels) and that the well documented gastric protective effects of sodium salicylate (200 mg/kg p.o. 30 min before indomethacin) upon gastric damage induced by indomethacin (2 mg/kg p.o.) was accompanied by a reduced inhibition of PG biosynthesis. This type of study, involving *ex vivo* RIA of eicosanoids, may be extended to other organs of predicted NSAID toxicity, e.g. kidney, enabling new agents with a favourable pharmacodynamic profile to be identified.

APPLICATION OF RIA TO CLINICAL STUDIES

The emphasis in this article has been on a mechanistic approach to new drug development, particularly the inhibition of PG and LT biosynthesis. Application of the techniques outlined enables the pharmacodynamics of selective 5-lipoxygenase and dual cyclo-oxygenase/5-lipoxygenase inhibitors to be monitored initially in human volunteer studies. Direct RIA of LTB_4 in human plasma does not suffer the problems encountered in rat and mouse plasma, and levels of LTB_4 are much higher than found in rodents. Fig. 2 illustrates an example of this application, showing the inhibition of LTB_4 due to nordihydroguiaretic acid when added *in vitro*.

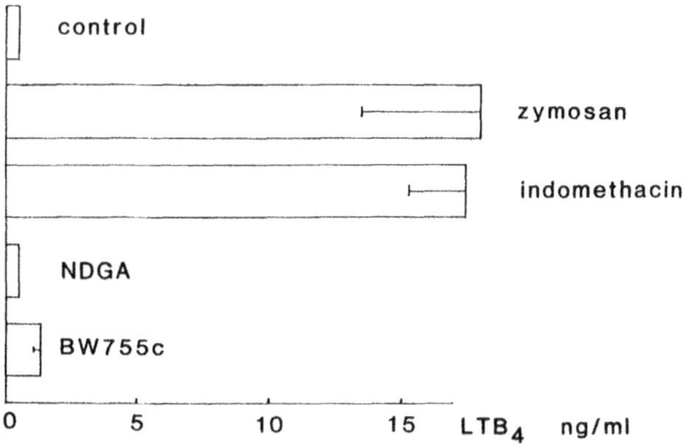

Fig. 2. Effect of indomethacin (5 µg/ml), nordihydroguiaretic acid (5 µg/ml) and BW755C (50 µg/ml) on zymosan-stimulated LTB_4 biosynthesis in human blood.

In the clinical setting the effects of 5-lipoxygenase and dual cyclo-oxygenase/5-lipoxygenase inhibitors can be measured *ex vivo*, using a range of stimulants to promote eicosanoid biosynthesis. This approach should prove particularly useful in studying the pharmacodynamic effects of new drugs in elderly patients in whom drug metabolism may be impaired. The procedures outlined may be used to titrate the therapeutic dose of a drug, to monitor compliance and to assess the effect of drug-drug interactions on eicosanoid biosynthesis. We can anticipate that new agents based on such a mechanistic approach will soon be undergoing clinical evaluation.

Acknowledgement

We thank Miss Heather Downie for preparing this manuscript.

References

1. Granström, E. & Kindahl, H. (1979) *Adv. Prostagl. Thrombox. Res.* 5, 119-210.
2. Carey, F., Forder, R.A., Gibson, K.H. & Haworth, D. (1988) *Prostaglandins Leukotrienes & Med.*, in press.
3. Young, R.N., Kakushima, M. & Rokach, J. (1982) *Prostaglandins* 23, 603-613.
4. Levine, L., Morgan, R.A., Lewis, R.A., Austen, K.F., Clark, D.A., Marfat, A & Corey, E.J. (1981) *Proc. Nat. Acad. Sci.* 78, 7692-7696.
5. Vermylen, J., Defreyn, G., Carreras, L.O., Machin, S.J., Van Schaeren, J. & Verstraete, M. (1981) *Lancet i*, 1073-1075.
6. Maguire, E.D. & Wallis, R.B. (1983) *Thrombosis Res.* 32, 15-27.
7. Randall, M.J. & Wilding, R.I.R. (1982) *Thrombosis Res.* 28, 607-616.
8. Carey, F. & Haworth, D. (1986) *Prostaglandins* 31, 47-59.
9. Carey, F., Forder, R.A. Haworth, D. & Woollard, P.M. (1987) *Biochem. Soc. Trans.* 15, 430-431.
10. Carey, F., Forder, R.A. & Haworth. D. (1987) *Br. J. Pharmacol.* 91, 404P.
11. McMillan, R.M., Millest, A.J., Proudman, K.E. & Taylor, K.B. (1986) *Br. J. Pharmacol.* 87, 53P.
12. Carey, F. & Forder, R.A. (1986) *Prostaglandins Leukotrienes Med.* 22, 57-60.
13. Carey, F., Haworth, D., Edmonds, A.E. & Forder, R.A. (1988) *J. Pharmacol. Meths.*, in press.
14. Haworth, D. & Carey, F. (1985) in *Inflammatory Mediators* (Higgs, G.A. & Williams, T., eds.), Macmillan, Basingstoke, pp. 37-45.

#NC(A)

NOTES and COMMENTS relating to

EICOSANOIDS AND RELATED ANALYTES

See pp. 60-63 for material that gives perspective,
and p. 61 for supplementary analytical refs.

Comments relating to particular contributions:

#A-1 to -3 & -6, and NC(A)-1: p. 59
#NC(A)-2: p. 60

#NC(A)-1

A Note on

DEVELOPING METHODS FOR A THROMBOXANE SYNTHETASE INHIBITOR AND A PROSTACYCLIN ANALOGUE IN BIOLOGICAL FLUIDS: PROBLEMS AND SOME SOLUTIONS

M.V. Doig and R.A. Clare

Analytical and Clinical Pharmacokinetic Services
Department, Wellcome Research Laboratories
Beckenham, Kent BR3 3BS, U.K.

PG's[⊗] and TX's have a wide range of biological activities including stimulation of muscles, dilation of small arteries, lowering of blood pressure, inhibition of gastric secretions and platelet aggregation. They are also implicated in imflammatory reactions, kidney function and immune responses [1]. Several articles in this vol. and in Vol. 16 deal with the determination of PG's and TX's and investigation of their biosynthesis. To define and elucidate their role in disease states, various inhibitors and analogues have been synthesized. These likewise have to be quantified at low levels without interference from endogenous compounds. Our experiences in developing two such methods are now described. In the first, solid-phase extraction problems arose.

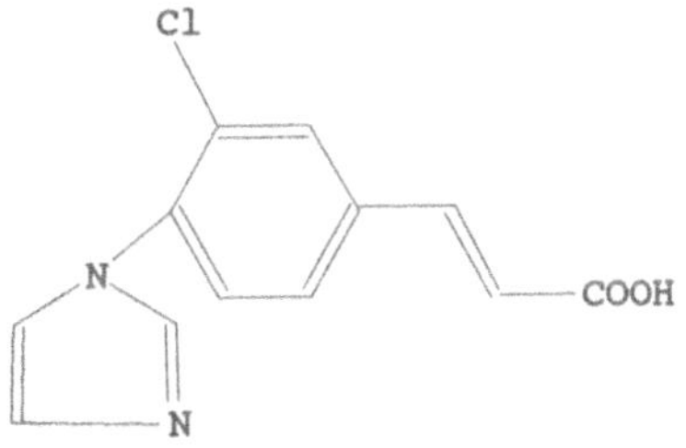

GC ASSAY FOR A TX-SYNTHETASE INHIBITOR

The inhibitor, A287, has the structure shown. The i.s. used, A491, was the Z form of A287 without the chlorine atom.

Problem 1: derivatization and chromatography.- The only derivatization that worked routinely was methylation using diazomethane.

[⊗]PG, prostaglandin; TX, thromboxane; i.s., internal standard; DEA, TEA: di-, tri-ethylamine; TMS, trimethylsilyl; WCOT, wall-coated open tubular.

Fig. 1. GC trace showing A287 (peak E287) after oral dosing (dog), extraction of plasma and derivatization (see text).

HP5730 GC assembly with NPD. Column: DB5, 15 m × 0.53 mm i.d.; oven at 210°, injection port 260°, detector 300°. Voltage 16 V. Carrier: N_2, 25 ml/min; detector gases: H_2, 3 ml/min; air, 30 ml/min.

Z287 is a metabolite, and Z491 is the i.s.

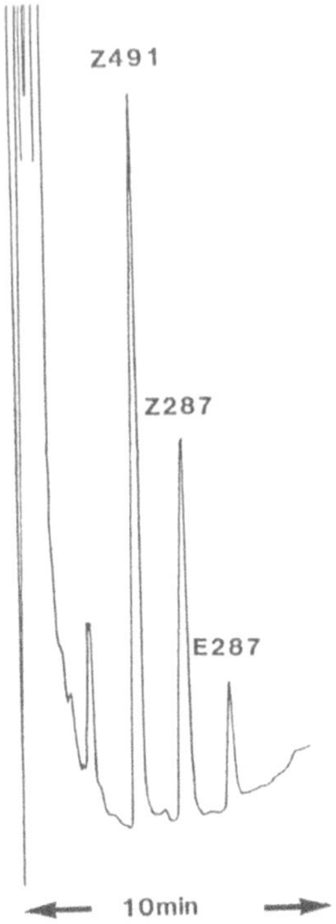

Even after derivatization, bad tailing effects were seen on packed GC columns even with commercial packings such as 3% SB2100DB (Supelco), specially deactivated for use with basic compounds. Tailing was suspected to be due to interactions between the imidazole group and the free silanols and acidic groups on the glass column and silica support. The use of cross-linked WCOT fused silica columns completely solved this problem (Fig. 1).

Problem 2: sample clean-up.- Since solvent extraction was not specific enough for this assay when using GC-NPD, solid-phase extraction was investigated with some interesting results. Initially a wide range of Bond-Elut cartridges were tried: C-18, C-8, C-2, phenyl, cyanopropyl and SAX. After i.s. addition the plasma was acidified to pH 3 by adding 0.6 M acetic acid. All cartridges retained A287 but no organic solvent would elute it with any acceptable level of recovery. Again ionic interactions were suspected, so we tried washing with 2% (v/v) DEA in methanol. This was successful, but the presence of DEA in the final extract was unacceptable for GC with NPD.

The final method developed involved loading the acidified plasma sample onto a phenyl Bond-Elut cartridge that had been pre-washed with methanol and water. The cartridge was then washed with the following solvents (3 ml of each): water, methanol, ethyl acetate, hexane, 2% (v/v) DEA in hexane, hexane, and ethyl acetate. The A287 was then eluted in 3 ml methanol with a recovery of >80% (Fig. 1).

Subsequent to our presentation of this method, other ways of masking the activity of the free silanols present in solid-phase extraction cartridges have been suggested (see 'Notes' on cartridge use in **#E**, this volume, also [2, 3]). These included replacement of the water in the pre-wash with 1% (v/v) aqueous TEA or with 0.1 M ammonium acetate, or elution of the compound with an acidified solvent. For readers who have also experienced such extraction problems, another pertinent article in this vol. is that by McDowall et al. (#D-5).

GS-MS ASSAY FOR A PROSTACYCLIN ANALOGUE

Sample preparation.- Acidified plasma, after addition of i.s., was loaded onto a pre-washed C-18 Bond-Elut cartridge. This was then washed with 1 ml water and 2 ml 10% (v/v) aqueous methanol and, after vacuum-drying, 2 ml '40/60' petroleum ether. Finally the compound was eluted in 1 ml methyl formate, and derivatized to form the methyl ester TMS-silyl ethers using diazomethane and bis(TMS)-trifluoroacetamide. Analysis was by GC-MS with selective ion monitoring.

The problem.- Blank plasma manifested interferences which limited the sensitivity of the assay and were identified as 'cholesterol-related'. After unsuccessful trial of different conditions for sample preparation and derivatization, improvement of the GC was attempted. With three different techniques as listed with the instrumentation employed, the following values were obtained for limit of sensitivity on-column (and per ml plasma):
- packed-column GC-MS with HP5730GC/VG16F MS, 500 pg (5 ng);
- capillary-column GC-MS with HP5890GC/HP Mass Selective Detector, 40 pg (1 ng);
- packed-trap-capillary or MDSS GC-MS with Packard 438AGC/VG16F MS, 40 pg (200 pg).

In the packed-trap capillary GC technique [4] as finally adopted, the first column was a 15 m × 0.53 mm DB-1 WCOT fused silica column, and the second a 15 m × 0.25 mm OV-17 WCOT fused silica column. Cold trapping between the two columns was achieved using liquid nitrogen. This technique was very successful, and is highly recommended for resolving co-eluting interferences present in extracts from biological samples.

References

1. Salmon, J.A. & Flower, R.J. (1983) in *Hormones in Blood*, 3rd edn., Vol. 2 (Gray, C.H. & James, V.H.T., eds.), Academic Press, New York, pp. 237-319.
2. McDowall, R.D., Pearce, J.C. & Murkitt, G S. (1986) *J.Pharm. Biomed. Anal. 4*, 3-21.
3. Vendrig, D.E.M.M., Holthuis, J.J.M., Erdelyi-Toth, V. & Hulshoff, A. (1987) *J. Chromatog. 414*, 91-100.
4. Deans, D.R. (1968) *Chromatographia 1*, 18-22.

#NC(A)-2

A Note on

THE STEREOCHEMICAL ANALYSIS OF MONOHYDROXYFATTY ACIDS, PARTICULARLY 12-HYDROXY-5,8,10,14-EICOSATETRAENOIC ACID

P.M. Woollard[®]

Institute of Dermatology
St. Thomas's Hospital, London SE1 7EH, U.K.

Monohydroxy acids are major lipoxygenase-like metabolites in extracts from the skin of psoriatic patients. An interest in the origin and therefore mechanism of formation of these compounds stimulated investigation of their stereochemistry. Chiral methods using HPLC have been developed for separation and quantification of the hydroxyl enantiomers as well as for their multi-mg preparation.

Diastereomeric dehydroabietyl (DHA) urethane derivatives of racemic hydroxyfatty acid methyl esters were prepared by reaction with DHA isocyanate (0.38 M) in the presence of 4-dimethylaminopyridine (0.26 M) at 60° for 65 h, and were partially purified using silic SepPak mini-columns eluted with hexane/diethyl ether mixtures. The DHA isocyanate reagent was synthesized by reaction of the amine hydrochloride with trichloromethylchloroformate.

Separation of the diastereomeric derivatives was carried out initially on a silica Spherisorb S5W column with hexane/isopropanol/methanol mixtures. However, the column was easily overloaded, and retention times were inconstant. These problems were eliminated, and improved separation achieved, by use of a Spherisorb Cyano column eluted with hexane/tert.-butyl methyl ether mixtures. Using this system it was possible to separate all unsaturated racemic hydroxyfatty acid derivatives tested. However, the presence of unsaturation appeared to be necessary for their resolution. Analysis of ng amounts of ^{14}C-labelled hydroxyfatty acids and accurate estimation of enantiomeric composition was also feasible. The use of semi-preparative columns facilitated larger-scale formation of the DHA urethane derivatives which, after deprotection, yielded parent compounds of >97% stereochemical purity. More recently, the use of HPLC employing chiral phases has simplified preparative and analytical procedures with consequent increases in yield and sensitivity.

[®]now at Wellcome Research Laboratories, Beckenham, Kent BR4 OPW.

Examination of 12-hydroxy-eicosatetraenoic acid from psoriatic skin has revealed that the major isomer present is the 12(*R*)- and not the 12(*S*)-hydroxyl enantiomer as previously thought.

Ref. added by Senior Editor to foregoing Forum Abstract:-

Woollard, P.M. (1986) *Biochem. Biophys. Res. Comm. 136*, 169-176.- "Stereochemical differences between 12-hydroxy-5,8,10,14-eicosatetraenoic acid in platelets and psoriatic lesions."

Comments on material in #A

Comments on #**A-1**, C. Robinson - HPLC STUDY OF PGD_2 METABOLISM

Remarks by J.C. Frölich.- On analysis of $PGF_{2\alpha}$ in urine by GC-MS we consistently saw a small peak which preceded it and was almost identical in mass spectrum; we wonder if it could be $PGF_{2\beta}$ (and ask whether Robinson would agree to **this** term rather than his term, $9\alpha,11\beta$-PGF_2). **Reply by Robinson.**- This compound could indeed be observed in human urine as a kidney product if PGD_2 11-ketoreductase is indeed present in human kidney and PGD_2 is generated by this organ.- Cf. a **remark by Frölich** that his group had found by GC-MS-MS some renal excretion of unchanged PGD_2 which, in view of the high enzyme levels found by Robinson, could be the source of urinary $PGF_{2\beta}$. **Robinson, answering M.V. Doig.**- The differences seen in metabolic profile between i.v. and intranasal administration of PGD_2 probably reflect merely the different doses actually administered. **Robinson, answering F. Carey.**- We have had no success in trial of several strategies to raise antisera to PGD_2, using a variety of carrier proteins and conjugating agents.

Comments on #**A-2,** J.C. Frölich - EICOSANOIDS BY GC-MS/MS
& #**A-3**, M. Höller - PG's AND TXB_2 IN PERFUSED RAT BRAIN

J.C. Frölich, replying to F. Carey.- We have insufficient data to indicate what are 'normal' values for 6,15-dioxo-$PGF_{1\alpha}$ in peripheral human venous blood. **Carey** mentioned having found, with an antibody to this compound, levels of <5 pg/ml, below the detection limit of the RIA. **Frölich, answering Carey,** said there was no evidence for the existence of 6-oxo-PGE_1 as a PGI_2 metabolite in man. **Höller, answering D. Wilkinson and D. Dell** who asked about extraction recoveries: the value of ~88% was obtained by adding standards to the freeze-dried powder, not to the tissue prior to freezing since metabolizing enzymes could have caused trouble. **Remark by Wilkinson.**- The measured levels may be very different from the true levels due to the time taken to isolate and prepare the brain for evaluation. I suggest that a better design of standardization would be helpful.

Comments on #**A-6,** F. Carey - RIA OF EICOSANOIDS
& #**NC(A)-1**. M.V. Doig - TX SYNTHETASE INHIBITOR & PG ANALOGUE

F. Carey was asked by Frölich about the non-effect of benoxaprofen on lipoxygenase pathway metabolites. **Response.**- Our data showed that at anti-hyperalgesic doses benoxaprofen failed to inhibit calcium-ionophore-stimulated LTB_4 formation in rat and mouse blood, in accordance with a report by Salmon and co-workers in 1985. **Carey** felt it would be interesting to pursue a suggestion put to him, to study different isomers of NSAID's for their effects on lipoxygenases.

Concerning the nature of the new LTB_4 metabolite that he had found **(question from Frölich)**, he reckoned there were two possible structures: 5(*R*),12(*S*)-dihydroxyeicosa-xxx,14*Z*-trienoic acid where xxx is 6*Z*,8*E* or 8*E*,10*E*. **Remark by R. Woestenborghs to M.V. Doig:** for your Compound A287 I would expect ECD to give a better response than NPD. **Reply:** no better, and no ECD-responsive i.s. was available.

Comments on #**NC**(**A**)**-2**, P.M. Woollard - HETE STEREOCHEMICAL ANALYSIS

Reply by P.M. Woollard to A.J. Hutt, who had asked which method he favoured for bioanalysis: I have not tried a chiral column (costly!) with samples of biological origin; with the profens we form diastereoisomers and separate them by achiral NP-HPLC. **F. Carey** wondered whether it is via lipoxygenase that 12(*R*)-HETE is formed. **Reply.-** This is possible, as is a cytochrome P-450 route; at present we lack decisive evidence. **Other responses by Woollard.-** (1) [in reply to **R. Heath**] Unlike the biologically active 12-HETE, 13-HETE (which is produced in ~5-fold larger amount) is not a neutrophil chemoattractant; it merely causes very slight erythrema when applied to skin. (2) We did not find any of the hydroperoxy compounds in scale extracts.

A Forum abstract which gives analytical perspective (cf. art. #A-5):

MEASUREMENT OF 5-LIPOXYGENASE PRODUCTS BY HPLC AND RIA

- J. Fauler and J.C. Frölich (Hannover Medical School)

HPLC and RIA are the two major methods used for the analysis of 5-lipoxygenase products. The enzyme catalyzes the dioxygenation of arachidonic acid. This initiates the transformation into a large number of structurally related compounds. Despite the fact that arachidonic acid is the common precursor, chemical properties of the different 5-lipoxygenase products vary considerably.

In the past, various HPLC systems have been used to separate structurally related compounds [*1*]. RP-HPLC remains the method of choice for the analysis of 5-lipoxygenase products because of its simplicity, its high recovery and its great versatility. Most 5-lipoxygenase products contain strong chromophores. On the other hand, immunoassay techniques have been developed for analysis of LTC_4 [*2*] and LTB_4 [*3*]. Their sensitivity is in the range 10-100 pg. However, the specificity of these assays is generally sufficient to permit direct measurement of these leukotrienes and their 11-trans isomers in biological samples. Consequently only HPLC analysis followed by an immunoassay allows a specific and sensitive quantitation of 5-lipoxygenase products in biological samples.

Refs.: [*1*] = W.S. Powell, & [*3*] = J.A. Salmon, *as cited on opposite p.*; [*2*] = L. Levine, *p. 339 in* [3] *that appears following opposite p.*

Table 1. Examples of chemical assay of eicosanoids in biological samples: **inc** = incubate, usually of leukocyes (typically macrophages). Usual abbreviations for eicosanoid types (pep = sulphidopeptide); *r* signifies radiotracers employed. LL = liquid-liquid extraction; deprot. = deproteinization; cpGC = capillary GC; C-18 or NP = HPLC (UV, absorbance detector). *Refs. are overleaf.*

Analyte(s) & sample type	Extraction/separation procedures	Ref.
6-keto-$PGF_{1\alpha}$ etc., **inc**:	LL; derivatize; cpGC-MS(NCI)	5
*r*Various PG's, **myocytes**	LL; TLC (study of synthase/EGF/aspirin)	6
PGE_2, **gastric mucosa**:	LL; derivatize; NP, UV	7
PG's, TXB_2, pepLT's, **inc**: (effective & quick procedures)	deprot.(EtOH); extract res.; C-18 (radial compr^n best) & grad. (1 run suffices), UV	8
*r*Prostacyclins, **perfused kid**.:	LL; TLC or C-18 (epoxidation study)	9
LTB_4, **inc**:	C-18; C-18 with ClO_4^- & acid; electrochem.	10
pepLT's, **inc**.:	reduce, cleave, derivatize; cpGC-MS	11
pepLT's, **inc**.:	C-18 [oxalic to 'deionize'], UV	12
*r*pepLT's, **inc (granulocytes)**:	LS - EDTA-washed C-18 (XAD disfavoured), then LL; C-18, UV	13
non-pep LT's, **spiked plasma**:	LS - XAD (C-18 poorer); C-18, UV	14

SOME ANALYTICAL APPROACHES FOR EICOSANOIDS

- Senior Editor's compilation, supplementing arts. in Sect. **A**

Eicosanoid assay is beset with difficulties and pitfalls, as is evident from Sect. #A. A general survey of methodology, including enzyme investigation, appeared in 1982 [1]; one pertinent article (J. Maclouf, p. 213) deals with RIA of 6-keto-$PGF_{1\alpha}$. Table 1 collates some recent non-RIA approaches, focused on cells rather than body fluids. Whilst some books in the area neglect analysis, useful 'snapshots' of analytical advances appear annually [2-4]. Body-fluid assays feature in [2], e.g.: solid-phase extraction, derivatization, capillary GC-MS(NCI) and TLC (C.T. Dollery & S.E. Barrow; p. 91); plasma TX levels by RIA (E. Granström et al., p. 67); LTB_4 analysis (GC-MS, e.g. disclosing serum levels of ~0.2 ng/ml: I.A. Blair et al., p. 61; RIA, e.g. for synovial fluid: J.A. Salmon, p. 25); HPLC approaches (W.S. Powell, p.53); assays on urine (C. Patrono, p. 71). Bioassay approaches (cf. K.A. Rainsford, #B-1 in this vol.) feature in [4] (S-E. Dahlen et al., p. 615).

The citations now given, including those in Table 1 (the extraction and HPLC conditions in [8] are noteworthy), will help readers with particular interests to search literature for continuing progress, in the context of fmol levels as coped with by GC-MS [11]. Confidence in assaying plasma will increase, one precaution being to have a cyclooxygenase inhibitor (e.g. indomethacin) present initially. Solvent extraction ('LL' in Table 1) may fall out of favour with the advent of cartridge methods, but the type evidently needs judicious choice of solid phase, the contenders being C-18 and an XAD-type resin (Table 1), also C-2 ('Amprep' mini-columns) which Amersham International advocate rather than C-18 or C-8 in their kit procedures for $PGF_{2\alpha}$ and TXB_2 in plasma. Appropriate pre-washing is vital.

References

1. Lands, W.E.M. & Smith, K.L., eds. (1982) *Methods in Enzymology 86 ('Prostaglandins and Arachidonic Acid Metabolites')*, 1-705 [Academic Press, New York].
2. Hayaishi, O. & Yamamoto, S., eds. (1985) *Advances in Prostaglandin, Thromboxane and Leukotriene Research 15:* analytical section, pp. 1-101 [Raven Press, New York].
3. Zor, U., et al., eds. (1986), *as for* 2., *16:* pp. 327-396.
4. Samuelsson, B., et al., eds. (1987) *as for* 2., *17B:* pp. 587-626.
5. Matsuda, H., Kuzuya, T., Kamada, T., Tada, M. & Matsuura, K. (1986) *J. Chromatog. 374*, 347-353.
6. Bailey, J.M., Muza, B., Hla, T. & Salata, K. (1985) *J. Lipid Res. 26*, 54-61.
7. Stein, T.A., Angus, L., Borrerd, E., Auguste, L.J. & Wise, L. (1987) *J. Chromatog. 385*, 377-382.
8. Henke, D.C., Kouzan, S. & Eling, T.E. (1984) *Anal. Biochem. 140*, 87-94.
9. Wong, P.Y., Malik, K.U., Taylor, B.M., Schneider, W.P., McGiff, J.C. & Sun, F.F. (1985) *J. Biol. Chem. 260*, 9150-9153.
10. Herrman, T., Steinhilber, D. & Roth, H.J. (1987) *J. Chromatog. 416*, 170-175.
11. Balazy, M. & Murphy, R.C. (1986) *Anal. Chem. 58*, 1098-1111.
12. Müller, M. & Sorrell, T.C. (1985) *J. Chromatog. 343*, 213-218.
13. Verhagen, J., Wassink, G.A., Kijne, G.M., Viëtor, R.J. & Bruynzeel, P.L.B. (1986) *J. Chromatog. 378*, 208-214.
14. Salari, H. & Steffenrud, S. (1986) *J. Chromatog. 378*, 35-44.

INHIBITION OF ARACHIDONIC ACID METABOLISM - Interface between #A and #B - J.A. Salmon: *from Biochemical Society Abstracts for a 1988 meeting**

"[Its] oxidative metabolism is increased in inflamed tissues.... [mainly involving, for PG's and LT's resp.] cyclo-oxygenase and 5-lipoxygenase... PG's, particularly PGE_2, mediate some cardinal signs of inflammation (erythrema, oedema, hyperalgesia).... NSAID's provide symptomatic relief by inhibiting PG synthesis. Experimental data also suggests that LT's are involved in mediating inflammatory responses (e.g. cell influx). As yet, no inhibitors of 5-lipoxygenase have been evaluated in clinical trials and therefore the effectiveness of such compounds in treating human disease has not been established.... [There are known] inhibitors of other arachidonic acid metabolizing enzymes."

* *with Senior Editor's acknowledgement to the author and the Society.*

Further perspective
- for readers not familiar with eicosanoids

DIAGRAM: Some products that arise from arachidonic acid
- adapted by Senior Editor from a diagram which M.V. Doig furnished for her article (with J.A. Salmon; pp. 117-120) in Vol. 16 of this series [(1985) *Bioactive Analytes......* (Reid, E., et al., eds.), Plenum]

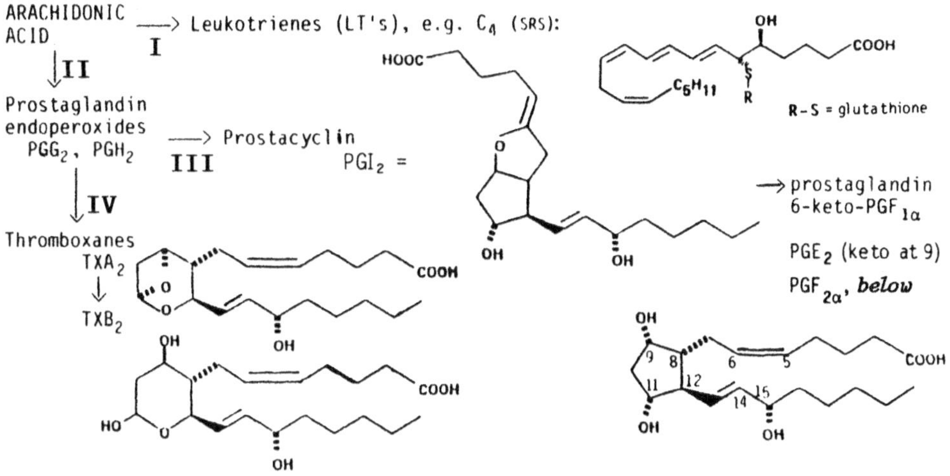

ENZYMES: **I**, lipoxygenase; **II**, fatty acid cyclo-oxygenase; **III**, prostacyclin synthetase; **IV**, thromboxane synthetase.

AMPLIFICATION
- with acknowledgement to J.A. Salmon as the author of an informative article (1986) in *Development of Drugs and Modern Medicines* (Gorrod, J.W., et al., eds.), Horwood/VCH, pp. 136-148

Arachidonic acid liberated from phospholipids can be acted on by enzyme complexes: **I**, giving rise (through an oxidation, then reduction) to hydroxy-eicosatetraenoic acids (HETE's) or, via LTA_4, to other LT's including peptido derivatives (SRS-A = 'slowly reacting substance of anaphylaxis'); **II** (microsomal), giving rise to unstable endoperoxides which, enzymatically [or (for $PGF_{2\alpha}$) non-enzymatically] yield various products including PGI_2 (unstable; via **III**), primary PG's (PGE_2 & PGD_2, besides $PGF_{2\alpha}$), and (further alternative, involving **IV**) TXA_2 which spontaneously furnishes TXB_2.

PG's are of ubiquitous origin and possess a cyclopentane ring, whereas TX's have an oxane ring. LT's possess a conjugated triene system (they were discovered in leucocytes; hence the name). TX's arise especially in blood platelets (thrombocytes). J.A. Salmon surveys the therapeutic potential of inhibitors of **I-IV** and phospholipase.

Section #B

ANTI-INFLAMMATORY DRUGS

#B-1

ANALYSIS OF THE ACTIONS OF ANTI-INFLAMMATORY DRUGS, WITH PARTICULAR REFERENCE TO EFFECTS ON INTERLEUKINS AND LEUCOCYTES

K.D. Rainsford

Anti-inflammatory Research Unit
Strangeway's Research Laboratory
Wort's Causeway, Cambridge CB1 4RN, U.K.

The importance of IL's and other cytokines in chronic inflammation has aroused interest in their assay in contexts such as inflammatory exudates and stimulated or unstimulated leucocytes. Assays developed for two main IL's, viz. IL-1 and IL-2, include cell responses such as production of other mediators, or reactions (e.g. proliferation, aggregation) whose specificity depends on the responsiveness of the cell type as well as the purity of the extracts, fluids or culture media being assayed, and on the assay conditions. Certain cell clones are more selective in the response to the IL's (e.g. the NOB-1 sub-clone of the EL-4 cell for IL-1), so improving the specificity of the LAF assay. Conditions are described for improving the specificity and responses of IL-1 in the CRA using the bovine nasal septum. The recent development of RIA's and ELISA's for assaying IL-1 and IL-2 is considered together with limitations and conditions in applying the methods. Appropriate animal models for studies of the effects of anti-rheumatic agents are discussed in relation to the use of the above-mentioned assays.*

During the past 1-2 decades the emphasis in determining the mode of action of AID's, especially the NSAID's, has concentrated on establishing their actions on the production of the PG's and other eicosanoids [1]. Ancillary studies on their mode of action have included defining their effects on (a) PMN and monocyte functions (emigration, chemotaxis, chemokinesis, superoxide and lysosomal enzyme release), (b) platelet aggregation and endothelial adhesion,

* *Abbreviations*.- HETES, hydroxyeicosatetraenoic acids; IL, interleukin; LA(F), lymphocyte activation (factor); LT, leukotriene; (NS)AID, (non-steroidal) anti-inflammatory drug; PG, prostaglandin. PMN, polymorphonuclear neutrophil leucocyte. CRA, *see text*.

(c) vascular permeability, and (d) connective tissue metabolism [1]. More recently, there has been evidence for the role of IL's and other cytokines in the development and maintenance of the chronic disease manifestations of severe arthritic conditions (e.g. rheumatoid arthritis) and other inflammatory states [2]. Also there is evidence that the abnormalities of lymphocyte function result from excesses or deficiencies in the production of IL's and other mediators as well as from alterations in intrinsic receptor status [2]. Hence there is now much interest in methods for the analysis of cytokines, especially IL-1 and IL-2, in inflammatory tissues and fluids (e.g. those from synovia) and in plasma samples from patients with rheumatoid arthritis, ulcerative colitis, etc., as well as from pertinent experimental animal and cellular systems.

ASSAYS FOR INTERLEUKIN 1 (IL-1)

Because of the lack of well described and convenient biochemical assays, reliance has been placed until recently on bioassays. For IL-1 (details in the cited refs.) these have included the following.-

(a) The classical thymocyte co-stimulated proliferation assay (known also as the LAF assay) as introduced by Gery et al. [3] (see Fig. 1) and developed further by Mizel and co-workers using special cell lines [4, 5] and further modified by others (see refs. [4,6 & 7], Fig. 1, and discussion below).

(b) The resorption of cartilage and bone [8-10], reflecting the previously described 'catabolin' activity [8, 10] (Fig. 2).

(c) The induction of fever following intracerebral or i.v. injection of the test substances or standards into rabbits or mice [10, 11].

(d) The production of PGE_2 and collagenase by fibroblasts, chondrocytes or other connective tissue cell types [9, 10].

(e) The fibroblast proliferation assay [10].

(f) The chemotaxis of lymphocytes [12] and PMN's [13].

Each of these assays represents a characteristic response to IL-1 which is believed to occur *in vivo*, although in some cases the best evidence has come from *ex vivo* studies. Collectively these responses, when being applied for identification to a single protein component or semi-purified extract, can be considered as denoting IL-1 or IL-2 activity by definition.

THE THYMOCYTE CO-STIMULATION (LAF) ASSAY

The LAF assay has formed the mainstay for defining IL-1 activity (details outlined in Fig. 1). As applied by many authors to biological fluids or tissue extracts, it has often been performed without adequate purification of the starting materials or controls to obviate effects of non-specific stimulators or inhibitors. The importance of

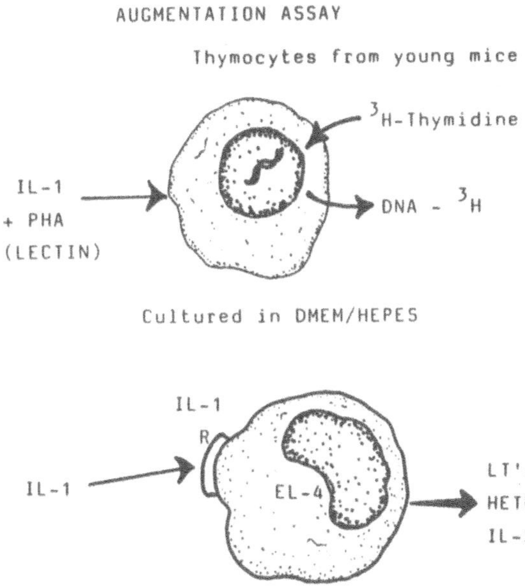

Fig. 1. The conventional thymocyte bioassay for IL-1: outline of the principles.

Upper diagram: the classical co-stimulation method in which the sample containing IL-1 is added to thymocytes from young mice (usually C3H HeJ) together with a mitogen such as the lectin phytohaemagglutinin (PHA) and radiolabelled thymidine. After incubation for 24 h in Dulbecco's modified Eagle's medium (DMEM) with HEPES buffer, the radioactivity incorporated into isolated DNA (in macromolecular precipitates) is determined. The IL-1 induces incorporation exceeding that with the lectin alone.

Lower diagram: an improved version of the assay. Using EL-4 or similar cells which respond to IL-1 without the need for a lectin, the IL-1 activity can be measured by measuring the IL-2 produced - either by the CT6 cell bioassay (see later in text, & Fig. 3), or by ELISA. Alternatively but less conveniently, the response to IL-1 can be measured by measuring eicosanoids such as the LT's or HETES. *Note by Ed.*- Incubations are at 37° in these and other assays.

performing serial dilutions in the LAF assay has been illustrated recently [14]: it was shown that the presence of a natural inhibitor in the media of lipopolysaccharide-stimulated monocytes interfered with the responsiveness of T-lymphocytes to the IL-1. One of the inhibitory substances was shown to be PGE_2 [14] and this and other eicosanoids have been shown to modulate T-lymphocyte responses to IL-1 induced proliferation [15], a feature which is obviously of importance for the actions of those AI agents that are known to affect eicosanoid production. Other naturally occurring inhibitors have been described [e.g. 16-18]. Evidently care must be taken to be aware of the presence of these inhibitors, and certainly serial dilution bioassays are the best means of recognizing the presence of putative inhibitors.

One obvious means of removing doubt about the role of inhibitors or activators affecting the T-cell response to IL-1 is to purify the fluids, extracts or media being assayed. HPLC on hydroxyapatite

[*continued on p. 71*

Fig. 2. Bioassay for IL-1-like activity (ILA), measured as cartilage-degrading or catabolin activity, released from synovial tissue (**1**) or from other cell or tissue systems.

For studying the effects of drugs on the release of synovial ILA, pig or human synovium is incubated in the presence or absence of drugs in DMEM containing 5% foetal calf serum (FCS) [19]. After 4-8 days the medium is harvested, serially diluted from 1:0 to 1:10 in DMEM + 5% FCS, and added to the bovine nasal cartilage system (**2**).

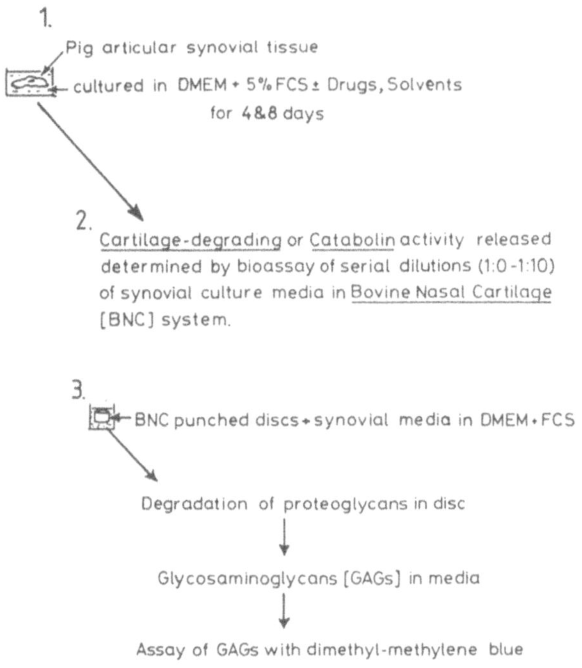

For this assay [10] small discs (~2 mm thick × ~3 mm diam.) are obtained under sterile conditions from the nasal septum and, after pre-incubation for 1-2 days [20], are incubated in 96-well Microtiter dishes with 200 µl of the above medium or a positive IL-1 standard (e.g. 5 units/ml human recombinant IL-1α, which is sufficient to give maximal degradation of cartilage). Media fractionated by gel-HPLC can also be added as a means of quantifying fractions with defined M_r, so improving assay precision [19]. After 3-6 days' incubation, the degradation of proteoglycans in the disc is determined by assaying the glycosamino-glycan (GAG) content of the medium and dividing this by that remaining in the papain-digested cartilage disc (dimethyl-methylene blue metachromatic assay for GAG [21]; 3). Besides, or instead, assay can be radiometric: the discs are pre-labelled with $^{35}SO_4^{2-}$ for 1-2 days (100 µCi/well) and thoroughly washed in buffered saline with cold 1 mM Na_2SO_4. After adding the test medium or standards and incubating (4 days), the medium and papain-digested discs are assayed for radioactivity (x) after precipitation with cetyl-pyridinium chloride (CPC, 2% w/v). Where y = radioactivity in the CPC-precipitated medium, % degradation = $x/(x+y)$. Advantageously, this assay gives the actual disc-proteoglycan degradation independently of drug- or IL-1-induced inhibition of proteoglycan synthesis. Also, induced changes in GAG turnover can be checked by determining the specific activity of the radiolabelled GAG's in the papain-digested disc's CPC-precipitable GAG's with that in the CPC-precipitated GAG's in the medium.

[22] or with C-18 RP columns [12] and affinity chromatography [23] have been reported. Hydroxyapatite HPLC has not, in the author's experience, been particularly successful for routine use as there appear to be considerable losses in bioactivity during this purification. IL-1 fractions can, however, be conveniently purified on conventional chromatographic grade hydroxyapatite (BioRad) using mini-columns (set in plastic; 1 cm high), eluted with a gradient of 10 to 500 μM pH 7.0 sodium phosphate buffer (but 1-5 mM $CaCl_2$ must be added to achieve column stability).

Some assay problems

Problems that may be encountered include the following.-

(a) Thiol-reactive drugs may interact with the mercaptoethanol added to the medium (the concentrations differing from one laboratory to another), maybe affecting the total reduced thiols in the medium or the stability of the drug.

(b) The lipopolysaccharide (derived from *Escherichia coli*) used in stimulating monocytes to produce IL-1 may carry through purification procedures and be associated with isolated protein fractions, so affecting the response of the lymphocytes. This can be overcome by adding polymyxin B, which binds to lipopolysaccharide. Alternatively, T-cells of a type unresponsive to the lipopolysaccharide (e.g. those from C3H/HeJ mice) can be used. Another approach can be to employ other stimuli of IL-1 production (e.g. A-23187, a calcium ionophore) which can be separated out from the proteins with IL-1 activity, or constitutive cell lines (e.g. the human monocytoid leukaemia cell lines, JOSK-I, -K, -M and -S [20]) which could be considered as a cellular source of IL-1 production mimicking that from human monocytes.

(c) The co-mitogens used to sensitize the T-cells to the actions of IL-1 (i.e. ConA and PHA [3]) may interact with the drugs being assayed for anti-IL-1 activity. A recently developed approach that avoids use of these mitogens is to employ the NOB-1 sub-clone of the EL-4 cell line, or other related cells, which constitutively produce IL-2 in response to IL-1 alone [7]. The IL-2 so produced can be assayed by the CTLL cell, using ^{3}H-thymidine incorporation as the index of proliferative response (similar to that in the LAF assay).

A useful procedure to avoid misinterpretation of responses in the thymocyte assay is to add antisera to IL-1 to act as a positive control for the assay.

THE CARTILAGE RESORPTION ASSAY (CRA)

While this assay has not been used to anywhere near the extent of that of the LAF assay, it has been used in the purification of pig IL-1 (ascribed to catabolin activity [10]), and does have particular utility, especially in respect of convenience, cost and specificity.

Although problems such as can arise in the LAF assay likewise beset the application of the CRA, especially with respect to the effects of inhibitors or stimulatory substances in the extracts or media being assayed, the CRA does have fewer problems overall. Thus, in the application of the system using cartilage derived from the bovine nasal septum ([8, 10]; details and refs. in Fig. 2 legend), we have found [19, 20, 24 & unpublished work] that this system is unresponsive to (a) all prostanoids at <500 ng/ml, (b) LT's and HETES at <50 ng/ml, (c) all cyclo-oxygenase or lipoxygenase inhibitory NSAID's (up to 100 μM), (d) corticosteroids, (e) carrageenan and a number of related inflammagens, and (f) protease inhibitors. It is, however, sensitive to the actions of quinoline anti-malarials, high concentrations of auranofin (but not other gold complexes), retinoids and lipopolysaccharide. Thus on the basis of these defined limits it is possible to be confident in employing the CRA assay on culture media and synovial fluids.

Particular care is required to perform serial dilution bioassays to reveal the presence of natural inhibitors or stimulators (other than the eicosanoids, of course). The same requirements for purified fractions can be extended to the CRA as stipulated above for the LAF assay, at least for verification of the methods, and accordingly the same solutions can be applied.

Automated assay of proteoglycan degradation products.- The methods for measuring the proteoglycan degradation of the bovine nasal cartilage (BNC) induced by IL-1 involve the assay of glycosaminoglycans (GAG's) in the medium compared with those remaining on the disc [10], as amplified in the legend to Fig. 2. The assays can be automated with considerable saving in labour and materials by use of a Microtiter plate photometric reader [25]. The reported method uses a Dynatech MR600 micro-plate reader which is relatively expensive (~£7,000, + computing equipment cost). We have recently adapted the method for use in a Cambridge Life Sciences model CLS 962 micro-plate reader, which retails at about one-fifth of the cost of the Dynatech instrument and is equally accurate and relatively fast. The manual control is the only non-automated component, but is quite speedy as plates can be read off within 10 min using the Epson P-40 printer attachment. Our adaptation involves (a) using full-strength dimethylmethylene blue dye (as in the original description of Farndale et al. [21]), (b) employing filters of wavelength 540 nm, which matches well the peak maximum for the dye-GAG complex (using shark chondroitin sulphate A, Sigma), which is at 535 nm, and (c) setting the machine at zero in air, not on the blank dye solution as might normally be contemplated with conventional spectrophotometry, this being necessary to ensure low error in the linear determination according to the above conditions.

IMMUNOASSAYS FOR IL-1

RIA's have only recently become available. The Cistron kit using ^{125}I-IL-1β (Cistron Biotechnology, Pine Brook, NJ 07058) is relatively expensive and of course measures only the human β form of IL-1. In our hands it gives satisfactory results, but the kit's lifetime should be reckoned as over-stated, and much care should be taken regarding the conditions for its transportation. It is as well to work with purified fractions (from HPLC [13] or AcA-54 gel chromatographic columns [10]), especially in view of the costliness of the assays and the paucity of the available information on the potential for interference from components other than IL-1 in the biological fluids being analyzed.

Some ELISA assays are under development (e.g. Cistron Biotechnology) and these will certainly improve the costs and application of specific immunoassays, besides the standard of assay methodologies. Bioassays will, however, still have an important place in quantifying responses, especially in the study of responses to AI agents.

ASSAYS FOR INTERLEUKIN 2 (IL-2)

Assays for IL-2, a growth factor for T-cells, again have been mainly those derived on the basis of the proliferation of lymphocytes, usually involving use of special cell lines (e.g. the IL-2-dependent CTLL-20, or EL-4 cells; see Fig. 3 for details) [26]. Recently, ELISA assays have become available (Genzyme Corp., Boston, MA 02111), and their application is likely to be extended considerably in the future.

GENERAL COMMENTS AND CONCLUSIONS

The availability of specialist cell lines has greatly facilitated the development of highly specific bioassays for cytokines and also, as in the case of the human pro-monocyte cell line U-937, for studies of the transcriptional control of cytokine production and the relationship to cell proliferation.

The success with determining the actions of drugs on leukocyte functions and the actions of monokines depends on the identification or confirmation of their actions *in vivo* and thus, together with the experimental design, on the choice of animal models or the status and type of disease state of patients from whom leukocytes or fluids are derived for study. In relation to the application to animal model systems it is clearly important to understand the roles of the cytokines in the various models [e.g. 26, 27]. The logical animal systems involve the established polyarthritis models developed in rats and mice, with the monoarticular arthritis (induced by antigen or IL-1) in rabbits as an additionally useful system.

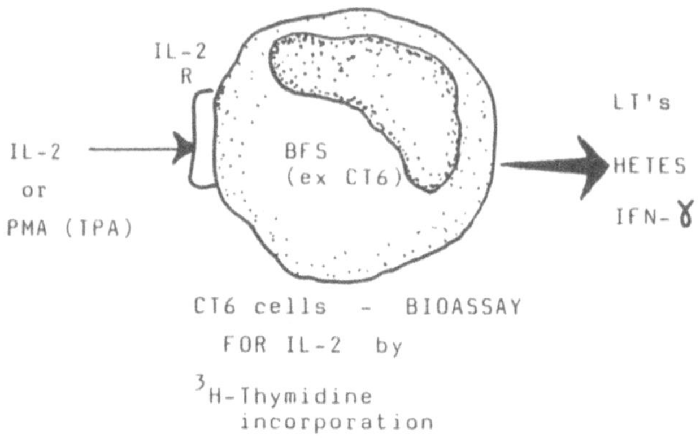

Fig. 3. Bioassay for IL-2, employing specific cell lines (e.g. the BFS line derived from CT6 cells) which are highly specific for IL-2 (as a result of the high density of the receptor, R, for this IL) and which also function independently of IL-1. The cellular response to IL-2 can be determined by either the incorporation of ^{3}H-thymidine into isolated DNA, or by assay of interferon (INF-γ) by conventional bioassay or by ELISA. As with the IL-1 assay (Fig. 1), it is also possible to measure the biological response by measurement of the HETES or LT's so produced, but this is less convenient. It should be noted, however, that in both the IL-1 and the IL-2 assays using cell lines the biological response of these cells can be modified by the released LT's, HETES and prostanoids.

Moreover, when it comes to discerning the actions of drugs in these animal models it is important to consider the dosage protocols and the relationship to effects on particular stages of the inflammatory process. Thus in the arthritis induced in rats by Freund's adjuvant, drugs which affect the progress of the disease when administered after it has developed (at 10-14 days) are more likely to resemble the actions of conventional inhibitors of PG synthesis (e.g. indomethacin). Those which are effective only when given from zero time (i.e. at the establishment of the disease) to 7 days post-induction are more likely to have influences on the lymphocyte manifestations of the disease, and therefore may be considered potentially 'disease-modifying' agents (though clearly such a term must be applied only cautiously). Clearly, the assays for cytokines in these systems will largely be performed on peripheral blood components and lymphocytes derived from spleen and lymph nodes [26, 27].

Synovial tissues can, of course, only be obtained from rodents more or less in an inflamed state, and the limits to the availability of such material and the inability to procure it in appreciable quantities from control animals will effectively limit the value of studies performed on this material. However, the rabbit monoarticular arthritis model using IL-1 or other inflammagens may prove useful in the study of synovial and cartilage material from inflamed tissues. The cartilage implanted in air pouches initially induced on the back or neck of rats and mice [28], especially in those animals with established polyarthritis, could prove an especially useful system to study cartilage-synovial interactions and the role of inflammatory cells in cartilage degradation *in vivo*.

References

1. Kitchen, E.A., Dawson, W., Rainsford, K.D. & Cawston, T. (1985) in *Anti-inflammatory and Anti-rheumatic Drugs*, Vol. I (Rainsford, K.D., ed.), CRC Press, Boca Raton (FL), pp. 21-87.
2. Billingham, M.E.J. (1987) *Br. Med. Bull. 43*, 350-370.
3. Gery, I., Gershon, R.K. & Wacksman, B.H. (1972) *J. Exp. Med. 136*, 128-142.
4. Mizel, S.B. (1981) in *Manual of Macrophage Methodology* (Herbscowitz, H.B., et al., eds.), Marcell-Dekker, New York, pp. 407-425.
5. Gillis, S. & Mizel, S.B. (1981) *Proc. Nat. Acad. Sci. 78*, 1133-1137.
6. Simon, P.L., Laydon, J.T. & Lee, J.C. (1985) *J. Immunol. Meth. 84*, 85-94.
7. Gearing, A.J.H., Bird, C.R., Bristow, A., Poole, S. & Thorpe, R. (1987) *J. Immunol. Meth. 99*, 7-11.
8. Dingle, A.T. (1983) *J. Rheumatol. (Suppl. 11), 10*, 38-42.
9. Mizel, S.B., Dayer, J-M., Krane, S.M. & Mergenhagen, S.E. (1981) *Proc. Nat. Acad. Sci. 78*, 2474-2477.
10. Saklatvala, J., Pilsworth, L.M.C., Sarsfield, S.J., Gavrilovic, J. & Heath, J.K. (1984) *Biochem. J. 224*, 461-466.
11. Fontana, A., Weber, E. & Dayer, J-M. (1984) *J. Immunol. 133*, 1696-1698.
12. Moissec, P., Yu, C-L. & Ziff, M. (1984) *J. Immunol. 133*, 2007-2011.
13. Camp, R.D.R., Fincham, N.J., Cunningham, F.M., Greaves, M.W., Morris, J. & Chu, A. (1986) *J. Immunol. 137*, 3469-3474.
14. Haynes, D.R., Whitehouse, M.W. & Vernon-Roberts, B. (1987) *10th Internat. Congr. Pharmacol.*, Sydney, Abstr. No. 0-372.
15. Brandwein, S.R. (1986) *J. Biol. Chem. 261*, 8624-8632.
16. Muchmore, A.V. & Decker, J.M. (1986) *J. Biol. Chem. 261*, 13404-13407.
17. Cannon, J.G., Tatro, J.B., Reichlin, S. & Dinarello, C.A. (1986) *J. Immunol. 137*, 2232-2236.
18. Roberts, N.J., Prill, A.H. & Mann, T.N. (1986) *J. Exp. Med. 163*, 511-519.
19. Rainsford, K.D. (1987) *Agents & Actions 21*, 337-340.
20. Rainsford, K.D. (1986) *J. Pharm. Pharmacol. 38*, 829-833.
21. Farndale, R.W., Sayers, C.A. & Barrett, A.J. (1982) *Conn. Tissue Res. 9*, 247-248.
22. Koeck, A. & Luger, T.A. (1984) *J. Chromatog.. 296*, 293-300.
23. Onoue, K., Sasaki, T., Yamamoto, T., Lin, B.H. & Matsuda, H. (1986) *Biochim. Biophys. Acta 881*, 437-445.

24. Rainsford, K.D. (1985) *Agents & Actions 16*, 55-57.
25. Meade, C.J., Twyholm, M., McMahon, A., Bodmer, J. & Swann, B.P. (1987) in *Proceedings 75th Anniversary of the Strangeways Research Laboratory - The Control of Tissue Damage*, Arthritis & Rheumatism Council for Research, London, pp. 30-34.
26. Lee, J.C., Rebar, L., Demuth, S. & Hanna, N. (1985) *J. Rheumatol. 12*, 885-891.
27. Phadke, K., Carlson, D.G., Gitter, B.D. & Butler, L.D. (1986) *J. Immunol. 136*, 4085-4091.
28. Sin, Y.M., Sedwick, A.D. & Willoughby, D.A. (1984) *J. Path. 142*, 23-30.

#B-2

IMMUNOASSAY OF BECLOMETHASONE 17,21-DIPROPIONATE AND METABOLITES

W.N. Jenner and D.J. Kirkham

Department of Biochemical Pharmacology
Glaxo Group Research Ltd.
Ware, Herts. SG12 ODJ, U.K.

A radioimmunoassay (RIA) has been developed for determining beclomethasone 17,21-dipropionate (BDP) in human body fluids, after solvent extraction, at levels down to 0.1 (plasma) or 1 (urine) ng/ml. The 3-O-carboxymethyloxime derivative of BDP was coupled by a mixed-anhydride reaction to bovine serum albumin (BSA) for raising a rabbit antiserum, or to [^{125}I]iodohistamine to provide a radiolabel. Known BDP metabolites and endogenous steroids had negligible cross-reactivity.

BDP given orally (2 mg) or intranasally (200 µg) to volunteers was undetectable in plasma and urine. If given i.v. (1 mg) BDP was detectable up to 1 h post-dose. It was eliminated rapidly, mainly by metabolism and/or excretion into the G-I tract. In assaying an active metabolite of BDP, its 17-monopropionate (B17MP), by a similarly developed RIA, the 21-propionate that may arise by transesterification cross-reacts extensively. The monopropionate ester pair was determined in plasma and urine from volunteers given BDP orally or i.v.

BDP is a potent synthetic corticosteroid administered by inhalation for the treatment of asthma (e.g. Becotide®) or by intranasal spray for the treatment of rhinitis (Beconase®). In each case a single inhalation or spray contains 50 µg of the drug, the typical daily dose being ~400 µg. Although BDP has been on the market since the early 1970's, its pharmacokinetics have never been investigated, in part due to the lack of an assay suitable for the low plasma concentrations expected following therapeutic doses. In order to rectify this situation, an assay has now been developed for studies in man. RIA was the technique of choice, because of the established success of this approach for the analysis of steroids, and of the high sensitivity potentially attainable.

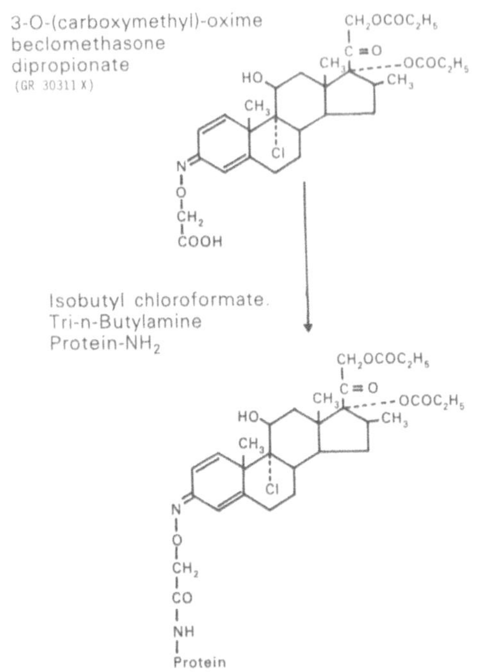

Fig. 1. Immunogen production.

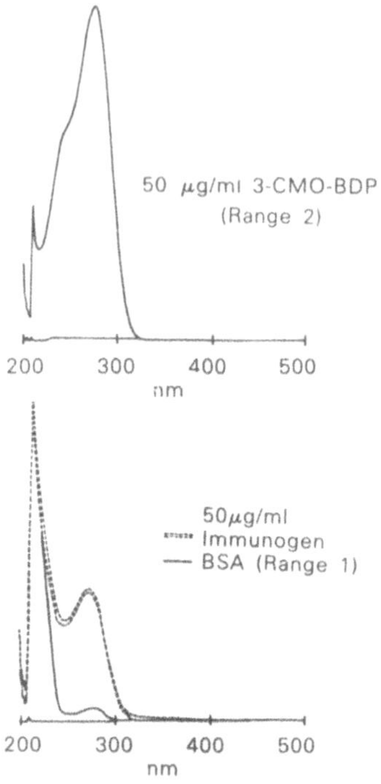

Fig. 2. Spectrophotometric analysis of BDP immunogen. (**1** and **2** are AUFSD settings.)

RIA DEVELOPMENT: BDP

Immunogen and antiserum preparation.- The 3-*O*-carboxymethoxime derivative of BDP was conjugated to BSA by the mixed anhydride method [1] using reaction conditions established by Kominami et al. [2] (Fig. 1). The product was purified by dialysis and gel filtration, and the extent of incorporation of hapten into the protein was determined by spectrophotometry (Fig. 2) to be 25 mol/mol. This high extent of incorporation was confirmed by elemental nitrogen microanalysis and by determination of. the number of unreacted lysine ε-groups in the immunogen compared with the unreacted protein [3]. The primary immunization dose of BDP-BSA conjugate in rabbits was 500 μg i.d, with booster doses in Week 11 (200 μg, i.m.), Week 20 (300 μg, i.m. and i.d.) and Week 64 (250 μg i.m. and s.c.), followed 7 and 13 days later by harvesting of the antiserum. The immunogen was injected in Non-ulcerative Freund's Adjuvant (Morris; Guildhay Antisera, Univ. of Surrey) containing BCG vaccine BP (Glaxo Labs.).

Antiserum characterization using ^{3}H-BDP.- ^{3}H-BDP with the radiolabel in the 17-propionate moiety was synthesized. Although, unfortunately, it was not possible to produce material with specific radioactivity >1 Ci/mmol, this was high enough to establish that the antiserum

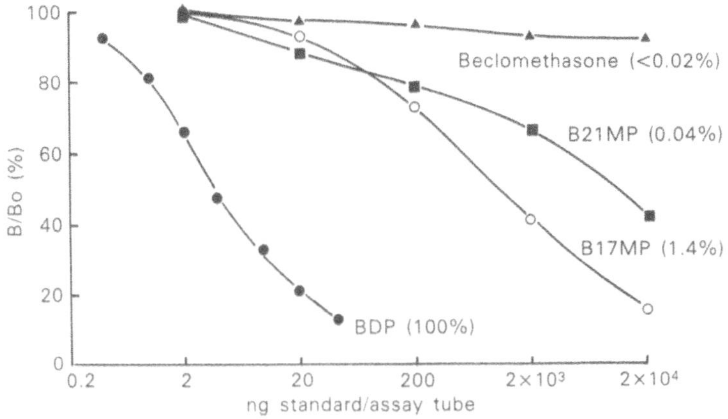

Fig. 3. Metabolite specificity of the RIA for BDP. The antiserum (Rabbit R5900) was used at a final dilution (titre) of 2400. The radiolabel was ^{3}H-BDP (1 Ci/mmol). The % values are cross-reactivity (% ratio of BDP concn. and that of possible metabolite at 50% B/BO).

had a potentially high titre and assay sensitivity and was specific with respect to BDP metabolite standards (Fig. 3). It was important to establish the cross-reactivity of these metabolites in the assay because it is known that the drug is rapidly de-esterified to B17MP and beclomethasone [4]. B17MP can also transesterify to the 21-propionate (B21MP) under certain conditions [5].

Steroid RIA's can be subject to interference by endogenous steroids in the sample. With BDP this did not appear to be a problem, as negligible cross-reactivity was found with testosterone, progesterone, cortisone, cortisol and corticosterone.

Although the basis of a specific assay had been obtained, the sensitivity of the standard curve in diluent (~5 ng BDP/ml) with the relatively low specific activity radiolabel would not have sufficed for determining the drug in body fluids in subjects receiving therapeutic doses. An alternative radiolabelled form of BDP was therefore sought.

Use of ^{125}I-labelled BDP.- [^{125}I]Iodohistamine was covalently coupled to 3-*O*-carboxymethyloxime BDP by the mixed anhydride reaction (Fig. 4), essentially according to Nars & Hunter [6]. The yield of radiolabelled product was ~35 μCi (11% radioisotopic yield). Standard curves prepared with the ^{3}H and ^{125}I radiolabels were compared under non-optimized conditions (Fig. 5). As expected, the very high specific radioactivity radio-iodine label (>2000 Ci/mmol) produced a far more sensitive standard curve (~50 pg BDP/ml) than the low-activity ^{3}H radiolabel (1 Ci/mmol).

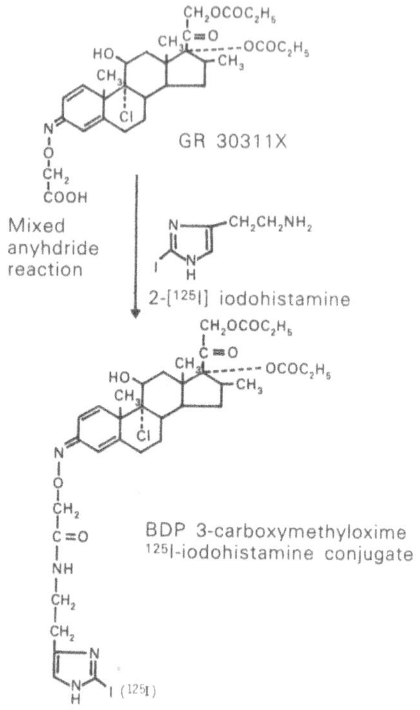

Fig. 4. Synthesis of BDP 3-carboxymethyloxime [^{125}-I]-iodohistamine conjugate.

Fig. 5 *(below)*, Standard curves for BDP using ^{3}H (■) and ^{125}I (●) radiolabel. Each curve was produced in assay buffer under standard conditions with antiserum R5900.⊗

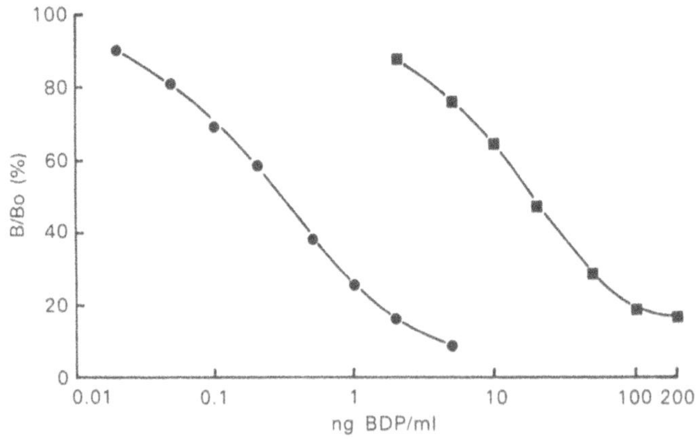

The previously established high specificity of the assay for BDP relative to BDP metabolites was confirmed using the ^{125}I-radiolabel. Furthermore, non-interference by endogenous steroids was confirmed by comparing standard curves in control human plasma, serum and 'charcoal-stripped' control serum. The charcoal treatment [7] would have removed all endogenous steroids from the sample. The standard curves in each case were essentially superimposable.

⊗ For Fig. 5 and also Figs. 3 & 6-10 the RIA conditions were essentially as shown later in Scheme 1.

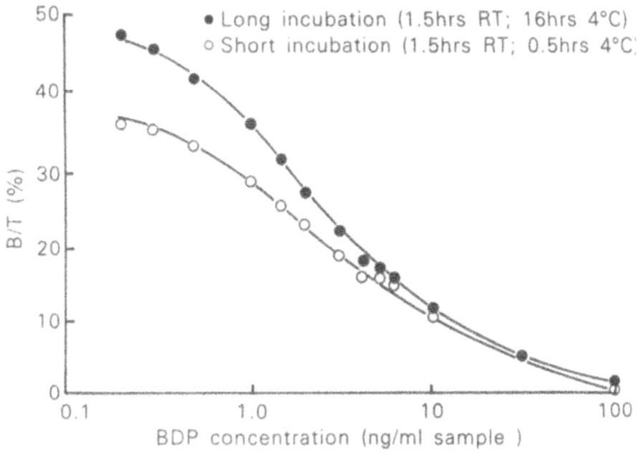

Fig. 6. Comparison of BDP standard curves after long and short incubations.

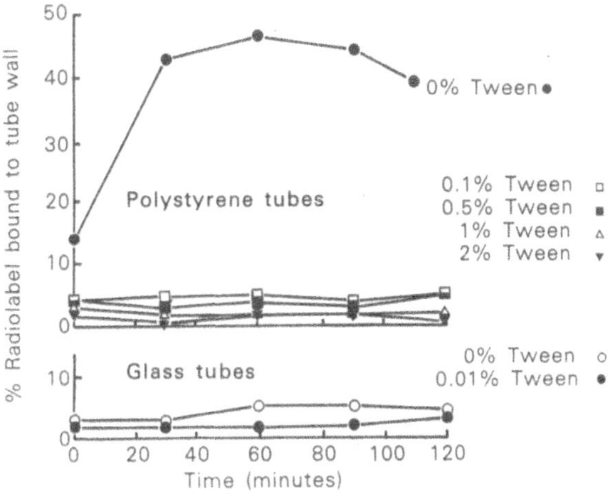

Fig. 7. Effect of Tween 20 on ^{125}I-BDP binding to polystyrene and glass assay tubes.
Results in an earlier experiment, without Tween: Up to 60 min the curves for glass and polystyrene were as in Fig. 7; beyond 60 min the rise continued. Binding to polypropylene was ~18% at 0 and 10 min; thereafter it was 80-85% of the value for polystyrene (~350% of that to glass).

Methodological problems

The work had so far been carried out under arbitrary, non-optimized conditions using charcoal phase-separation. When an experiment was set up to examine the effect of incubation on the standard curve in assay diluent, it was noticed that the assay unexpectedly failed to reach equilibrium rapidly: there were marked differences in the extent of antibody (Ab) binding between short and long incubations at 4° (Fig. 6). This appeared to reflect loss of radiolabel from solution as the incubation progressed.

This phenomenon was examined further (summary in Fig. 7 legend) by incubating solutions of radiolabel in diluent in the normal polystyrene assay tubes, in polypropylene tubes and in glass tubes. Whereas with glass tubes <15% of the radiolabel bound to the tube walls in 3 h, the plastic tubes showed progressive loss, attaining ~50%. This non-specific binding of radiolabel could be suppressed (Fig. 7) by addition of a detergent, Tween 20, to the assay diluent.

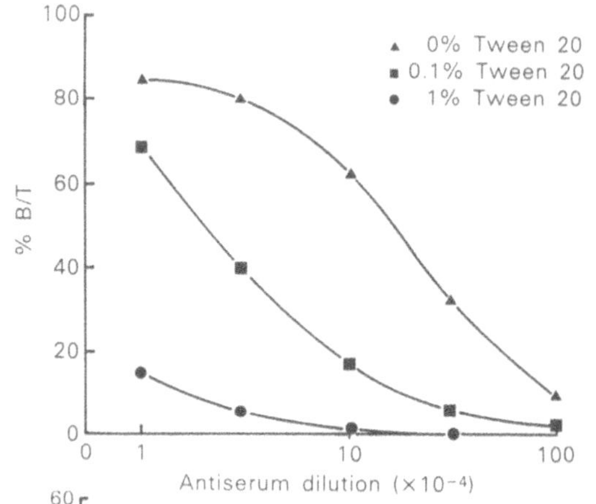

Fig. 8. The effect of Tween 20 on antiserum dilution profiles in polystyrene tubes.

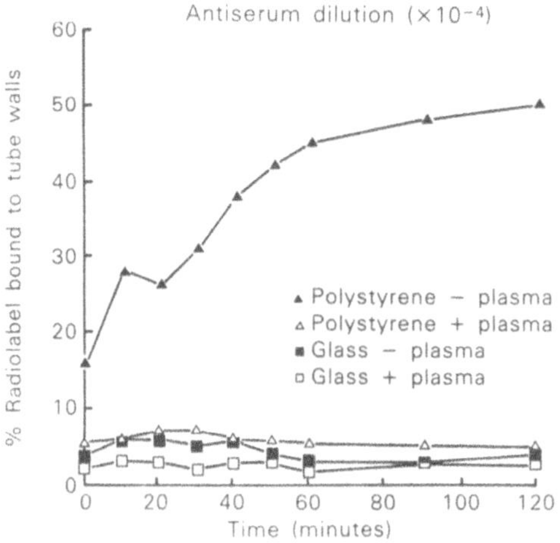

Fig. 9. The effect of plasma on ^{125}I-BDP binding to assay tubes (100 µl of plasma).

However, Tween itself had an adverse effect on Ab binding and could not be used as a routine addition to the diluent (Fig. 8).

Plasma (20% v/v) or BSA was also effective in minimizing the binding of radiolabel to tube walls (Fig. 9). Therefore, in the analysis of plasma samples, with standards made up in control plasma, non-specific binding problems would be expected to be minimal. Unfortunately, however, the presence of plasma also caused problems in the assay by reducing considerably (~50%) the amount of radiolabel bound to Ab in both glass and polystyrene tubes.

It was considered that the presence of corticosteroid-binding globulin (BCG) in plasma could contribute to the plasma matrix effect by binding BDP. Many RIA's for corticoids require the presence of a blocking agent in the assay diluent to counteract the effect of this protein. Three standard 'blocking agents' were investigated, viz. danazol, cortisol and 8-anilino-1-naphthalene sulphonic acid (ANS). These would be expected to reduce the binding of BDP to BCG by competing for binding sites on the protein. However, none of these agents had any appreciable effect on the BDP standard curve in plasma.

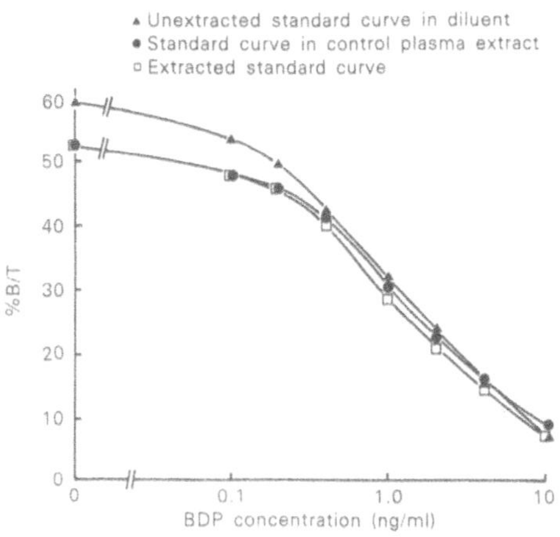

Fig. 10. The effect of plasma extracts on the standard curve.

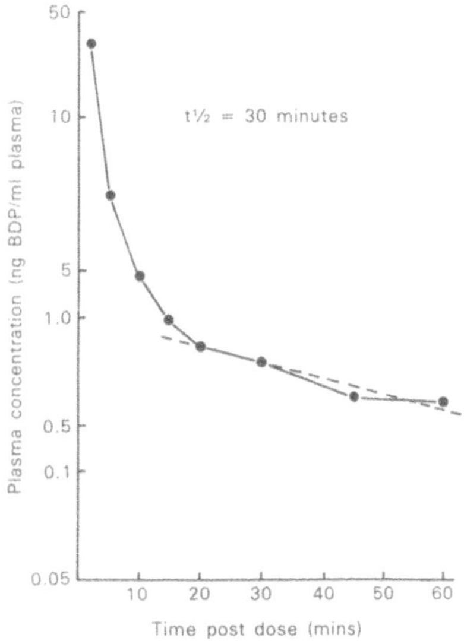

Fig. 11. Plasma BDP in a volunteer given 1 mg BDP i.v.

Extraction of samples.- Because of the considerable plasma-matrix effect it was clear that the drug would have to be extracted from samples prior to analysis. A number of extraction solvents were investigated and it was concluded that diethyl ether provided the best combination of extraction efficiency (>90%) and convenience. The ether extract was evaporated under a N_2 stream and the drug residue was reconstituted in ethanol/water (1:1 by vol.) for assay. The extraction largely eliminated the plasma-matrix effect, as shown in Fig. 10 by the similarity of the standard curves prepared by spiking into diluent and into extracts of control plasma. The latter curve was, moreover, virtually identical with one obtained by extracting spiked plasma samples. This confirmed that the ether extraction process had a high efficiency.

Stability of BDP in plasma.- Now that a satisfactory analytical procedure had been obtained, BDP stability was of concern, particularly because human serum contains an esterase that can hydrolyze BDP to B17MP [8]. Fresh heparinized control blood and plasma samples were spiked with BDP (1 and 10 ng/ml) and incubated for up to 3 h at room temperature or 1 h at 37°, with removal of aliquots at intervals for extraction and assay. No significant loss of drug was found. Evidently BDP is relatively stable in whole blood and plasma, allowing pharmacokinetic studies with no special precautions in handling blood samples. Similar experiments with spiked plasma indicated that BDP is stable indefinitely in plasma stored at -20°.

PLASMA (1 ml), incl. BDP-spiked samples for std. curve (0-10 ng/ml) and quality control (0.1, 0.4, 1.0 & 5.0 ng/ml)

↓ *Extract in glass tubes with 3 ml ether; dry down under N_2 at 50°*

Residue

↓ *Dissolve in 1 ml 1:1 ethanol/water*

Processed sample ⟶ RIA medium

For RIA: *Diluent (300 µl), + 20,000 dpm ^{125}I-BDP (50 µl in ethanol) + antiserum (100 µl in diluent), giving*

URINE

↓ *Dilute × 10 in 1:1 (by vol.) ethanol/water; take 1 ml*

Sample ⟶ RIA medium

↓ *1.5 h at room temp. & 0.5 h at 4° in 5 ml borosilicate tubes; then + 0.5 ml 1% dextran-coated charcoal in diluent; 20 min spin, 4°*

Supernatant: *take 0.5 ml for counting (10 min)*

Scheme 1. BDP assay (50-100 pg/ml plasma detectable). Diluent: 0.05 M sodium phosphate buffer pH 7.4 containing 0.6% NaCl & 1% gelatin.

FINALIZED ASSAY PROCEDURE, AND PHARMACOKINETIC STUDIES

Scheme 1 shows the assay procedure adopted. It was applied to subjects given BDP orally, intranasally or (Fig. 11; see below) i.v. In plasma and urine samples from two pairs of volunteers, given 2 mg BDP as a suspension in orange juice or 200 µg intranasally, in no case was BDP detectable (assay sensitivity: 100 pg/ml plasma, 1 ng/ml urine). In another male volunteer given 1 mg i.v., the drug was detectable in plasma up to 1 h post-dose (Fig. 11; $t_{\frac{1}{2}}$ shown); <0.5% of the dose was excreted as unchanged drug in the urine in 24 h. Preliminary values were as follows: k_{el}, 0.023 min^{-1}; AUC (trapezoidal method, extrapolating concentrations to zero time; $0 \rightarrow \infty$), 261 ng/ml min; V_D [namely: dose/(AUC × k_{el})] = 166.7 l; clearances (ml/min): plasma, 3834; non-renal, 3817; renal, 17. The very high non-renal clearance, far exceeding the hepatic blood flow (~1500 ml/min), indicates notable extra-hepatic clearance.

RIA DEVELOPMENT: B17MP

As BDP is known [4] to rapidly furnish B17MP, an active metabolite [9], a method for assaying it was sought. Antiserum and radiolabel production and RIA performance were essentially as for BDP, although sensitivity was lower: the detection limit in the absence of plasma was ~1 ng/ml (Fig. 12, *left*). Cross-reactivity was extensive (1600%) for the 21-monoester, which could be assayed down to <0.1 ng/ml, but negligible for BDP, beclomethasone and especially cortisol (Fig. 12, *right*).

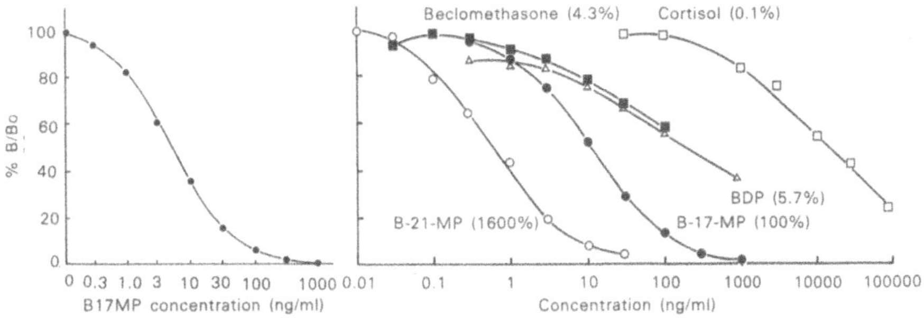

Fig. 12. BMP ('B17MP') assay. *Left:* example of a standard curve in assay diluent produced under standard assay conditions, for B17MP. *Right:* cross-reactivities in the 'B17MP' assay (expressed as in Fig. 3).

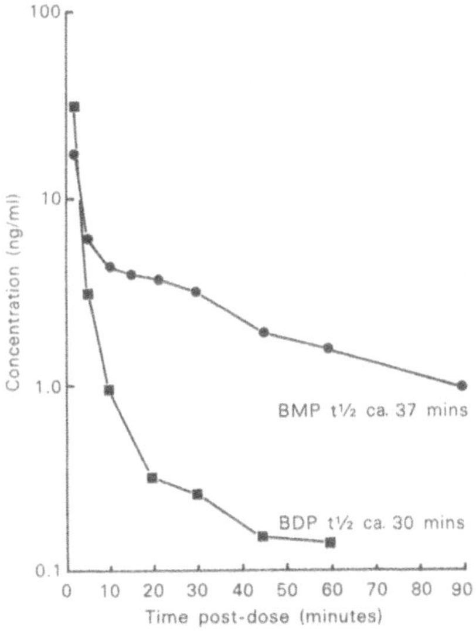

Fig. 13. BMP and BDP concentrations following i.v. administration of 1 mg of BDP to a volunteer.

Transesterification of the 17-propionate moiety to the 21-position during immunogen synthesis or after immunization could account for the 21-monoester (having given rise to Ab's) being the preferred ligand. Possibly the 21-monoester could exist *in vivo* due to trans-esterification from B17MP or direct hydrolysis of BDP; hence the assay cannot be used for specifically determining B17MP. However, in the pharmacokinetic study where 1 mg BDP was given i.v. the assay has served to indicate semi-quantitatively the combined concentration of the monoesters (BMP) in plasma: they were detected up to 90 min post-dose (Fig. 13).

Except at the first two times, 2 and 5 min, the BMP levels were considerably higher than the BDP levels, confirming that the drug is rapidly and extensively hydrolyzed *in vivo*. The atypical plasma concentration-time profile of 'beclomethasone monopropionate', the highest concentration of metabolites being found in the first sample post-dose, probably indicates rapid 'first-pass' metabolism by the lung, which is known to have considerable BDP esterase activity [8]. Results not yet confirmed indicate that BMP is likewise present in plasma following oral administration of BDP (1 mg).

Acknowledgements

The authors wish to acknowledge the valuable contributions to the work of Ian Fellows, Isobel Brown and Christopher Lear.

References

1. Erlanger, B.F., Borek, F., Beiser, S.M. & Lieberman, S. (1959) *J. Biol. Chem. 234*, 1090-1094.
2. Kominami, G., Yamauchi, A., Ishihara, S. & Kono, M. (1981) *Steroids 37*, 303-314.
3. Habeeb, A.F.S.A. (1966) *Anal. Biochem. 14*, 328-336.
4. Martin, L.E., Tanner, R.J.N., Clark, T.J.H. & Cochrane, G.M. (1974) *Clin. Pharmacol. Ther. 15*, 267-275.
5. Gardi, R., Vitali, R. & Ercoli, A. (1963) *Gazz. Chim. Ital. 93*, 431-450.
6. Nars, P.W. & Hunter, W.M. (1973) *J. Endocrinol. 57*, xvii-xviii.
7. Stockhill, C. (1979) *Annals Clin. Biochem. 16*, 275.
8. Andersson, P. & Ryrfeldt, A. (1984) *J. Pharm. Pharmacol. 36*, 763-765.
9. Harris, D.M. (1975) *J. Steroid Biochem. 6*, 711-716.

#B-3

HPLC ANALYSIS OF ANTI-INFLAMMATORY AGENTS (NON-STEROIDAL AND STEROIDAL)

J.C. McElnay

Deparment of Pharmacy
The Queen's University of Belfast
Medical Biology Centre, 97 Lisburn Road
Belfast BT9 7BI, N. Ireland

AI agents in biological fluids are commonly analyzed chromatographically, maybe by GC but usually by HPLC - typically RP - with UV detection (sometimes fluorescence or EC). HPLC is also convenient in percutaneous absorption studies on AI agents in gels and creams. Solvent extraction is often not necessary if the drug concentration, e.g. in plasma, is relatively high: precipitation of plasma proteins is often sufficient, provided that a pre-column is used.*

We have used RP-HPLC for mefenamic acid, ibuprofen, indomethacin, fluocinolone and (in percutaneous absorption studies) benzydamine. Straight-phase HPLC was preferable for prednisone, prednisolone, hydrocortisone, dexamethasone and chloroquine. Solvent extraction was avoided wherever possible; usually an i.s. was used, to improve accuracy. UV or (chloroquine) fluorescence detection was used. Basic HPLC equipment usually suffices for these drugs in biological fluids, with no need for gradient elution or specialized columns and detectors.

Therapeutic drug monitoring (e.g. using immunoassay techniques) is not routine for drugs used in the treatment of inflammatory conditions. It is mainly for pharmacokinetic research that the methods used for analyzing biological fluids (usually serum, urine and synovial fluid) for such drugs have been developed. The methods are usually chromatographic, for the sake of sensitivity and specificity. Although GC methods are available for a number of AI agents, e.g. aspirin [1], a review of all published methods as in Clarke's compilation [2] shows that HPLC is the most commonly used method. HPLC is also

* *Abbreviations.*- AI, anti-inflammatory; EC, electrochemical; Fa, fluocinolone acetonide; i.s., internal standard; r, correlation coefficient; RP, reversed phase; SDS, sodium dodecyl sulphate.

convenient for examining the diffusion of AI agents from gels and creams during percutaneous absorption studies.

Depending on the chemical characteristics of the individual AI drugs, a range of detection systems has been used, including UV, fluorescence and EC. RP-HPLC with UV detection is the most widely used method due to its more general application; only a few compounds will fluoresce or be easily oxidizable or reducible. For most methods, simple organic-solvent extraction procedures which decrease solvent-front size, increase analytical column life and concentrate the drug prior to injection have been described. Extraction is, however, time-consuming and is often unnecessary if the drug is present at relatively high concentration. Thus, for plasma, deproteinization prior to analysis is often a sufficient preparation for chromatography provided that a pre-column is used to protect the analytical column.

A number of non-steroidal AI drugs (NSAID'S), corticosteroids and drugs that modify inflammatory disease have been quantiified using HPLC in our laboratories during studies in the areas of pharmacology (pharmacokinetics) and biopharmaceutics (e.g. percutaneous absorption; buccal partitioning; tablet dissolution). This short review aims to describe the HPLC methods and highlight the difficulties experienced during work with the main drugs studied, namely indomethacin, prednisolone, chloroquine, benzydamine and Fa. Details of the sources of columns, detectors and data-processing modules are included for each assay since it is these and not the pumping or injection systems that contribute to most inter-laboratory variability in assay procedures. Routinely the mobile phase was initially filtered through a suitable membrane filter and degassed for 10 min in an ultrasonic bath.

ASSAY OF INDOMETHACIN

This potent non-steroid AI agent is available as both solid and liquid preparations for oral administration and, for rectal administration, as a suppository formulation. Such formulations differ in drug-release rate *in vitro* [3], and hence in clinical efficacy at a given dose. We have examined the pharmacokinetics of indomethacin after night-time administration of the two most popular U.K. brands of suppository [4], using a sensitive specific HPLC method to measure indomethacin in the plasma and urine of 12 volunteers in a double-blind crossover study.

The chromatographic conditions were based on those used by Skellern & Salole [5], while sample preparation was based on a 1977 protocol by E. Lin & L.Z. Benet (Pharmacy School, Univ. of California at San Francisco). For precipitation of plasma proteins, acetonitrile (0.8 ml) - containing 1 μg/ml mefenamic acid as i.s. - was added to 0.4 ml of each plasma sample. After vortexing for 30 sec and centrifuging at 4000 **g** for 10 min, 100 µl of the supernatant was injected onto a 37-53 µm Pellicular ODS guard column (Millipore; 5 cm) linked to a 5 µm Hypersil ODS column (Shandon; 150 × 4.6 mm). The mobile phase, delivered at 2 ml/min, was a buffer/methanol mixture (30:70 by vol.), the buffer being 0.1 M Na acetate adjusted to pH 3.2 with glacial acetic acid. Integrated peak areas were measured (Hewlett Packard 3390A) at 254 nm (Kontron 750), and the peak-area ratio was utilized for estimating unknown sample concentrations by reference to a standard calibration curve. Urine samples were spiked with i.s. and 100 µl aliquots injected directly onto the column. The assay method was reproducible: r = 0.9998 over the range 0-5 µg/ml, and C.V. 0.0366 for 10 plasma samples each containing 0.625 µg/ml. Fig. 1 shows a typical chromatogram for plasma. (The two suppository formulations turned out to be similar in plasma profiles and urinary excretion [4].)

No major difficulties arose in running the assay. Occasionally the mobile phase pressure increased to 4000 psi, due to progressive blocking of the in-line filter or the inlet frit of the pre-column by insoluble matter from the samples. When the pressure reached 4000 psi the analysis was stopped and the frits removed for sonic cleaning. With continuous (automatic) sampling the cleaning was required on alternate days. More recently (unpublished) we have added a block heater to the chromatography system used for indomethacin. Keeping the columns at 30° tends to prevent a drop, during an overnight run, in the UV absorbance baseline, occurring together with an increase in retention times when room temperature decreases.

ASSAY OF PREDNISOLONE

There is controversy in the literature concerning the bioavailability of plain and enteric-coated prednisolone tablets, notably in patients with renal transplants, G-I or respiratory disorders, or rheumatoid arthritis [e.g. 6-8]. We have compared the two types in volunteers [9] and in patients with chronic obstructive airways disease; endogenous cortisol (hydrocortisone) was one of the pharmacodynamic parameters measured.

A normal-phase HPLC system developed by Delargy [10]⊗ was found most suitable for the assay of plasma samples. In preparing calibration samples for plasma cortisol, pooled plasma from volunteers had to be stripped of its endogenous cortisol, using activated charcoal which was separated from the stripped plasma by centrifugation (2000 **g**, 30 min). The treatment, for 2 h at room temperature with magnetic stirring, was with 4 g charcoal/100 ml plasma. The charcoal reduces endogenous cortisol to below the level detectable in the assay [11].⁺

⊗ based on work by Rose & Jusko [11]

⁺ Other uses of charcoal, especially 'coated', are indicated, with a review ref. [E. Reid, *Analyst 101*, 1-18], in Vol. 5, this series: art. by A.A.A. Aziz et al.- *Ed.*

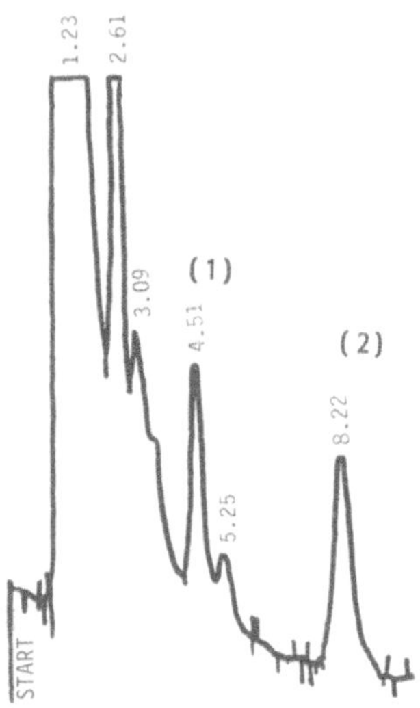

Fig. 1. HPLC of indomethacin (1) and spiked-in mefenamic acid as i.s. (2) in plasma.

Fig. 2, *right.* HPLC of prednisolone (3), cortisol (2) and, as i.s., dexamethasone (1) in plasma.

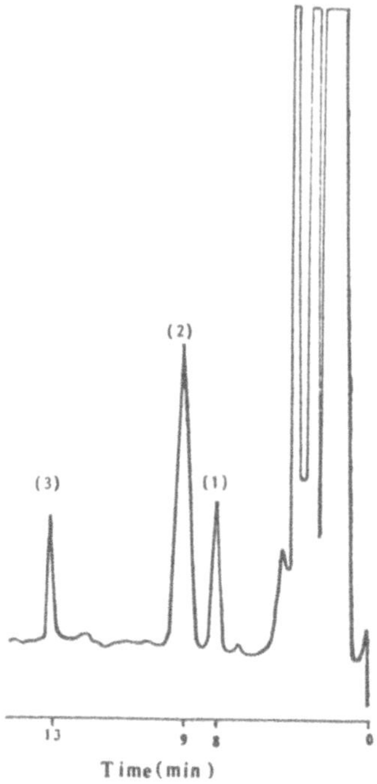

A double extraction procedure was used to extract the steroids from calibrator and patient plasma samples. The samples (1.9 ml), spiked with 0.1 ml dexamethasone (1 µg/ml solution) as i.s., were extracted into 6 ml of diethyl ether/dichloromethane (DCM; 60:40 by vol.) by vortexing for 1 min. The tubes were centrifuged at 2,000 **g** for 5 min, and 5 ml of the organic phase aspirated off into a second tube containing 1.0 ml 0.1 M NaOH. After vortexing (30 sec) and recentrifuging, 4 ml of the washed organic phase was aspirated into a third tube and dried down with an air stream at 45°. The residue was reconstituted in 250 µl of the mobile phase, namely DCM/water-saturated DCM/methanol/tetrahydrofuran/glacial acetic acid (66.45: 30.0:2.5:1.0:0.05 by vol.).

The volume injected was 100 µl. The analytical column was 5 µm Hypersil (Shandon; 250 × 4.6 mm), with a 30-38 µm HC Pellosil pre-column (Beckman; 50 × 4.6 mm). The flow rate was 2.0 ml/min. For the steroids the detector (Shimadzu SPD-6A) was set at 240 nm. Fig. 2 shows a typical chromatogram. (In respect of pharmacokinetic and pharmacodynamic parameters the two tablet types appeared to differ, significantly so for the volunteers but not for the few patients so far studied [9].)

Prednisone can also be assayed by this method. Optimal resolution was obtained at room temperatures between 23° and 25°. When unacceptable baseline drift occurred the system was flushed with isopropanol, whereby this periodic problem was overcome. The steroids gave linear calibrations using the peak-height ratio technique (Hewlett Packard 3390A). The C.V.'s for prednisolone (n = 10; 250 ng/ml) and hydrocortisone (n =10; 100 ng/ml) were 0.049 and 0.047, and the two r's were 0.999 over the concentration ranges 0-200 and 0-500 ng/ml respectively. The assay appeared to perform best if the mobile phase was re-cycled (in a batch mode, up to 4 times), i.e. contained trace amounts of the steroids being analyzed. [Cf. 'HPLC...recycling' Index entries in Vols. 12 & 14, this series.- *Ed.*]

ASSAY OF CHLOROQUINE

Besides its primary use in malaria prophylaxis and treatment, chloroquine serves to modify inflammatory conditions, e.g. rheumatoid arthritis. Aspects which have interested us are its interaction with antacids in the gut [12] and its binding to glass and plastic surfaces [13, 14], studied by fluorescence spectrophotometry, and - studied with HPLC - its pharmacokinetics in humans (unpublished), and its pharmacokinetic interaction with digoxin [15]. HPLC has the advantage in pharmacokinetic studies that desethylchloroquine, the major metabolite, can be measured simultaneously with the parent drug. Although several RP-HPLC methods have been reported for chloroquine [e.g. 16], in our hands a normal-phase method based on that of Alvan et al. [17] has proved more satisfactory for the two analytes in plasma, due to improved sensitivity.

They were extractable from plasma using a slight modification of the method of Staiger et al. [18]. A 1 ml aliquot of plasma was placed in a teflon-lined screw-capped test tube. Borosilicate glassware must be used throughout, since chloroquine binds to soda-glass [13]. Care should also be taken in the use of plastic tips, since we have shown that chloroquine binds to diverse plastic materials [14]. Pre-soaking the plastic material (24 h) in a solution of chloroquine followed by thorough washing with deionized water tends to prevent further binding from taking place. Cellulose acetate filters should be avoided in filtering or sterilizing weak chloroquine solutions since binding to this material is extensive [14].

To the plasma was added, as i.s., 50 µl of 6,8-dichloro-4-(1-methyl-4-diethylamino-butylamino)-quinoline; the amount of i.s. added should approximately correspond to the expected amount of chloroquine in the samples being analyzed. To each tube was added 0.5 ml 0.5 M NaOH. After gentle shaking by hand, 3 ml n-hexane was added and the tube contents were mixed in a rotating extractor for 30 min. Then 50 µl n-octanol was added to aid layer separation. After centrifugation, 2 ml aliquots of the hexane layer were evaporated to dryness at room temperature under air. The residue was dissolved in 200 µl of mobile phase, and 50 µl was injected onto the column.

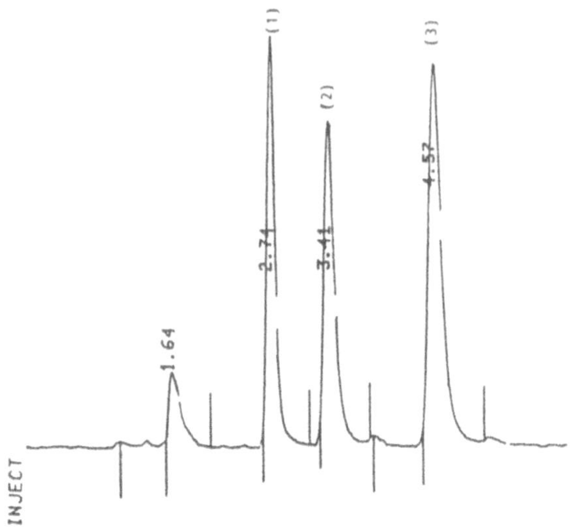

Fig. 3. HPLC of chloroquine (2), desethylchloroquine (3) and the i.s. *(see text)* (1) in plasma.

The mobile phase consisted of acetonitrile/methanol/diethylamine (80:19.6:0.4 by vol.) and was delivered at 1.5 ml/min. The analytical column, 5 μm Ultrasphere silica (Beckman; 150 × 4.6 mm), was protected by a 10 μm Ultrapack silica pre-column (Beckman; 45 × 4.6 mm). During assay development a variable wavelength UV detector was used at 344 nm; but to increase sensitivity with study samples a fluorescence detector was used (Waters model 420; 338 nm excitation, 400 nm emission). Peak areas (Waters Data Module 730) were utilized in preparing calibration curves and in determining unknown chloroquine and desethylchloroquine concentrations. Fig. 3 shows a typical chromatogram. For chloroquine and desethylchloroquine (0-100 ng/ml) the calibration r's were 0.983 and 0.986, both improving to 0.991 over the concentration range 0-1000 ng/ml. The respective C.V.'s (1 μg/ml; n = 10) were 0.0389 and 0.0147 respectively. The method allowed their concentrations in plasma to be followed readily over a 150-h study period in volunteers after a single 1 g dose of chloroquine phosphate.

ASSAY OF BENZYDAMINE

This non-steroidal AI drug, used orally in some countries but available only as a topical preparation (Difflam® cream) in the U.K., is used to relieve painful inflammatory conditions of the musculoskeletal system. A gel preparation, not yet commercially available, containing 3% benzydamine hydrochloride, has also been formulated for specific purposes, e.g. use with ultrasound in phonophoresis (enhancement of percutaneous absorption of drugs through ultrasound perturbation). The few methods described for benzydamine assay include the use of ^{14}C-labelled drug [19], fluorescence [20] and HPLC with fluorescence detection [21]. Since we were interested in assaying gel and cream formulations and buffer samples used in phonophoresis studies, a new method based on HPLC with UV detection was devised [22].

Dissolution of benzydamine gel samples was carried out by vortexing for 30 sec in 10 ml of acetonitrile/water (1:1 by vol.), followed by direct injection onto the column. Indomethacin (0.1 ml of 1 mg/ml soln.) was added initially to the 10 ml. Difflam® cream samples were weighed into centrifuge tubes and dissolved by vortexing for 5 min in 10 ml of tetrahydrofuran/isopropanol (30:60), with indomethacin addition as above. Direct injection of 20 µl onto the column was performed after centrifuging (1100 **g**, 15 min).

The mobile phase, delivered at 0.9 ml/min, was acetonitrile/water/acetic acid (62:37.5:0.5) containing 5 mM SDS; the acetic acid maintained the pH at 4.0. The 5 µm Novapak C-18 analytical column (Waters; 150 × 3.9 mm) and the 37-53 µm Pellicular ODS guard column (Millipore; 50 × 4.6 mm) were kept at 30° within a block heater.

Detection was at 305 nm (Spectroflow 773; Kratos), and peak-area ratios (Shimadzu Chromatopac C-R3A recording data processor) was used for calibration and for unknown samples. Fig. 4 shows a typical chromatogram. Gel and cream samples each equivalent to 18.75-600 µg benzydamine hydrochloride/ml both gave r = 0.999, with C.V.'s (at 600 µg/ml; n = 10) of 0.015 and 0.016 respectively. Using the method we have shown for the gel preparation, by measuring loss of drug from the vehicle as it penetrates into the skin, the ineffectiveness of phonophoresis [22]. The assay, and the following assay, should be readily adaptable to (e.g.) saliva, plasma, synovial fluid or urine.

ASSAY OF FLUOCINOLONE ACETONIDE

Fa [*not* author's abbreviation - *Ed.*] is a steroidal agent used primarily in treating inflammatory skin disorders. Ultrasound, studied in steroid treatment of musculo-skeletal inflammatory conditions [23], has been shown by us, using the skin blanching test, to cause a small but statistically significant increase in percutaneous absorption of Fa in volunteers [24]. To quantify this effect pharmacokinetically, the influence of ultrasound on the *in vitro* Fa penetration through lipid membranes including skin is being studied. An assay for determining Fa in aqueous buffers has therefore been developed.

In these diffusion experiments, methanol was added (to 4% v/v) to the phosphate buffer (pH 7.4) because of the poor aqueous solubility of the drug. An aliquot (100 µl) of this solution was injected, without prior extraction, onto the 37-53 µm Pellicular ODS pre-column (Waters; 50 × 4.6 mm) linked to the 5 µm Novapak C-18 analytical column (Millipore; 150 × 3.9 mm). The mobile phase consisted of water/acetonitrile/acetic acid (65:35:0.5) containing 5 mM SDS, flowing at 0.8 ml/min. No i.s. was used, the 238 nm Fa peaks as shown in Fig. 5 being quantified by peak-area integration; otherwise the above description for benzydamine including the 30° operation is applicable. There was linearity with 0-100 µg Fa/ml; the C.V. (2.5 µg/ml; n = 10) was 0.0043. The membrane tests are still in progress.

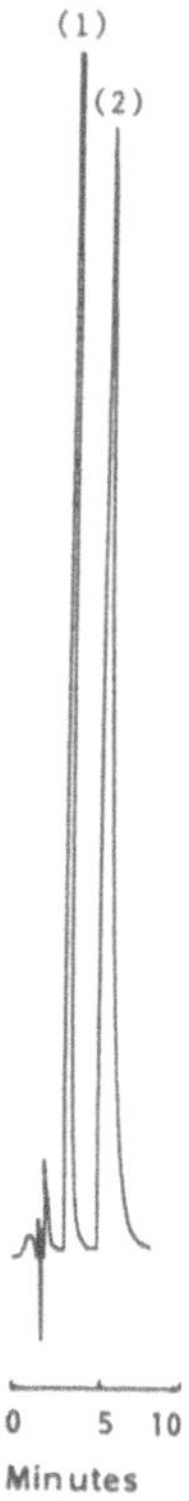

Fig. 4, *left*. HPLC of benzydamine (2) in a cream, with indomethacin (1) as i.s. Gels gave similar patterns.

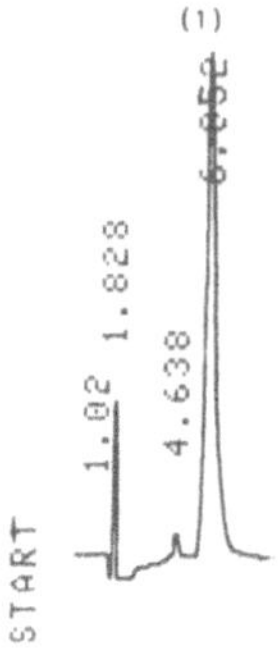

Fig. 5, *above*. HPLC of fluocinolone acetonide (1) in pH 7.4 phosphate buffer containing 4% v/v methanol.

CONCLUSION

The five assays detailed above indicate that a wide range of non-steroidal and steroidal AI drugs in biological fluids and aqueous buffers are amenable to isocratic HPLC with ordinary equipment. For most AI compounds analytical data are available in the literature (notably [2]). Where specific assay details are not available for a particular drug and an assay has to be developed from first principles, useful review articles may be consulted, e.g. on technical considerations [25] and on optimizing peak resolution [26].

References

1. Rance, M.J., Jordan, B.J. & Nichols, J.D. (1975) *J. Pharm. Pharmacol.* *27*, 425-429.
2. Moffat, A.C., Jackson, J.V., Moss, M.S. & Widdop, B., eds. (1986) *Clarke's Isolation and Identification of Drugs*, 2nd edn., The Pharmaceutical Press, London.
3. McElnay, J.C. & Nicol, A.C. (1984) *Int. J. Pharm.* *19*, 89-96.
4. McElnay, J.C., Taggart, A.J., Kerr, B. & Passmore, P. (1986) *Int. J. Pharm.* *33*, 195-199.

5. Skellern, G.G. & Salole, E.G. (1975) *J. Chromatog. 114*, 483-485.
6. Shaffer, J.A., Williams, S.E., Turnberg, L.A., Houston, J.B. & Rowland, M. (1983) *Gut 24*, 182-186.
7. Olivesi, A. (1985) *Therapie 40*, 5-7.
8. Hayes, M., Alam, A.F., Bruckner, F.E., Doherty, S.M., Myles, A., English, J., Marks, V. & Chakraborty, J. (1983) *Annals Rheum. Dis. 42*, 151-154.
9. McCann, J.P., McElnay, J.C., Nicholls, D.P., Scott, M.G. & Stanford, C.F. (1987) *Br. J. Clin. Pharmacol. 23*, 652P.
10. Delargy, H. (1981) *Ph.D. Thesis*, Dept. of Pharmacy, The Queen's University of Belfast.
11. Rose, J.Q. & Jusko, W.J. (1979) *J. Chromatog. 162*, 273-280.
12. McElnay, J.C., Mukhtar, H.A., D'Arcy, P.F. & Temple, D.T. (1982) *J. Trop. Med. Hyg. 85*, 153-158 (& see 159-163).
13. Yahya, A.M., McElnay, J.C. & D'Arcy, P.F. (1985) *Int. J. Pharm. 25*, 217-223.
14. Yahya, A.M., McElnay, J.C. & D'Arcy, P.F. (1986) *Int. J. Pharm. 34*, 137-143.
15. McElnay, J.C., Sidahmed, A.M., D'Arcy, P.F. & McQuade, R.D. (1985) *Int. J. Pharm. 26*, 267-274.
16. Bergqvist, Y. & Olin, A. (1982) *Acta Pharmaceutica Suecica 19*, 161-174.
17. Alvan, G., Ekman, L. & Lindstrom, B. (1982) *J. Chromatog. 229*, 241-247.
18. Staiger, M.A., Nguyen-Dinh, P. & Churchill, F.C. (1981) *J. Chromatog. 225*, 139-149.
19. Anderson, K. & Larsson, H. (1974) *Arzneim.-Forsch./Drug Res. 24*, 1686-1688.
20. Giacalone, E. & Valzelli, L. (1966) *Med. Pharmacol. Exp. 16*, 102-106.
21. Catanesse, B., Lagana, A., Marino, A., Picollo, R. & Rotatori, M. (1986) *Pharmacol. Res. Comm. 18*, 385-403.
22. Benson, H.A.E. & McElnay, J.C. (1987) *J. Chromatog. 394*, 395-399.
23. Skauen, D.M. & Zentner, G.M. (1984) *Int. J. Pharm. 20*, 234-235.
24. McElnay, J.C., Kennedy, T.A. & Harland, R. (1987) *Int. J. Pharm. 40*, 105-110.
25. Giese, R.W. (1983) *Clin. Chem. 29*, 1331-1343.
26. Drayer, D.E. (1984) in *Proceedings of the Second World Conference on Clinical Pharmacology and Therapeutics* (Lemberger, L. & Reidenberg, M., eds.), Am. Soc. for Pharmacology & Experimental Therapeutics, Bethesda, MD., pp. 809-819.

#B-4

ANALYTICAL APPROACHES ADOPTED DURING THE DEVELOPMENT OF FLURBIPROFEN

A. Bye*, W. Adams and D. Kaiser

Upjohn Ltd., Fleming Way, Crawley RH10 2NJ, U.K.
and The Upjohn Company, Kalamazoo, MI 49001, U.S.A.

Brief mention is made of the assay and stereoselective metabolism of ibuprofen. The NSAID[ø] *given attention in this article is flurbiprofen, the development of which entailed successive assay methods. That outlined below was used when the number of samples to be assayed was escalating.*

Requirement (flurbiprofen)	*A simple, specific and sensitive method applicable to biological samples (here described for serum) without interference from known metabolites; stereospecificity was not sought.*
End-step	*RP-HPLC with detection by UV (or, in the automated version, by fluorescence).*
Sample preparation	*Solvent extraction after acidification, or C-18 cartridge extraction (but mere protein precipitation in the automated version). Residue from drying down extract or eluate dissolved in mobile phase for HPLC.*
Comments	*Streamlining of the method, allowing cost-saving automation, was achieved through knowledge of metabolism and use of discriminating detection.*

In any drug development programme analytical support is needed from the beginning. Often the original methodology becomes either too cumbersome or obsolete as the needs of the development project change. Initially, precise measurements in relatively small numbers of samples are needed, without much idea of potential problems. With experience, methods evolve to overcome problems, to become more cost-effective and to satisfy new needs.

---

* addressee for any correspondence (at the Crawley address)

ø *Abbreviations:* NSAID, non-steroidal anti-inflammatory drug; PFB, pentafluorobenzyl; RP, reversed phase; i.s., internal standard.

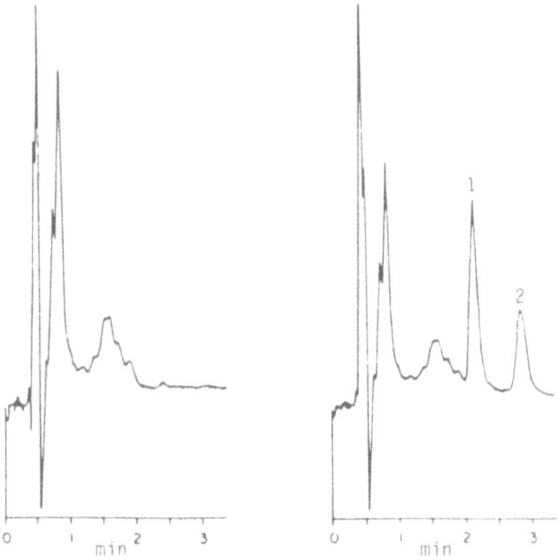

Fig. 1. Typical chromatograms for *(left)* 'blank' human plasma, and *(right)* human plasma containing ibuprofen (2) and, as internal standard, ibufenac (1); each 1 μg/ml. Column: 10 μm C-18 Radial-Pak, 10 × 0.5 (i.d.) cm. Mobile phase: 20 mM phosphoric acid/acetonitrile, 50:50 by vol., at 3 ml/min. Detection at 220 nm.

These analytical problems are related to increasing sample numbers and ever-shortening turn-round times. Eventually when methods become 'routine' a new compound is introduced into the overall development programme and usually introduces new analytical challenges although it can resemble the lead compound in pharmacology and chemical structure. To exemplify the problems associated with analytical support for drug development, the NSAID's ibuprofen and especially flurbiprofen are now considered.

THE BACKGROUND: IBUPROFEN

Ibuprofen

Ibuprofen [(*R*,*S*-2-(4-isobutylphenyl)propanoic acid] was one of the first NSAID's to come into general use. It is 99% bound to plasma proteins. The maximum daily dose, in the U.K., is 2400 mg, in divided portions. One reported method [1] is typical. In solvent comparisons with acidified plasma, only iso-octane/2-propanol (85:15 by vol.) gave complete extraction, 106.6 ±4.7% (S.D.); hexane gave 48.2 ±1.4%, petroleum ether 55.3 ±1.2%, chloroform 65.0 ±3.0%, dichloromethane 68.6 ±7.9%, and toluene 69.5 ±2.7%. RP-HPLC with UV detection (Fig. 1) gave a detection limit of 0.04 μg/ml plasma.

Current interest in ibuprofen relates more to its stereoselective inversion (*R*→ *S*) in man.* Because this need has introduced many specialist techniques (isotope work, MS, etc.) we have chosen to describe how we dealt with the analysis of flurbiprofen [*R*,*S*-2-(fluoro-4-biphenyl)propanoic acid (**F** in Fig. 2), another NSAID introduced into our drug development programme. It too is extensively bound (99%);

* See #B-6 (by A.J. Hutt & J. Caldwell)

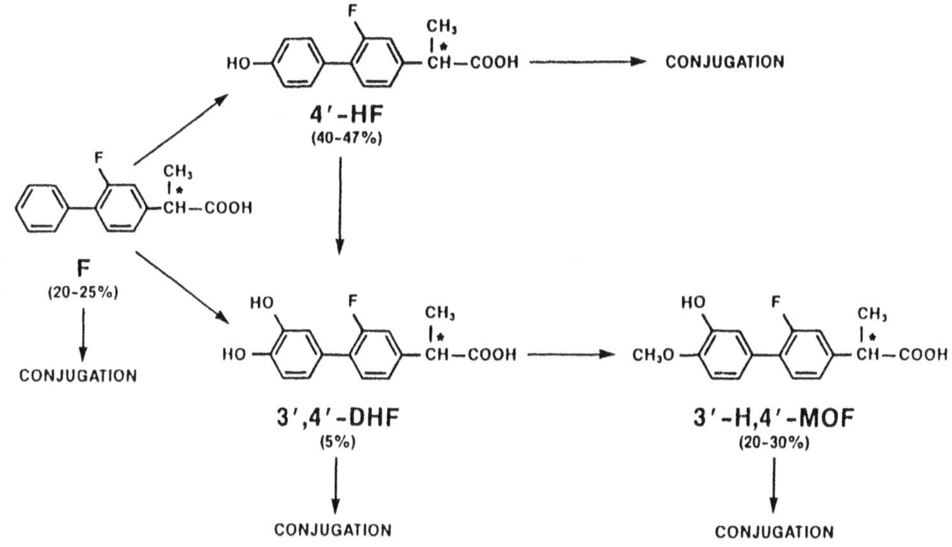

Fig. 2. Flurbiprofen (F) and its metabolism, studied with radio-label at position *.

the maximum daily dose is 300 mg in divided portions. It has useful fluorescence properties but has more known urinary metabolites (Fig. 2) than ibuprofen. Following standard doses (150-200 mg/day) peak drug levels in plasma are ~7 µg/ml compared with 40 µg/ml for ibuprofen after standard doses (1200-1600 mg/day).

INITIAL AND SUBSEQUENT METHODS FOR FLURBIPROFEN

A highly sensitive method was needed to study the absorption, metabolism and excretion of flurbiprofen in a limited number of samples from animals and humans at low doses (<10 mg) giving peak plasma levels of ~1 µg/ml. Such studies largely generate urine samples.

Choice of method.- Taking into account [2] the experience with ibuprofen and earlier investigations [3], GC (OV-17 column) in conjunction with electron-capture detection of the PFB esters was chosen. However, a TLC step had to be added to separate substances which interfered with the GC measurement of flurbiprofen [4]. Scheme 1 shows the pre-GC steps.

Comments.- Although useful information was gathered, enabling single-dose administration of flurbiprofen to be set at 50 or 100 mg, the method was too cumbersome for the growing numbers of analytical samples. It was reserved for metabolic investigation work.

Choice of method for multi-sample use.- A reliable, sensitive and specific method was needed to support the analysis of the growing

Scheme 1. Pre-GC steps.

Plasma (1 ml), *added to tubes already containing dried-down internal standard*

Add 0.25 ml M H_2SO_4, extract with benzene & centrifuge

Solvent extract

Evaporate down, wash tube with $CHCl_3$ and concentrate; → silica gel TLC plate

TLC zone of Rf 0.31

Scrape off; elute (methanol); add 25 mg K_2CO_3 & PFB reagent, react at 60°, 90 min, & evaporate to dryness

Derivatized analyte

Shake in cyclohexane/water (50:50) for GC (on 2 µl of organic layer)

number of plasma samples from animals and humans receiving single oral doses of >50 mg. By this time, parallel studies [5] had shown the absence of plasma metabolites.

RP-HPLC with UV detection was adopted, with ibuprofen as internal standard. For the requisite clean-up steps, extraction was by solvent (Scheme 2) or cartridge (Scheme 3). The choice of extraction solvent was critical if maximum sensitivity was required. Also, the mobile phase had to have sufficiently low pH (<5) or contain an ion-pairing agent to give acceptable peak shapes and workable retention times (Table 1). With these provisos good chromatography was feasible (Fig. 3) although many peaks unrelated to the drug were apparent. Sensitivity (0.2 µg/ml) was adequate for most purposes.

Comments on this method.- The greatly increased numbers of samples could now be processed. Experience with the UV-detection approach showed little interference from drug-related or endogenous substances in plasma. Also metabolites, especially urinary, were identified by classical radiolabel methods and were now available for further study. Sample preparation was now becoming the rate-limiting step.

Table 1. Effect of mobile phase pH and % acetonitrile (v/v) on R_T.

Flow, ml/min	Acetonitrile %	pH	Flurbiprofen R_T	Ibuprofen (i.s.) R_T
1	60	3	6.3	7.8
1	55	3	6.9	8.7
1	50	3	9.6	12.6
1	50	4	7.8	11.1
1	50	5	5.4	8.7
2	40	3	17.0	22.0

Serum (100 µl) spiked with i.s.

Add 500 µl 0.25 M H_2SO_4 & 4 ml diEt ether; cap, mix, centrifuge

Solvent extract

Evaporate to dryness & reconstitute in mobile phase

HPLC load solution

C-18 column; methanol/pH 5 acetate (65:35 by vol.); 1.5 ml/min; 229 nm

Scheme 2. HPLC preceded by solvent extraction. Ibuprofen as i.s. (in Scheme 3 also). *Plasma (fibrin clots!) can replace serum.*

Serum (1 ml) spiked with i.s.

Load into Prep-1 (duPont) cartridge, mix, add 0.2 ml 0.5 M H_2SO_4 & mix; load into processor, wash with 1 ml water & elute with 2 ml methanol

Eluate

Dry down & make up in mobile phase

HPLC load solution

C-18 column; acetonitrile/0.05 M acetic acid (40:60); 1.2 ml/min; 230 nm

Scheme 3. HPLC preceded by cartridge extraction.

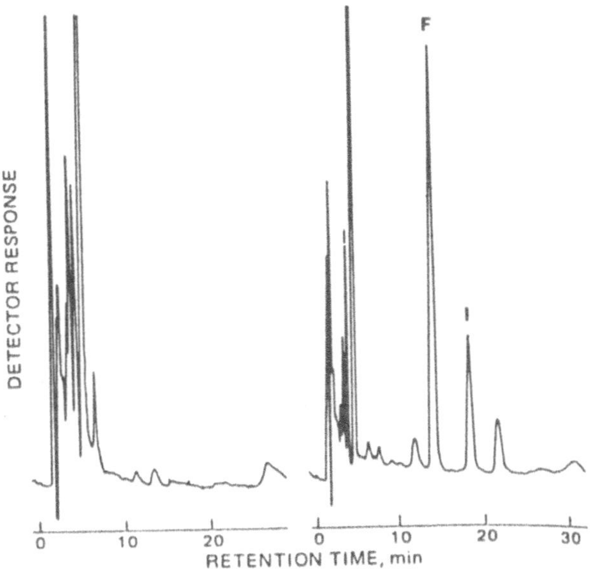

Fig. 3. Typical HPLC chromatograms following Scheme 3, for human plasma: *left*, 'blank'; *right*, flurbiprofen & ibuprofen (F, I) present.

FINAL SIMPLIFIED METHOD FOR FLURBIPROFEN IN PLASMA

The need now was for a simple, sensitive and specific assay which could be fully automated or readily sub-contracted.

Choice of method.- Now that drug, metabolites and structural analogues were available, RP-HPLC with fluorescence detection became the method of choice. The fluorescence characteristics, as shown

Scheme 4. HPLC preceded by protein precipitation.

Plasma/serum (0.1 ml)

Add 0.1 ml 0.5 M NaCl, 2 ml 0.05 M K phosphate buffer pH 2.6, and 1 ml acetonitrile containing the 4'-methoxy analogue as i.s.; mix & centrifuge

Supernatant, for HPLC (load 0.1 ml)

C-18 column; pH 2.6 buffer (as above)/tetrahydro-furan (THF), 55:45 by vol.; 1.9 ml/min; settings 260_{ex} & 320_{em} nm on the detector (Perkin Elmer 650-105).

in Fig. 4 for flurbiprofen and its 4'-hydroxy metabolite, warranted adoption of λ_{ex} 260 nm and λ_{em} 320 nm as optimal settings for the detector. Sample preparation entailed merely protein precipitation with acetonitrile at pH 2.6 (Scheme 4), with good recovery of the drug in the supernatant. Peak shape with the tetrahydrofuran-containing eluent was best at low pH (2.6), and a good i.s. was the methoxy analogue of the drug [2-(2-methoxy-4-biphenyl)propanoic acid]. The resulting chromatograms were remarkably clean (Fig. 5), and the major metabolites could be identified and quantified [6].

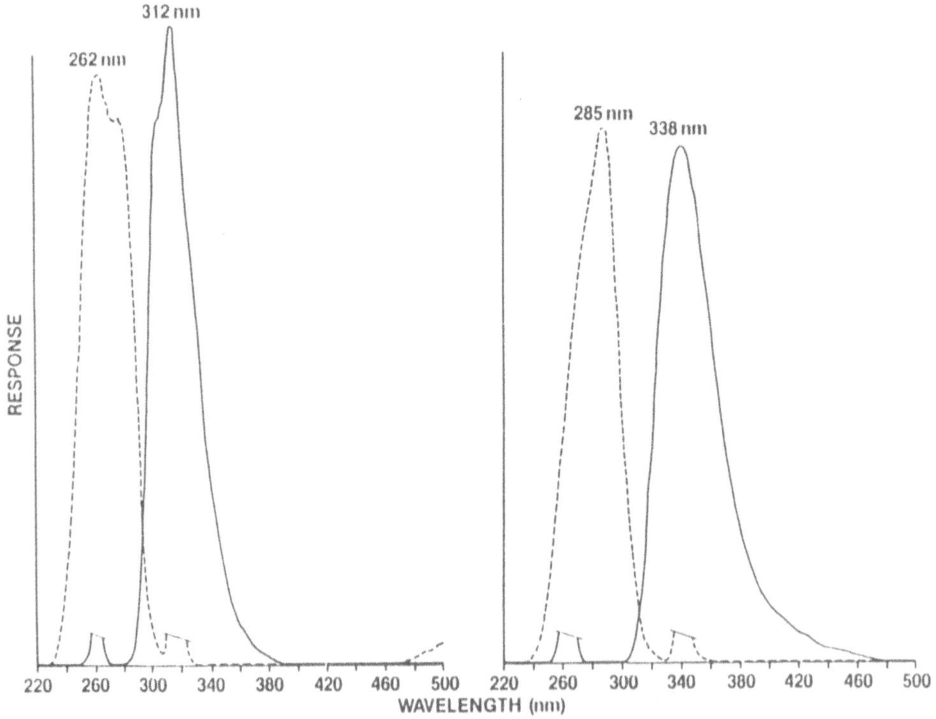

Fig. 4. Absorption (----) and emission (——) spectra: *left*, flurbiprofen; *right*, 4'-hydroxyflurbiprofen. Evidently 260 nm is suitable for excitation of either; 320 nm emission was chosen to give maximum sensitivity for drug, still adequate for metabolite.

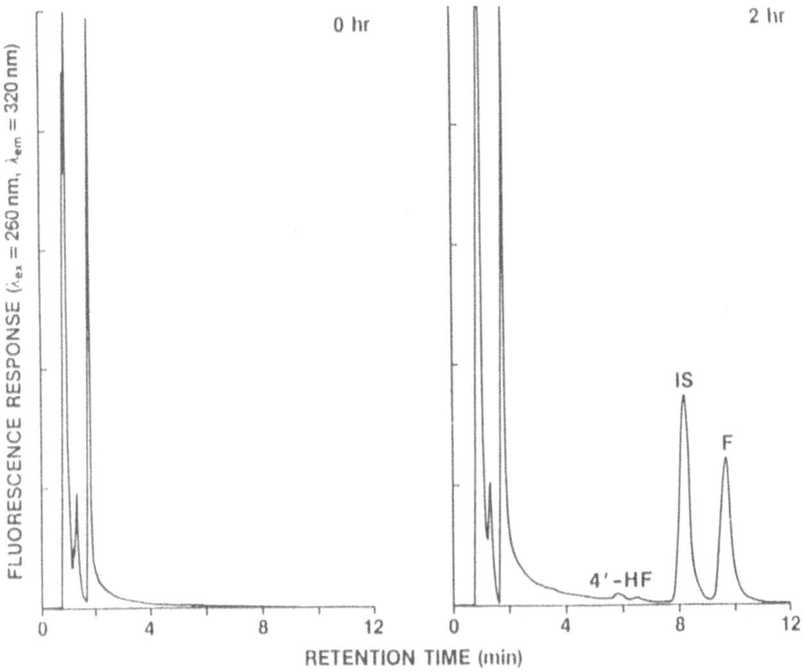

Fig. 5. Typical chromatogram (procedure as in Scheme 4) for *(left)* blank plasma, and *(right)* plasma 2 h after a dose of 50 mg flurbiprofen in man. Full quantitation of circulating hydroxy metabolite (cf. the trace amount seen) needs initial conjugate hydrolysis.

Because of the simplicity of the method, full automation was possible. The main considerations in introducing robotics are as follows.-

Operator actions, irrespective of introducing robotics:
Thaw samples. Pipette serum, then calibration standards and appropriate blank additions.
Operator actions that become first robotic steps:
Pipette precipitating solvent containing i.s. Vortex-mix.
Remaining steps, if done by operator:
Centrifuge. Transfer supernatant to vials. Add buffer. Load autosampler, which injects aliquots onto HPLC column.
ALTERNATIVELY, remaining steps if done by robotics:
Pipette buffer into the tube. Vortex-mix. Centrifuge. Inject aliquots onto HPLC column.

Comments.- A variant of the above fluorescence method [7] proved robust enough for contract-laboratory work and, particularly with robotics, has become the 'work-horse' method within the drug development programme. However, the method was made possible only by constant review of the accruing analytical information and by the drug develop-

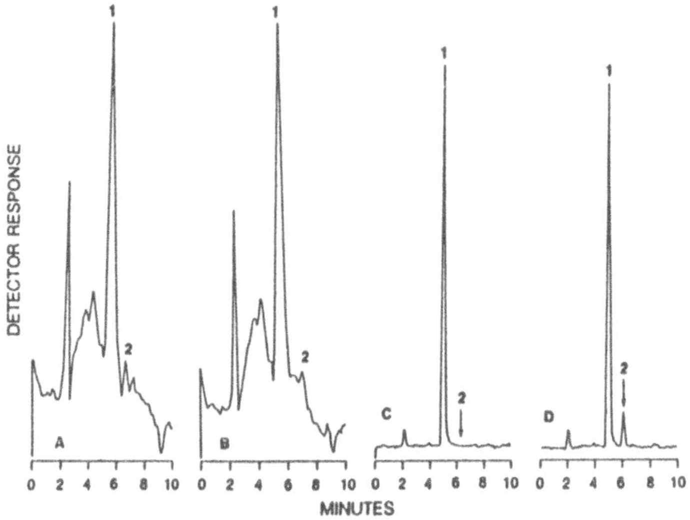

Fig. 6. Typical chromatograms (as in [7]) for human serum using UV (**A**, **B**) and fluorescence (**C**, **D**) detection. All samples contained 0.1 μg/ml i.s. (**1**), and **B** and **D** also contained 0.1 μg/ml flurbiprofen (**2**). Differentiation from the blank was achievable by the fluorescence method but not by the UV method at this sensitivity.

ment team's integrated approach. The clean simplicity of the fluorescence method together with a good baseline are the prime requirements of any routine assay. Fig. 6 exemplifies within-laboratory method comparison, with the lowest level measurable by the fluorescence assay: the blank in the UV assay showed an artefactual 'flurbiprofen' peak.

OVERALL METHOD DEVELOPMENT, METABOLITES, AND SAMPLES OTHER THAN BLOOD

We have outlined the analytical approach that was adopted for the compound. Although any approach must be compound-specific we stress that methodology should evolve with the needs placed on it. It is always desirable to streamline methodology, but the absolute requirement is for specificity, with the attendant sensitivity - which, in the final method, allowed 0.1 μg/ml of the drug to be measured in plasma (or serum).

The occurrence of metabolites was established largely by the analysis of urine samples. For urine, in which the drug levels were such that the aliquot sizes could be similar to those for plasma, a preliminary splitting of conjugates was performed, by alkaline hydrolysis followed by neutralization [6]. Other analytical steps were similar to those outlined for plasma in Schemes 1-3 and - protein removal being inapplicable - in Scheme 4, where a hyrolysate was obtained as follows.- Instead of the 0.1ml 0.5 M NaCl addition, 0.05 ml 1 M NaOH was added, and then, after 20 min at room temperature to effect hydrolysis, 0.05 ml 1 M HCl.

Although metabolites at first seemed irrelevant to plasma, further work showed trace amounts (Fig. 5). This influenced the evolving design of HPLC mobile-phase composition that is evident from the successive Schemes outlined above; thus Scheme 4 used THF rather than acetonitrile. Throughout this method development the prime chromatographic requisite (cf. Table 1) was for an acidic aqueous buffer with organic modifier on a C-18 RP column. For good selectivity of the compounds of interest, high concentrations of modifier were needed (40-65% v/v); hence 'fine tuning' was difficult. Indeed we found that changing solvents (methanol, acetonitrile, THF) was the most effective way of changing selectivity. In our hands THF gave best selectivity and the conditions described in Scheme 4 suited our needs best.

Our assays have been mainly on plasma or serum (see legend to Scheme 2) and urine. However, from time to time the methods (mainly Schemes 1 and 4) have been found useful for bile, faeces and toxicology samples.

References

1. Litowitz, H., Olanoff, L. & Hoppe, C.L. (1984) *J. Chromatog.* *311*, 443-448.
2. Kaiser, D.G. & Martin, R.S. (1978) *J. Pharm. Sci.* *67*, 627-630.
3. Kaiser, D.G. & Van Giessen, G.J. (1974) *J. Pharm. Sci.* *63*, 219-221.
4. Kaiser, D.G., Show, S.R. & Van Giessen, G.J. (1974) *J. Pharm. Sci.* *63*, 567-570.
5. Risdall, P.C., Adams, S.S., Crampton, E.L. & Marchant, B. (1978) *Xenobiotica* *8*, 691-704.
6. Adams, W.J., Bothwell, B.E., Bothwell, W.M., Van Giessen, G.J. & Kaiser, D.G. (1987) *Anal. Chem.* *59*, 1504-1509.
7. Albert, K.S., Gillespie, W.R., Raabe, A. & Garry, M. (1984) *J. Pharm. Sci.* *73*, 1823-1825.

#B-5

THE ASSAY OF DICLOFENAC AND METABOLITES: A REVIEW

P.H. Degen

Research and Development, Pharmaceuticals Division
Ciba-Geigy Ltd., CH-4002 Basle, Switzerland

Diclofenac sodium, o-(2,6-dichlorophenyl)aminophenylacetic acid sodium salt, is a well-established potent NSAID, given in low doses (25-100 mg/day). In man, diclofenac is extensively metabolized mainly by oxidative pathways followed by glucuronide conjugation. A small percentage of the original drug is glucuronidated directly. As diclofenac has a high total clearance, the plasma levels are low, necessitating a specific and very sensitive assay for pharmacokinetic studies.*

As diclofenac contains two halogen atoms, the first choice was GC-ECD, after derivatizing the carboxyl group by esterification. However, diclofenac esters tend to partially cyclize to an indolinone, particularly at acid pH or high temperature as in a GC injection block. The remedy is either complete conversion to a cyclized form or, without pre-chromatographic derivatization, use of HPLC. Amongst the available assays (GC, GC-MS, HPLC, TLC), only a few are suitable for routine determinations of low concentrations. Recent developments include assays for the mono- and di-hydroxy metabolites, whose levels can help demonstrate absorption in biopharmaceutical studies.

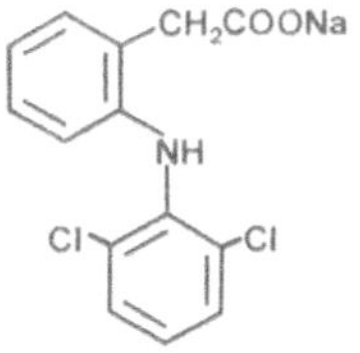

$C_{14}H_{10}Cl_2NO_2Na$ MW 318.13

DICLOFENAC SODIUM

VOLTAREN (VOLTAROL)

For the potent NSAID diclofenac sodium, the active ingredient of VOLTAREN, oral administration results in ~60-70% being excreted in urine as conjugates of hydroxylated metabolites (3'-hydroxy-; 4'-hydroxy-; 5-hydroxy-; 4',5-dihydroxy-; 3'-hydroxy-4'-methoxy-diclofenac) which analytical method design, especially for urine, should comprehend. With HPLC methodology, diclofenac may be measured directly

**Abbreviations*.- NSAID, non-steroidal anti-inflammatory drug; i.s., internal standard; ECD, electron-capture detector (similarly NPD, FID).

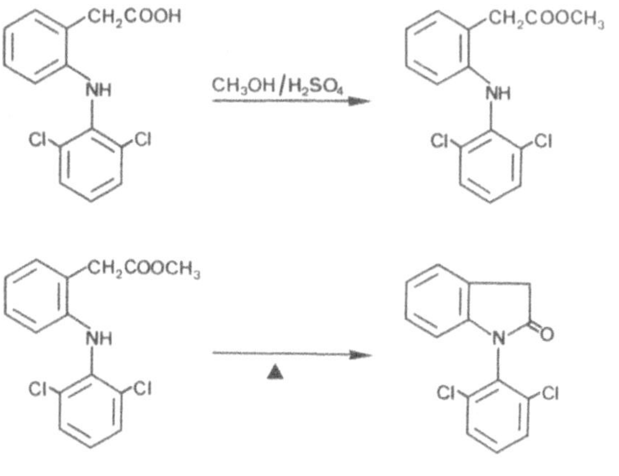

Fig. 1. Esterification and cyclization to the indolinone.

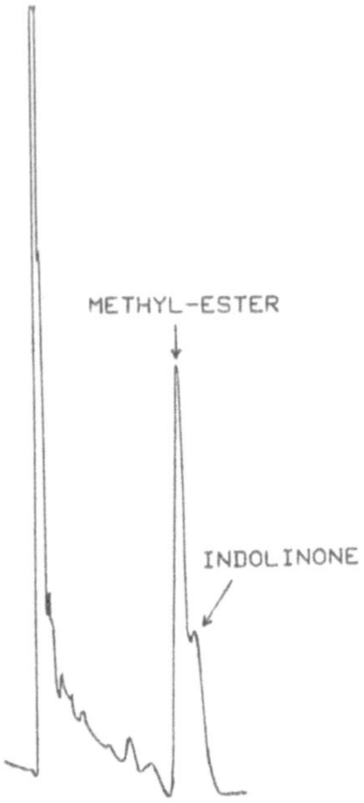

Fig. 2 *(right)*. GC chromatogram of diclofenac methyl ester.

when extracted from biological material, using UV detection. However, the sensitivity of these procedures is insufficient for all types of pharmacokinetic studies. For quantification by GC, diclofenac and its metabolites must be derivatized. This may be achieved by esterification after liquid-liquid extraction, or by cyclization to an indolinone or by extractive alkylation. Fortunately, diclofenac already contains two chlorine atoms that facilitate its ECD detection. Quantification by GC-MS is the method of choice when low levels of diclofenac are anticipated, as in studies of its percutaneous absorption.

STABILITY

Diclofenac sodium itself is stable in weakly alkaline solutions. Some of the extracted metabolites are susceptible to oxidative degradation, which can be reduced by adding ascorbic acid. In urine, conjugates of otherwise unchanged diclofenac hydrolyze at room temperature and during thawing of frozen samples. However, this is not a major problem since in urine the sum of free and conjugated compounds is measured, the strong alkali in the extractive alkylation being hydrolytic.

METHODOLOGY

The published assay methods are listed in Table 1. Diclofenac is easily extracted from biological fluids at acid pH. Esterification of the carboxyl group for subsequent GC gives an ester which may partly cyclize (Fig. 1) in the hot GC injection block, producing varying ratios of methyl ester and indolinone (Fig. 2).

Table 1. Methods for the assay of diclofenac in biological fluids. D = diclofenac; hydroxylated metabolites denoted (e.g.) 5, 4'5 (dihydroxy). P = plasma, S = serum, U = urine*. Formation of an indolinone denoted →ind. Extractive alkylation denoted ex-alk. For UV detection: nm value stated. LOQ = limit of quantification, ng (n) or µg (µ)/ml (for nmol/l, n × 3.14). DCM = dichloromethane; Me = methyl; i = iso; Et_2O = diethyl ether.

Assayed	Extraction (ex.), derivatization and end-step (pck = packed, cpy = capillary)		LOQ	Ref.
D; U, hyd.	Benzene ex.; →ind.;	pck-col. GC-ECD	2 n	1
D; P, S	DCM ex.; acylation with acetyl chloride;	pck-col. GC-FID	100 n	2
D, 4', 5, & 4'5; U ± hyd.	ex-alk. with MeI/DCM →ind.	pck-col. GC-ECD	10 n	3
D; P	C-18 SepPak ex.;	HPTLC; 290 nm	<100 n	4
D + other NSAID's; P	Deproteinization with acetonitrile;	HPLC; 254 nm	40 n	5
D; P	Benzene ex.; → Me ester	pck-col. GC-ECD	2.5 n[Ø]	6
D, 4', 5, 3' & 4',5; U, hyd.	ex-alk. with MeI/DCM →ind.;	cpy-col. GC-ECD	30 n	7
D; S, U	Chloroform ex.;	HPLC; 254 nm	1 µ	8
D; P	Hexane/iPrOH (90:10) ex. (freeze aq. phase);	HPLC; 215 nm	5 n	9
D; S	unstated;	direct UV	0.6 µ	10
D; blood, U, tissue	Chloroform ex.; react with Na periodate & H_2SO_4;	colorimetry, 550 nm	5 µ	11
D; P, U	Toluene ex.; →Me ester	pck-col. GC-MS	0.2 n	12
D; P	Benzene ex.; →ind.;	pck-col. GC-MS	0.2 n[+]	13
D; P	Chloroform ex.; → amide	pck-col. GC-NPD	100 n	14
D; P	Benzene ex.;	HPLC-electrochem. det.	<5 n	15
D + other NSAID's; U	Et_2O ex.;	HPLC; 254 nm	?	16
D, P. D + 3', 4' & 5; U.	Hexane/iPrOH (90:10) ex.;	HPLC; 282 nm	P: 10 n U: 200 n	17
D + 3', 4', 5, 4'5, 3',4'-methoxy; P	Et_2O/DCM (2:1) ex.; ex-alk. with MeI;	cpy-col. GC-ECD	10 n[+]	18
D + other NSAID's; P	Deproteinization with perchloric acid;	HPLC; 254 nm	400 n[+]	19

*hyd. denotes hydrolysis, ±hyd. before and after hydrolysis.
Ø LOQ much higher (100 n) in this group's earlier use (1980) of this approach.
+ limit of detection (LOD), not LOQ.

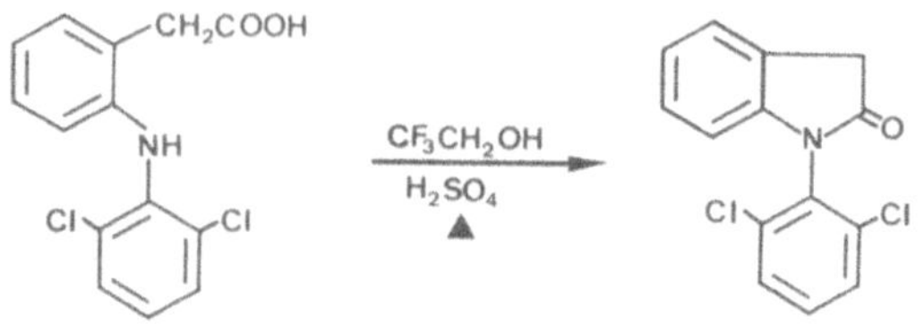

Fig. 3. Direct cyclization of diclofenac to the indolinone with trifluoroethanol/H_2SO_4.

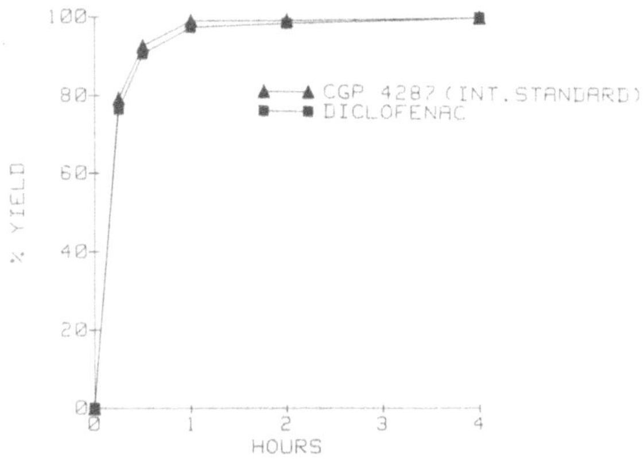

Fig. 4. Kinetics of the formation of indolinone as in Fig. 3, at 80°.

Therefore an ester was sought which was easier to cyclize than the methyl ester. The reaction with trifluoroethanol/H_2SO_4 directly produces the indolinone (Figs. 3 & 4) and led to the first GC method [1] using packed columns and ECD. Despite the fairly complicated work-up, the procedure was used in most pharmacokinetic and bioavailability studies [e.g. 20-25].

Ikeda et al. [6] reported improved sensitivity (× 2-3) when using the methyl ester method instead of indolinone formation; but in our hands the improvement was only ~1.6-fold. Moreover, the chromatograms showed considerably more interfering peaks than after indolinone formation.

A procedure involving GC-MS was reported [13] for determining very low plasma levels (0.6 nmol/l detectable) after percutaneous application of a topical formulation of diclofenac to humans. After extraction, diclofenac was converted to its indolinone by treatment with pentafluoropropionic anhydride in hexane at room temperature (30 min). The indolinone was found to be stable for at least 4 days at room temperature.

Other procedures, based on GC-FID [2], HPTLC [4], HPLC [5, 8, 15, 19], GC-MS [12], spectrophotometry [10, 11] or GC-NDP [14] are somewhat insensitive. For routine determinations of unchanged diclofenac in biological fluids after oral, rectal or parenteral applications, two HPLC procedures proved to be very effective and reliable [9, 17].

Fig. 5. Structures of diclofenac and 5 metabolites:

I, diclofenac;
II, 4'-OH-diclofenac;
III, 5-OH-diclofenac;
IV, 3'-OH-diclofenac;
V, 4',5-di-OH-diclofenac;
VI, 3'-OH-4'-methoxy-diclofenac.

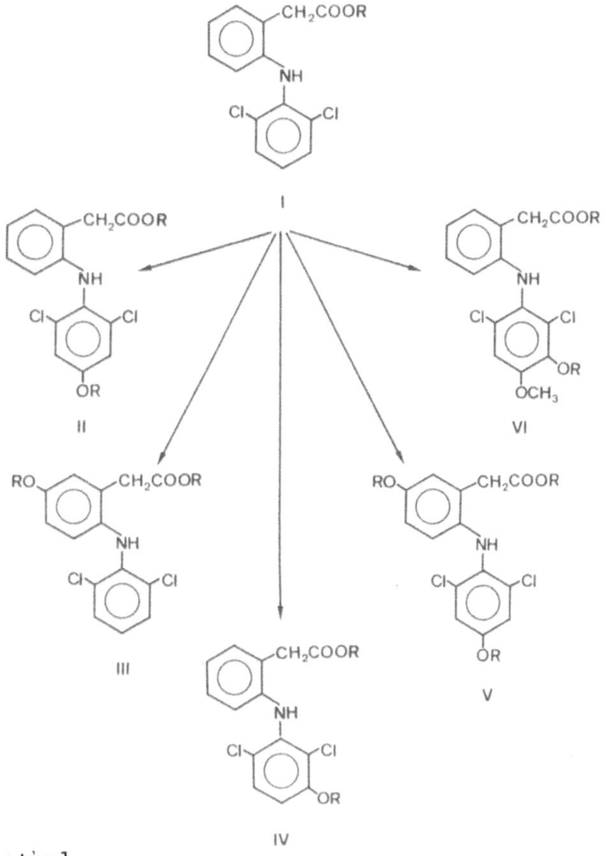

Fig. 6. Formation of dimethyl-indolinone derivatives by extractive alkylation. I-VI as in Fig. 5 legend. (CH_3J = methyl iodide.)

	R	R'	R'' *in deriv.*
I	H	H	H
II	H	H	OCH_3
III	OCH_3	H	H
IV	H	OCH_3	H
V	OCH_3	H	OCH_3
VI	H	OCH_3	OCH_3

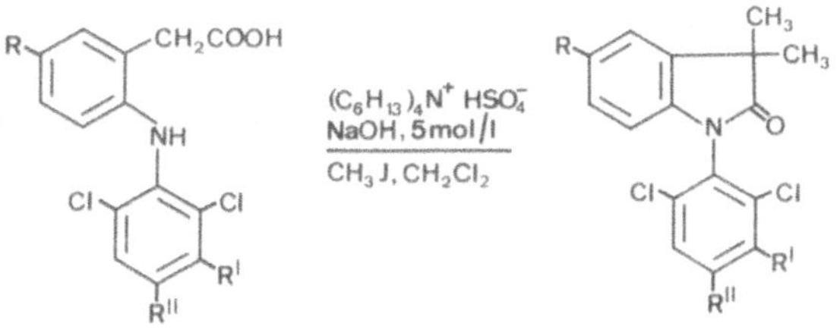

Methods that allowed the measurement of metabolites (Fig. 5) were based on GC-ECD with packed columns [3], HPLC [17] or GC-ECD with capillary columns - the only approach that allows all known metabolites of diclofenac to be resolved. The simultaneous determination of diclofenac and 5 metabolites is achievable by extractive alkylation (Fig. 6) followed by capillary-column GC-ECD (Fig. 7). The products of the extractive alkylation with methyl iodide are

Fig. 7. Chromatogram of a plasma (1 ml) extract from a volunteer collected 6 h after a single oral dose of diclofenac sodium as a slow-release formulation. The aliquot put onto the GC column was 2 µl. The nmol amount in the 1 ml sample was 1.44 for the i.s. and, for the analytes in Fig. 5 legend, was reckoned to be: I, 1.21; II, 0.65; III, 0.15; IV, 0.35; V, 0.15; VI, 0.32.

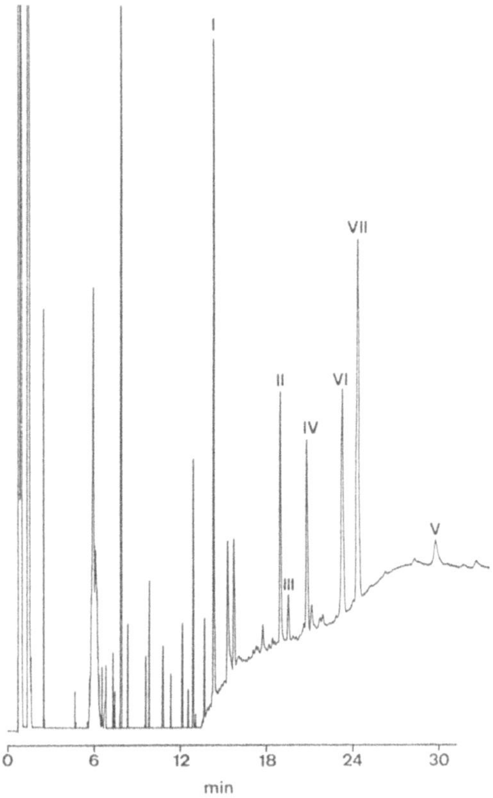

indolinone with two methyl groups on the α-carbon of the phenylpropionic acid moiety. The aromatic hydroxyl groups (of the metabolites) are converted into methoxy derivatives, which have excellent GC properties such as stability and ECD sensitivity. The measurement of diclofenac and the 5 metabolites serves for biopharmaceutical studies (e.g. [26] & Fig. 8). The total (free and conjugated) amount of diclofenac excreted after an oral dose is only ~7%, whereas the sum total of free and conjugated diclofenac and metabolites accounts for ~36%.

The new extractive alkylation procedure may replace the standard GC method, being simpler and faster. Following the solvent extraction at acid pH, the steps in the respective methods are as follows.-
Standard GC method [1] for unchanged drug (LOQ 6 nmol/l):-
Back-extract, & re-extract into organic phase; evaporate down, derivatize; perform a clean-up extraction; evaporate down; —> GC.
Improved GC method [18] for drug & 5 metabolites (LOQ 21 nmol/l):-
Evaporate down; perform extractive alkylation; evaporate down; perform a clean-up extraction, freezing to recover org. phase; —> GC.
Standard HPLC method [17] for drug & some mono-OH met's (LOQ 31 nmol/l):-
Recover the org. phase by freezing; —> HPLC.

Routinely the C.V.'s are ~10% and the LOQ's as stated.

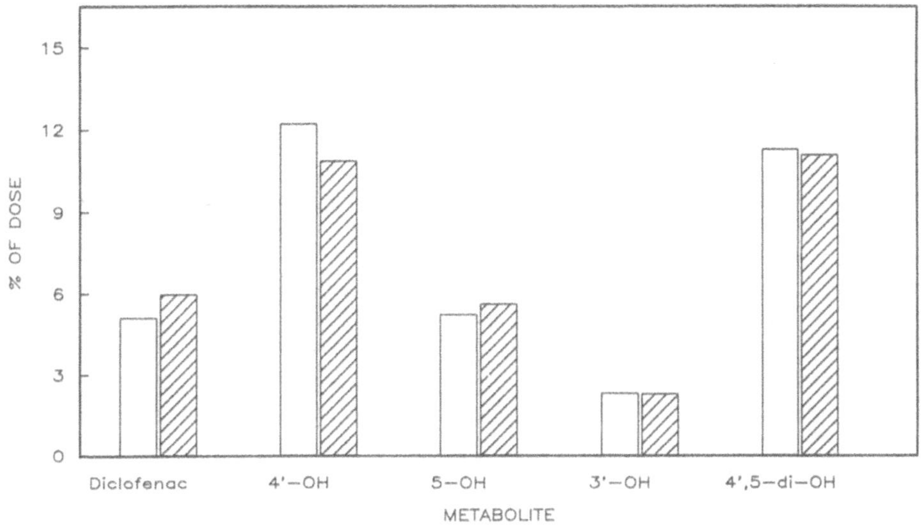

Fig. 8. Mean urinary excretion (0-48 h) of diclofenac and 4 metabolites in 6 healthy volunteers given 4 × 25 mg VOLTAREN enteric coated tablets *(open bars)* and, 2 weeks later, a 100 mg VOLTAREN slow-release tablet *(hatched bars)*. Assays by GC-ECD after extractive alkylation.

CONCLUSIONS

The choice of method for the quantitative determination of diclofenac depends on the type of study to be performed. Plasma determinations of unchanged drug are best done, with speed and reliability, by either the standard GC method or one of the HPLC methods. HPLC is particularly useful when large sample numbers are to be processed. For very low levels in plasma or in synovial fluid, particularly after percutaneous administrations of diclofenac, the GC-MS approach is best; but GC without MS can, by careful optimization of the chromatographic column, measure down to as little as ~6.3 nmol/l (2 ng/ml). For determining low levels of diclofenac together with the individual 5 metabolites in plasma or urine, the choice falls on the capillary-column GC method.

References

1. Geiger, U.P., Degen, P.H. & Sioufi, A. (1975) *J. Chromatog. 111*, 293-298.
2. Brombacher, P.J., Cremers, H., Verheesen, P.E. & Quantjel-Schreurs, R.A. (1977) *Arzneim.-Forsch. 27*, 1597-1599.
3. Schweizer, A., Willis, J.V., Jack, D.B. & Kendall, M.J. (1980) *J. Chromatog. 195*, 421-424.
4. Schumacher, A., Geissler, H.E. & Mutschler, E. (1980) *J. Chromatog. 181*, 512-515.

5. Nielsen-Kudsk, F. (1980) *Acta Pharmacol. Toxicol. 47*, 267-273.
6. Ikeda, M., Kawase, M., Kishie, T. & Okmori, S. (1981) *J. Chromatog. 223*, 486-491.
7. Schneider, W. & Degen, P.H. (1981) *J. Chromatog. 217*, 263-271.
8. Said, S.A. & Sharaf, A.A. (1981) *Arzneim.-Forsch. 31*, 2089-2092.
9. Chan, K., Vyas, K.H. & Wnuck, K. (1982) *Anal. Lett. 15*, 1649-1663.
10. Jakovljevic, V. (1983) *Med. Pregl. (YU) 36*, 145-148.
11. Fartushny, A.F., Muzhanosvsky, E.B., Sedov, A.I. & Kvasor, E.V. (1984) *Farm. Zh. (KIEV) No. 4*, 46-49.
12. Möller, H., Stüber, W. & Ding, R. (1984) *Pharmazeut. Ztg. 129*, 2387-2392.
13. Kadowaki, H., Shiino, M., Uemura, I. & Kobayashi, K. (1984) *J. Chromatog. 308*, 329-333.
14. Lingemann, H., Haan, H.B. & Hulshoff, A. (1984) *J. Chromatog. 336*, 241-248.
15. Plavsic, F. & Culig, J. (1985) *Human Toxicol. 4*, 317-322.
16. Battista, H.J., Whinger, G. & Henn, R. (1985) *J. Chromatog. 345*, 77-89.
17. Godbillon, J., Gauron, S. & Metayer, J.P. (1985) *J. Chromatog. 338*, 151-159.
18. Schneider, W. & Degen, P.H. (1986) *J. Chromatog. 383*, 412-418.
19. Owen, S.G., Roberts, M.S. & Friesen, W.T. (1987) *J. Chromatog. 416*, 293-302.
20. Riess, W., Stierlin, H., Degen, P.H., Faigle, J.W., Gérardin, A., Moppert, J., Sallmann, A., Schmid, K., Schweizer, A., Sulc, M., Theobald, W. & Wagner, J. (1978) *Scand. J. Rheumatol., Suppl. 22*, 17-29.
21. Tsuchiya, T., Terekawa, M., Ishibashi, K., Noguchi, H. & Kato, R. (1980) *Arzneim.-Forsch. 30*, 1650-1653.
22. Zimmerer, J., Tittor, W. & Degen, P.H. (1982) *Fortschr. Med. 100*, 1683-1688.
23. Crook, P.R., Willis, J.V., Kendall, M.J., Jack, D.B. & Fowler, P.D. (1982) *Eur. J. Clin. Pharmacol. 21*, 331-334.
24. Haapasaari, J. Wuolijoki, E. & Ylijoki, H. (1983) *Scand. J. Rheumatol. 12*, 325-330.
25. Naito, S.I. & Tominga, H. (1985) *Int. J. Pharmaceutics 24*, 115-124.
26. Riess, W., Brechbühler, S., Degen, P.H., Dieterle, W., Dörhöfer, G. & Feldmann, K.F. (1983) *Int. J. Clin. Pharm. Res. III (6)*, 495-510.

#B-6

ENANTIOMERIC ANALYSIS OF 2-PHENYLPROPIONIC ACID NSAID'S IN BIOLOGICAL FLUIDS BY HPLC

[1]A.J. Hutt and [2]J. Caldwell

[1]Department of Pharmacy
Brighton Polytechnic
Brighton BN2 4GJ
U.K.

[2]Department of Pharmacology and
Toxicology, St. Mary's Hospital
Medical School
London W2 1PG, U.K.

The importance of stereochemistry in the disposition of the 2-Ø-propionic acid NSAID's is outlined. General approaches for the chromatographic resolution of these agents by both direct and indirect methods are briefly reviewed. Problems associated with the choice of HPLC mobile phase for the resolution of, and derivatization methods for the formation of, stable diastereoisomeric derivatives are illustrated with reference to pirprofen. The application of chiral mobile phases and commercially available HPLC CSP's to the analysis of these agents is examined, and problems which may arise during the 'work-up' of samples of biological origin are discussed.*

The 2-Ø-propionic acids (the 'profens') are an important NSAID group used for treating a wide variety of rheumatic diseases. They possess a chiral centre in the propionic acid moiety, and their pharmacological activity resides in the enantiomers of the *S* absolute configuration (generally dextrorotatory), the *R* enantiomers being either weakly active or inactive *in vitro* [1]. These differences in activity may be obscured *in vivo* due to the metabolic chiral inversion of the *R* enantiomers to their active *S* antipodes [2, 3]:

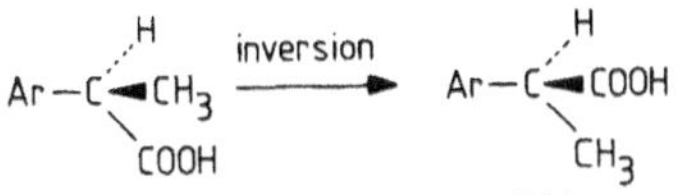

R(-) S(+)

* *Abbreviations, besides Editor's* t_r *for* retention time, Ø *for* phenyl.- NSAID, non-steroidal anti-inflammatory drug; THF, tetrahydrofuran; DMOA, *N,N*-dimethyloctylamine; CSP, chiral stationary phase; RP, reversed phase; MS, mass spectrometry (EI, electron impact).

The extent of the inversion depends on the structure of the acid and the animal species under investigation [3]. Besides the stereospecific chiral inversion, many of these agents undergo other stereoselective metabolic transformations, e.g. conjugation with glucuronic acid [4] and oxidation [2]. The overall result is more rapid elimination of the *R*- than of the *S*-enantiomers, an exception being the enantiomers of tiaprofenic acid which are eliminated at similar rates [5]. The majority of the 'profens' are racemic mixtures as bought and administered therapeutically, the exceptions being naproxen and the recently introduced flunoxaprofen [6] which are used as their *S*-enantiomers. Hence the determination of the enantiomeric composition of the 'profens' in biological fluids is important in both pharmacology and toxicology so that exposure to the active agent may be assessed particularly in studies attempting to relate biological activity to plasma drug levels.

GENERAL METHODOLOGY

Published general methods for enantiomeric HPLC analysis of the 'profens' involve either diastereoisomer formation followed by achiral chromatography (indirect resolutions) [7, 8], or formation of labile diastereoisomer complexes by interactions with a chiral solvent (e.g. quinine as a chiral counter-ion [9]) or surface, i.e. a CSP [10, 11] (direct resolutions). Many of these methods require prior derivatization of the 'profen' with either a chiral [7, 8] or an achiral [10] reagent to obtain useful chromatographic resolution.

Reaction of enantiomeric analytes with a second chiral compound (Fig. 1 gives examples) to yield a pair of diastereoisomer derivatives may give rise to problems associated with poor product yield, the optical purity of the reagent, racemization of the 'profen' and/or the chiral reagent and stereoselective derivatization - the so-called kinetic resolution. Many of these problems are associated principally with the indirect method of resolution and are less important when direct methods of resolution are employed, e.g. if derivatization of the 'profen' is required the reagent need not be chiral ([10, 12]; but see later and ref. [13]). Many of the chiral systems are sensitive to the presence of trace organic impurities often found in samples of biological origin and, to obviate problems, necessitating a sample 'clean-up' before enantiomeric analysis is carried out.

Both chromatographic approaches yield information on the absolute configuration of an analyte if analogues of known absolute configuration are available for comparison. This is of particular significance in the case of the 'profens' as many studies rely on chromatographic methods to demonstrate that chiral inversion has taken place. In addition, the elution order of the enantiomers or their diastereoisomeric derivatives may be reversed by either reversing the chirality of the derivatizing agent (indirect methods) or that of either the

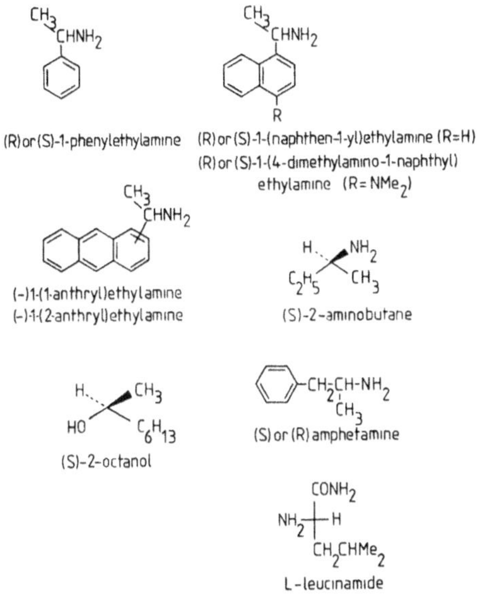

Fig. 1. Examples of chiral amines and alcohols used to derivatize the 'profens' prior to chromatography of the diastereoisomers formed.

Fig. 2. Metabolic and chemical oxidation of pirprofen - 2-[3-chloro-4-(3-pyrrolin-1-yl)phenyl]-propionic acid - to its corresponding pyrrole derivative - 2-[3-chloro-4-(1-pyrrolyl)-phenyl]propionic acid.

mobile phase additive or the CSP (direct methods). This may be analytically useful as it is possible to arrange for the enantiomer present in lower concentrations to be eluted from the column first, thereby somewhat increasing the sensitivity of the assay.

Pirprofen

Problems which may arise from indirect methods of resolution may be illustrated by our studies on pirprofen (Fig. 2) [14], the (+)-enantiomer of which is ~6 times more active than the (-)-isomer as an inhibitor of PGE_2 synthesis *in vitro* [15]. (**N.B.** The absolute configuration of its isomers has not been determined and the designation of optical rotation given refers to the free acid.)

The initial step was to prepare the two diastereoisomeric amides of pirprofen by reaction with *S*-1-(naphthen-1-yl)ethylamine via the mixed anhydride method (Fig. 3, eqn. 2) using ethylchloroformate. The resulting derivatives were characterized by NMR, MS and elemental analysis. In anticipation of HPLC using a radial-compression column and a non-aqueous solvent system, the resolution of the two synthesized diastereoisomers was examined by TLC using various solvent systems. The choice of solvents was based on a literature survey, e.g. combinations of iso-octane, dichloromethane, methanol, THF and hexane and methanol in dichloromethane.

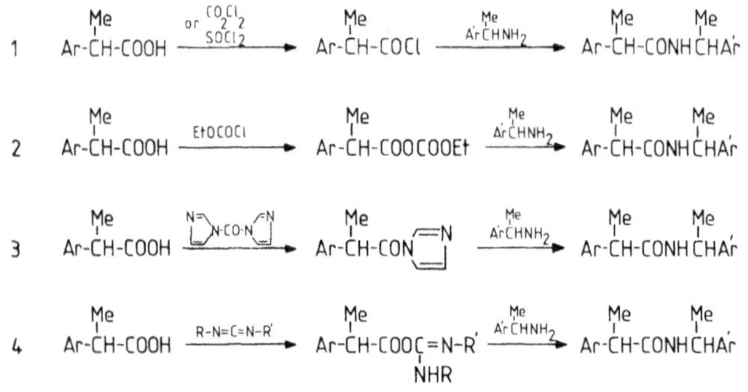

Fig. 3. Methods for derivatizing the 'profens' to yield diastereoisomeric amides. Various reagents have been used to activate the carboxyl goup, e.g. thionyl and oxalyl chlorides to yield acyl chlorides (**1**), ethylchloroformate to yield a mixed anhydride (**2**), 1,1'-carbonyldiimidazole to yield an imidazolide (**3**), and a carbodiimide to yield an intermediate adduct (**4**): e.g. R = R' = cyclohexyl : dicyclohexylcarbodiimide; R = ethyl, R' = 3-dimethylaminopropyl : 1-(3-dimethylaminopropyl)-3-ethylcarbodiimide.

For these diastereoisomers HPLC using hexane/THF (5:1 by vol.) offered promise, but baseline resolution was not obtained and the [(+)-acid : *S*-amine] diastereoisomer consistently gave tailing (Fig. 4) which was not improved by altering the solvent composition. Dichloromethane containing 0.5% (v/v) methanol was then found to give reasonable resolution, but this and the t_r's depended strongly on the source of the dichloromethane (Fig. 4). Different suppliers use different alcohols in different quantities as stabilizer, and the resolution was found to be extremely sensitive to alcohol content. The finally chosen system, ethyl acetate/hexane (1:4 by vol.), was found to be much more robust and less sensitive to minor changes in composition (Fig. 5). Even with this system, the resolution was markedly altered if the diastereoisomers were dissolved in ethanol before injection onto the column (Fig. 5).

DERIVATIZATION

All the techniques for resolving the 'profens' by the formation of stable diastereoisomeric derivatives, and some of those based on the direct methods, require derivatization, generally via chemical activation of the carboxyl group and reaction with the chiral or achiral derivatizing agent. Amongst the favoured methods (Fig. 3), one of the most common is coupling of the 'profen' with 1,1'-carbonyldiimidazole to yield the corresponding imidazolide (eqn. 3) [7, 16-18]. It may entail problems due to formation of ***N,N'***-disubstituted urea derivatives unless acetic acid is added to the reaction mixture *before* the chiral amine [16]. Other peaks, of unknown origin, have been

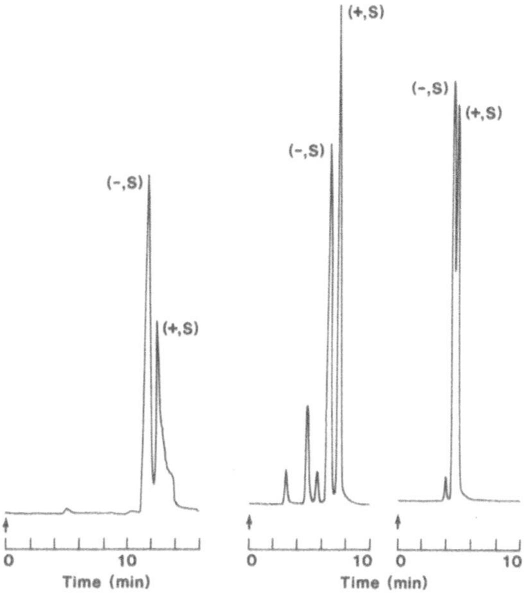

Fig. 4. Resolution of pirprofen enantiomers as *S*-1-(naphthen-1-yl)ethylamides, using: ***left:*** hexane/THF (5:1 by vol.): ***centre*** and ***right:*** methanol (0.5% v/v) in dichloromethane from two suppliers. Column: Radial-Pak cartridge (100 × 5 mm i.d.) containing 10 μm Porasil housed in a Z-module (Waters); M45 pump; M440 UV detector; Rheodyne 7251 injector (20 μl loop); 0.5 ml/min flow-rate.
Peak designation: (-,*S*) and (+,*S*) refer to the diastereoisomers prepared by reaction of either the (-)- or the (+)-acid and the *S*-amine.

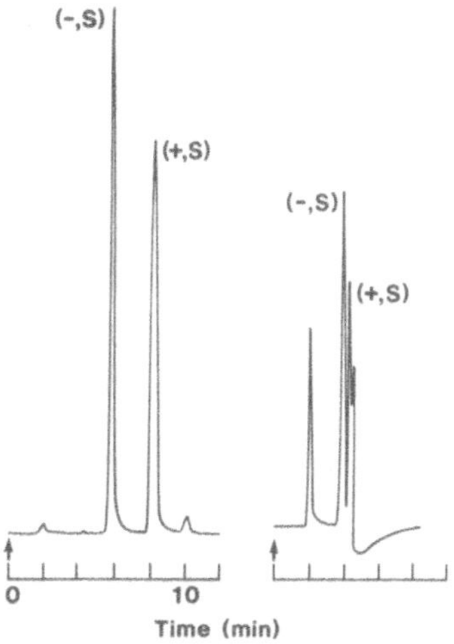

Fig. 5. HPLC of pirprofen *S*-1-(naphthen-1-yl)ethylamide derivatives following injection onto the column dissolved in either HPLC solvent *(**left**)* or ethanol *(**right**)*. HPLC conditions and peak designation as for Fig. 4. Solvent: hexane/ethyl acetate (4:1 by vol.); flow rate 0.8 ml/min.

observed in chromatograms when high concentrations of the coupling reagent are used [17], and its optimal concentration remains debatable [7, 17].

Using (±)-pirprofen, we examined the application of carbodiimides as coupling reagents (Fig. 3, eqn. 4). With dicyclohexylcarbodiimide the chromatogram showed new peaks, not due to the derivatizing agents, viz. the diimide and the amine [*S*-1-(naphthen-1-yl)ethylamine], nor to the diastereoisomer products. The major culprit was isolated and, by direct-insertion MS, identified as the adduct (Fig. 6) formed on reaction of the acid with the diimide (Fig. 3, eqn. 4). Changing to 1-(3-dimethylaminopropyl)-3-ethylcarbodiimide furnished 'clean' chromatograms.

Reaction rate and product yield were next investigated. Use of the chiral amine and the diimide coupling reagent gave, after 1.5 h,

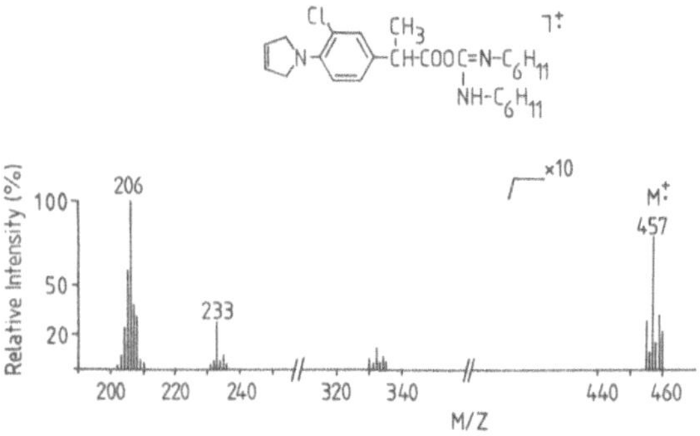

Fig. 6. Direct-insertion MS (EI, 70 eV) of the adduct isolated from reaction of racemic pirprofen with *S*-1-(naphthen-1-yl)-ethylamine using dicyclohexylcarbodiimide as coupling agent. Structure assignment: the odd mol. wt. (m/z 457) indicates the presence of an odd no. of N's; the ratio $M^{+\cdot} : [M+2]^{+\cdot} = 3:1$ and the loss of 2 protons to yield the ion $[M-2]^{+\cdot}$ (corresponding to formation of the pyrrole derivative) show presence of a pirprofen residue, and the ratio $M^{+\cdot} : [M+1]^{+\cdot}$ indicates the presence of 26 C's in the molecule.

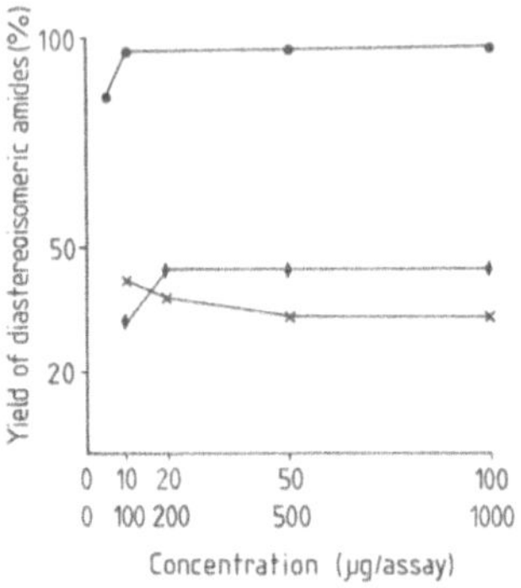

Fig. 7 Effect of concentration of 1-(3-diMe-aminopropyl)-3-Et-carbodiimide (◆) and *S*-1-(naphthen-1-yl)ethylamine (×) *(both lower scale)* and *(upper scale)* 1-OH-benzotriazole (1-HOBT; ●) on the yield of pirprofen diastereoisomeric amides. Drug racemate and other reagents: 100 µg/assay; for ◆ and ×, 1-HOBT omitted; 1.5 h reaction in each case. Constancy of peak ht. ratios for the 2 diastereoisomers indicates no stereoselective derivatization.

~40% yield, the HPLC reference material being the pure diastereoisomers. Published accounts of derivatization methods rarely state the overall yield; data are generally reported as 'final values' or as compound recovery on a % basis, which may be highly misleading in an analysis of the derivatization step. Fig. 7 shows the effect of increasing the concentration of the chiral amine or the diimide. Raising the diimide above 200 µg/assay had little effect, and raising the amine concentration decreased the product yield. Addition of 1-hydroxybenzotriazole (1-HOBT) to the reaction mixture, as little as 5 µg/assay, increased the yield appreciably (Fig. 7). Table 1 shows the effect of 1-HOBT on the derivatization time course.

Pirprofen undergoes both metabolic and chemical oxidation to yield its 'pyrrole' derivative, 2-[3-chloro-4-(1-pyrrolyl)phenyl]propionic acid [19] (Fig. 2), which is known to account for ~10% of the drug-related material in blood [19]. Fig. 8 shows the resolution of this compound's enantiomers as their diastereoisomeric amides. The elution order of the peaks corresponded to that of the products obtained by oxidation of pirprofen enantiomers. It is worthy of note that the resolution factor of the oxidation-product diastereoisomers exceeds that of the pirprofen diastereoisomers (Table 2).

Table 1. Time course of the reaction for the formation of the diastereoisomeric amides formed on reaction of the enantiomers of pirprofen with *S*-1-(naphthen-1-yl)ethylamine in the presence and absence of 1-HOBT - 10 μg, along with 100 μg each of the amine, racemic pirprofen and diimide (see text) in 1 ml CH_2Cl_2 at room temp.[⊗] Yields (%) relate to peak heights obtained for a synthetic mixture of the two diastereoisomeric amides. Peaks: 1st, [(-)-acid:*S*-amine]; 2nd (eluting later), [(+)-acid:*S*-amine]. The 1st/2nd height ratios were constant, indicating absence of stereoselective derivatization.

Time, h	***No*** 1-HOBT: peak yields		1-HOBT added: peak yields	
	1st	2nd	1st	2nd
0.0	-	-	-	-
0.25	29	28	73	71
0.50	35	33	91	89
0.75	37	36	-	-
1.00	38	38	90	88
1.50	41	40	91	91
2.00	39	39	95	94

Table 2. Resolution (R_S) and separation (α) factors for pirprofen and related 2-phenylpropionic acid enantiomers as their *S*-1-(naphthen-1-yl)ethylamides. Chromatographic conditions as for Fig. 5. The diastereoisomeric amide [*R*-acid:*S*-amine] (peak 1) precedes the [*S*-acid:*S*-amine] amide (peak 2). Distinguishing the peaks as $_1$ and $_2$, then where K' = capacity factor, t = retention time and w = peak width (min) at base: $\alpha = K_2'/K_1'$ and $R_S = 2(t_2 - t_1)/(w_2 + w_1)$.

Compound	α	R_S
2-Phenylpropionic acid	1.84	4.69
Ibuprofen	2.13	5.14
Carprofen	1.11	5.53
Pirprofen	1.60	4.67
Pirprofen 'pyrrole'	2.29	6.67

Attempts to separate the 'pyrrole' and pirprofen diastereoisomers by chromatography were unsuccessful. Furthermore, using 'spiked' plasma samples, decomposition of pirprofen occurred during extraction. As the parent drug was known to predominate over its oxidation product in plasma, it was decided to convert the pirprofen to its 'pyrrole' using 2,3-dicyano-5,6-dichlorobenzoquinone and then determine the enantiomeric composition of the oxidation product. This proved somewhat difficult as the reagents used in the derivatization and oxidation interfered with one another. Attempts then made to purify the products using a short column were unsuccessful.

[⊗] Finally the residue from rotary evaporation of the solvent was taken up in HPLC solvent (0.5 ml; 10 μl injected).

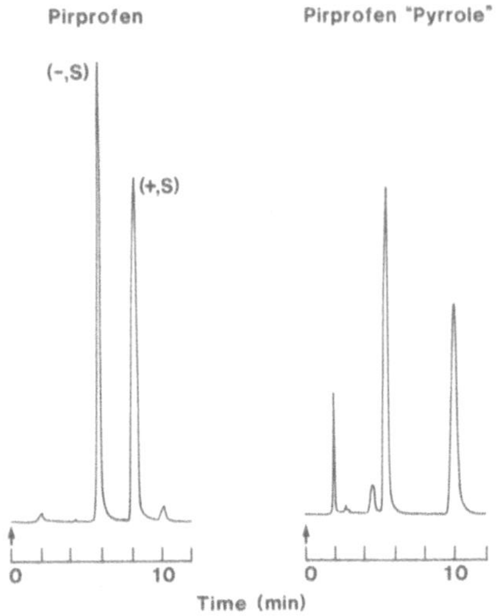

Fig. 8. HPLC of pirprofen *S*-1-(naphthen-1-yl)ethylamides and the equivalents from the 'pyrrole' oxidation product prepared by reaction with 2,3-dicyano-5,6-dichlorobenzoquinone. HPLC as for Fig. 5. The first 'pyrrole' peak corresponds to the (-)-pirprofen peak, i.e. same elution order for the respective diastereoisomers.

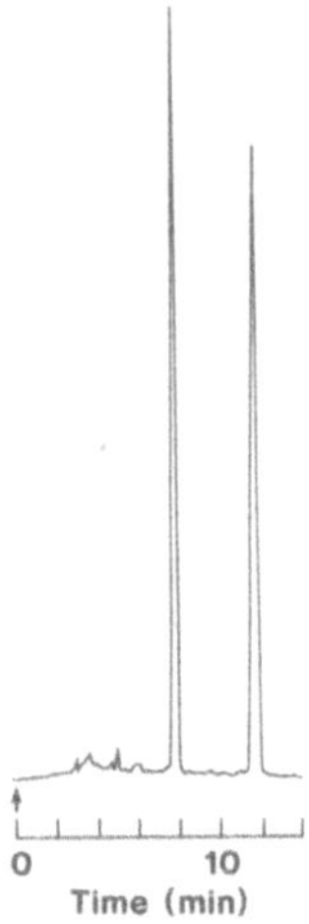

Fig. 9. HPLC of ketoprofen *S*-(1-naphthen-1-yl)ethylamides, prepared by reaction of racemic ketoprofen using conditions already described. The respective peak areas agree within 1%. Column: Hichrom 5 µm silica, 250 × 4 mm i.d., at 25° (water jacket). Solvent: hexane/ethyl acetate (7:3 by vol.); 1.0 ml/min. Pump, Spectra-Physics 8700; detector set at 254 nm (Spectroflow 747).

Other 'profens'.- The above analytical technique has been applied to the resolution of a number of other 'profen' NSAID's (Table 2), and similar results were obtained on transferring the system to a conventional stainless steel packed column (Fig. 9). The method has also been used in examining the stereochemical aspects of the metabolism of 2-Øpropionic acid in the rat, mouse and rabbit [8, 20].

CHIRAL CHROMATOGRAPHIC SYSTEMS

There have been few reports of the application of chiral chromatographic systems to the analysis of the 'profens' in metabolic studies. However, some general methods for their resolution have been reported.

Chiral mobile phases

The resolution of naproxen [2-(6-methoxynaphth-2-yl)propionic acid] using quinine as a chiral counter-ion with pentan-1-ol/dichloromethane (1:99 by vol.) as solvent has been reported [9]. Resolution was achieved using LiChrospher Si 500 and Nucleosil CN stationary phases but not with a LiChrosorb DIOL column [9]. A similar technique achieved resolution of indoprofen enantiomers, but the detection of the analytes was hindered by large background signals due to quinine, which rendered the method unsuitable for metabolic studies [21].

The resolution and subsequent analysis of the enantiomers, in material extracted from soil, of the herbicide fluazifop, viz. *R*,*S*-2-[4-(5-trifluoromethyl-2-pyridyloxy)phenoxy]propionic acid, has been reported [22] using a chiral mobile phase containing a Nickel II complex of L-prolyl-***N***-octylamine [22]. However, this system has not been applied to the analysis of the 'profens'.

We have attempted to resolve the enantiomers of 2-Øpropionic acid, ibuprofen and pirprofen using β-cyclodextrin (0.5% w/v) as a mobile phase additive, but without success although partial resolution was achievable for mandelic acid enantiomers, albeit with too long t_r's [23].

Chiral stationary phases (CSP's)

The resolution of ibuprofen, ketoprofen and naproxen on an α_1-acid glycoprotein column (Enantio Pac) has been reported [11] using DMOA as a mobile phase additive to the phosphate buffer solvent system (pH 5-7) containing propan-2-ol (1% v/v). The retention of the 'profens' was observed to depend on the DMOA concentration: an increase (from 0 to 10 mM) raised the K' of one 'profen' enantiomer whilst its antipode was unaffected. The elution order of the enantiomers was reported for naproxen, the *R*-enantiomer being the first eluted peak [11].

Using the alternative commercially available protein-based stationary phase, bovine serum albumin (Resolvosil-7), and various solvent systems based on phosphate buffer (0.1 M, pH 6-7.8) with differing concentrations of propan-2-ol (2-5% v/v), the mandelic acid and warfarin enantiomers could be resolved [23] but not those of 2-Øpropionic acid or pirprofen.

The application of a Pirkle column, (*R*)-***N***-(3,5-dinitrobenzyl)-phenylglycine, to the resolution of the 'profens' was first reported by Wainer & Doyle ([10]; cf. #D-1 and, in Vol. 16, #C-3). Derivatization of the 'profens' was found to be a prerequisite for enantiomeric resolution and a variety of ester and amide derivatives of ibuprofen were prepared. Optimal resolution was obtained using either benzyl or 1-naphthalene-methyl amide derivatives. A model for the interaction between the ibuprofen amide derivatives and the CSP was proposed, as in Fig. 10 [10].

The application of the above system to a metabolic study, in combination with MS, has been described by Crowther et al. [12]. The composition of ibuprofen excreted in horse urine, during 2 h after administration of the racemic compound, was shown to consist predominantly of the *S*-enantiomer. The ibuprofen was isolated from

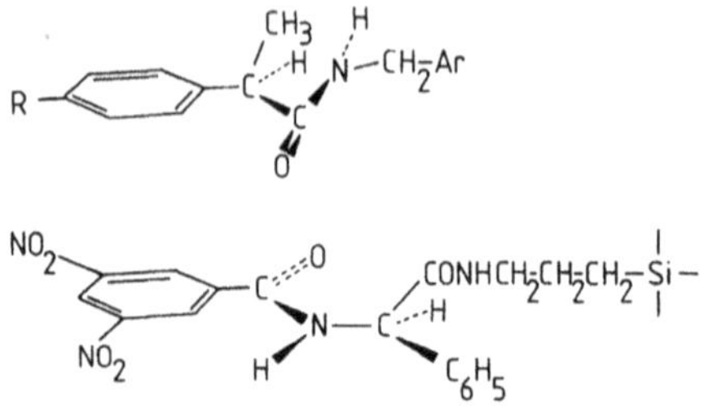

Fig. 10. Model proposed by Wainer & Doyle *(adapted from [10])* to describe the interaction between the (*R*)-ibuprofen 1-naphthalenemethyl amide (R = isoBu, Ar = 1-naphthalene) and the (*R*)-*N*-(3,5,-dinitrobenzoyl)Øglycine CSP. *Possible interaction sites:* amide dipoles, amide H bonding, π — π bonding between the electron-deficient 3,5-dinitrobenzoyl ring and the aromatic ring of the profen, interactions with the glycine residue's carbonyl and Ø groups (e.g. H-bonding, steric &/or π interactions).

urine by solvent extraction, followed by a TLC purification step, before derivatization to yield the benzylamide [12]. On attempting to carry out a similar analysis using crude extracts, an unknown material co-eluted from the column with the *R*-enantiomer derivative.

Recently the same CSP has been used to develop an enantioselective assay for pirprofen in plasma [13]. Problems alluded to above, concerning the separation of the enantiomers of the drug from those of the 'pyrrole' oxidation product, necessitated use of the CSP and formation of stable diastereoisomers (*R*-methylbenzamides) conjointly, to achieve both separation of compounds and resolution of enantiomers [13].

'Clean-up' methods prior to chiral analysis

Many of the methods described for the enantiomeric analysis of the 'profens' require an initial sample 'clean-up' step before the enantiomeric composition of the drug may be determined. This may be due to the presence of trace organic contaminants (the direct resolution methods being particularly sensitive) or to problems associated with other metabolic products of the drug. Most of these methods employ a TLC isolation step [12, 16, 22]. However, the method of Sallustio et al. [24] for 2-Øpropionic acid, ketoprofen and fenoprofen involves RP-HPLC followed by isolation of the chromatographic peak corresponding to the 'profen', extraction, derivatization and normal-phase chromatographic resolution. Whatever work-up procedures are chosen, it must not be forgotten that achiral processes may sometimes cause enantiomeric enrichment of a chiral compound [25]. All procedures must be validated with reference to constancy of enantiomeric composition: if standard enantiomers are not available, then it is necessary to use the racemate, ensuring that its 50:50 *S*/*R* composition remains unaltered [26].

CONCLUDING COMMENTS

Interest has revived in the pharmacology and toxicology of the enantiomers of chiral drug molecules, notably the 'profen' NSAID's in respect of action, chiral inversion and metabolism: since most are used as racemates, we do not even know how much active drug the individual is getting. Deep understanding of 'profen' actions, and hence the development of less toxic compounds, needs awareness of their pharmacology *vs.* their stereochemistry, which in turn depends absolutely on analytical methodologies as here described.

Acknowledgement

The work was supported in part by Ciba-Geigy, Horsham.

References

1. Hutt, A.J. & Caldwell, J. (1984) *Clin. Pharmacokin. 9*, 371-373.
2. Kaiser, D.G., Van Giessen, G.J., Reischer, R.J. & Wechter, W.J. (1976) *J. Pharm. Sci. 65*, 269-273.
3. Hutt, A.J. & Caldwell, J. (1983) *J. Pharm. Pharmacol. 35*, 693-704.
4. Lee, E.J.D., Williams, K., Day, R., Graham, G. & Champion, D. (1985) *Br. J. Clin. Pharmacol. 19*, 669-674.
5. Singh, N.N., Jamali, F., Pasutto, F.M., Russell, A.S., Coutts, R.T. & Drader, K.S. (1986) *J. Pharm. Sci. 75*, 439-442.
6. Lampa, E., Romano, A.R., Berrino, L., Tortora, G., DiGuglielmo, R., Filippelli, A., Gentile, B. & Marmo, E. (1985) *Drug Exp. Clin. Res. 11*, 501-509.
7. Maitre, J-M., Boss, G. & Testa, B. (1984) *J. Chromatog. 299*, 397-403.
8. Hutt, A.J., Fournel, S. & Caldwell, J. (1986) *J. Chromatog. 378*, 409-418.
9. Pettersson, C. (1984) *J. Chromatog. 316*, 553-567.
10. Wainer, I.W. & Doyle, T.D. (1984) *J. Chromatog. 284*, 117-124.
11. Hermansson, J. & Eriksson, M. (1986) *J. Liq. Chromatog. 9*, 621-639.
12. Crowther, J.B., Covey, T.R., Dewey, E.A. & Henion, J.D. (1984) *Anal. Chem. 56*, 2921-2926.
13. Sioufi, A., Colussi, D., Marfil, F. & Dubois, J.P. (1987) *J. Chromatog. 414*, 131-137.
14. Carney, R.W.J., Chart, J.J., Goldstein, R., Howie, N. & Wojtkunski, J. (1973) *Experientia 29*, 938.
15. Ku, E.C. & Wasvary, J.M. (1975) *Biochim. Biophys. Acta 384*, 360-368.
16. Van Giessen, G.J. & Kaiser, D.G. (1975) *J. Pharm. Sci. 64*, 798-801.
17. Singh, N.N., Pasutto, F.M., Coutts, R.T. & Jamali, F. (1986) *J. Chromatog. 378*, 125-135.
18. Pedrazzini, S., Zanoboni-Muciaccia, W., Sacchi, C. & Forgione, A. (1987) *J. Chromatog. 415*, 214-220.
19. Degen, P.H., Schweizer, A. & Sioufi, A. (1984) *J. Chromatog. 290*, 33-43.
20. Fournel, S. & Caldwell, J. (1986) *Biochem. Pharmacol. 35*, 4153-4159.
21. Bjorkman, S. (1985) *J. Chromatog. 339*, 339-346.
22. Bewick, D.W. (1986) *Pestic. Sci. 17*, 349-356. 349-356.
23. Hutt, A.J. & Caldwell, J. (1988) in *Metabolism of Xenobiotics* (Gorrod, J.W., Oelsbacher, H. & Caldwell, J., eds.), Taylor & Francis, London, pp. 335-344.
24. Sallustio, B.C., Abas, A., Hayball, P.J., Purdie, Y.J. & Meffin, P.J. (1986) *J. Chromatog. 374*, 329-337.
25. Tsai, W.L., Hermann, K., Hug, E., Rohde, B. & Dreiding, A.S. (1985) *Helv. Chim. Acta 68*, 2238-2243.
26. Caldwell, J. & Testa, B. (1987) *Drug Metab. Disp. 15*, 587-588.

#NC(B)

NOTES and COMMENTS relating to ANTI-INFLAMMATORY DRUGS

Comments relating to particular contributions: p. 131

#NC(B)-1

A Note on

THE DETERMINATION OF SUPROFEN IN BIOLOGICAL FLUIDS

C.J. Dyde, M.F. Barkworth and K.D. Rehm

iphar Institute for Clinical Pharmacology
Arnikastrasse 4
8011 Höhenkirchen-Siegertsbrunn, W. Germany

Requirement	A simple and specific method for determining suprofen in biological fluids generated from pharmacokinetic studies.
End-step	RP-HPLC (C-18) with UV detection at 290 nm.
Sample preparation	Suprofen and internal standard are extracted from acidified plasma/urine into chloroform. After separation of the phases, the organic layer is evaporated to dryness, and the residue re-dissolved.
Comments	The simple liquid-liquid extraction followed by automated HPLC allows the determination of up to 96 samples and standards per day. The assay shows excellent linearity up to at least 37.5 µg/ml; the detection limit is 0.1 µg/ml. The inter-assay precision (reproducibility) is <5%. Already ~3000 plasma and urine samples from pharmacokinetic studies have been assayed.

Ref. noted by Editor, relevant to the above Forum abstract:-

Michos, N., Zulliger, H.W., Barkworth, M.F., Johnson, K.J., Rehm, K.D., Toberiach, H. & Klein, G. (1986) *Arzneim.-Forsch. 36*, 941-948.

Comments on material in #B

Comments on #**B-1,** K.D. Rainsford - ANTI-INFLAMMATORIES & INTERLEUKINS
& #**B-2,** W.N. Jenner - BECLOMETHASONE ESTER IMMUNOASSAY

K.D. Rainsford, replying to J.C. McElnay.- In studies with chloroquine the presence of foetal calf serum minimized binding to the apparatus: only ~4-8%, depending on the drug concentration. **F. Carey** wondered what mechanism - maybe lysosomal stabilization or an effect on protease activity - underlies the action of anti-malarial drugs in the bovine nasal cartilage system.

W.N. Jenner answered R. Heath, who asked about the purpose of the 0.25 h period at 4° in the main incubation for assaying BDP: it served to temperature-equilibrate for the DCC separation step. **S.A. Wood asked** whether the time taken to raise the antisera was usually as long as 66 weeks. **Jenner's reply:** there was no hurry; boosters were given when the titre fell off. **In reply to G.P. Mould:** trial of different ways of iodinating beclomethasone to reduce the plasma effects gave no evident improvement.

Comments on #**B-3,** J.C. McElnay - HPLC OF ANTI-INFLAMMATORIES
& #**B-4,** A. Bye - APPROACHES TO FLURBIPROFEN ASSAY

Replies by J.C. McElnay to K.D. Rainsford.- The previously reported effects of ultrasound on the absorption through skin of benzydamine and of other drugs was indeed assessed by subjective procedures, e.g. pain relief. Ultrasound seemingly may have some effect unconnected with an influence on pharmacokinetics. **Reply to G.P. Mould:** suppositories are normally given at night, so daytime use for indomethacin was not tried. **A. Bye, answering A.J. Hutt:** we did not, and probably will not, investigate the stereochemistry of flurbiprofen disposition.

Comments on #**B-5,** P.H. Degen - DICLOFENAC ASSAY
#**B-6,** A.J. Hutt - ENANTIOMERS OF NSAID's BY HPLC
& #**NC(B)-1,** M.F. Barkworth - SUPROFEN ASSAY

P.H. Degen, replying to McElnay: a topical formulation of diclofenac is available commercially in some European countries including Switzerland. **Hutt, answering K.G. Feitsma:** for bioanalysis of NSAID's, done by chiral derivatization, we would have welcomed an opportunity to try a Pirkle column. **K.D. Rainsford asked M.F. Barkworth** whether any information is available concerning the relationship of plasma levels of suprofen to the reported flank-pain side-effects of the drug, i.e. whether there is any evidence of elevated levels in patients receiving this drug. **Reply:** we have no experience of this situation and in fact have never seen this side-effect in our Institute. **McElnay** asked whether he had tried analysis of the drug without solvent extraction, merely injecting directly after protein precipitation. **Reply:** not tried, but there is a literature report of such an assay involving column switching. **Comment by R. Woestenborghs:** a false impression of non-bioequivalence can be gained if absorption curves (e.g. syrup compared with capsules) are presented without specification of the doses used.

Supplementation by Senior Editor

Perspective on the **eicosanoid/anti-inflammatory drug 'interface'** concludes the 'NC' pages of Sect. **A.** Comparison of centrifugal TLC and preparative HPLC for separating 'profen' **enantiomers** is cited in the 'NC' pages that follow Sect. **E.** Various **assay methods for anti-inflammatory drugs,** including enantiomers, appear as an 'NC' Table at the end of Sect. **D,** supplementing the foregoing articles. Art. #**B-4** is now reinforced by citation of a C-8 HPLC assay, preceded by perchloric acid precipitation, of **flurbiprofen** (50 ng/ml detectable) in plasma and breast milk with UV detection:

Johnson, V.A. & Wilson, J.T. (1986) *J. Chromatog. 382*, 367-371.

The second entry in that Table (ref. [4], p. 220) relates closely to art.#B-6: for a range of profens, interfering GC peaks were minimized by using amphetamine enantiomers (with 1,1'-carbonyldiimidazole present) to form the diastereoisomers. For assay of biological samples, as demonstrated with etodolic and tiaprofenic acids, the alkalinized sample, if plasma or synovial fluid rather than urine, had to be ether-washed to remove basic impurities.

Section #C

ACE INHIBITORS AND CALCIUM ANTAGONISTS

#C-1

ANALYSIS OF ANGIOTENSIN-CONVERTING ENZYME INHIBITORS IN BIOLOGICAL FLUIDS

A. Rakhit

Clinical Biology, Research Department
Pharmaceuticals Division
CIBA-GEIGY Corp., Summit, NJ 07901, U.S.A.

During the last 10 years of development of several ACE inhibitors, diverse analytical methods including HPLC, GC, GC-MS, REI and RIA have been reported. The different analytical methods are reviewed here with respect to their principles, procedures, strengths and weaknesses, except for RIA (considered in the following article). HPLC detection approaches, some entailing appropriate pre-derivatization, include fluorescence, electrochemical and even UV. Whatever the approach, captopril is detectable down to ~10-20 ng/ml, and pentopril to ~50 ng/ml. For GC measurement the approach, depending on the analyte, may be ECD, FID or NPD. Initial reaction steps followed by GC-MS with SIM have also been used for captopril and its mixed disulphides and S-methyl metabolites in different biological fluids. The sensitivity limit ranges between 1 and 20 ng/ml depending on the clean-up method and the detector used.*

More recently the radioenzymatic approach, REI, has been found more convenient and universally applicable to monitor plasma levels of all active ACE inhibitors. REI has several advantages: it is simple and rapid; it eliminates the need for derivatization; it requires <0.1 ml of sample due to its high sensitivity; furthermore, it does not require sophisticated analytical equipment and hence is suitable for multi-sample analysis in clinical laboratories.

ACE (dipeptidyl carboxypeptidase, EC 3.4.15.1), also termed Kininase II, is known to play an important role in the regulation of fluid balance as well as blood pressure. The synthesis of inhibitors

**Abbreviations (besides obvious ones).*- ACE, angiotensin-converting enzyme. ECD, electron-capture detection; FID, flame-ionization & NPD, nitrogen-phosphorus. MS, mass spectrometry; SIM, selected ion monitoring. REI, radioenzymatic inhibition; RIA, radioimmunoassay. *Other abbreviations overleaf.*

which block the active site of the ACE system has led recently to the design of a new class of antihypertensive agents. Captopril, 1-[2(*S*)-3-mercapto-2-methyl-1-oxopropyl]-L-proline, was the first orally active ACE inhibitor to be marketed for the treatment of hypertension. Some of the side-effects observed in early clinical trials were believed to be due to the presence of the reactive thiol group in captopril This hypothesis led to the synthesis of several other ACE inhibitors lacking the -SH moiety. Of these only enalapril has so far been introduced (in the U.S. market); several others, e.g. ramipril, cilazapril, benazepril, pentopril and lisinopril are currently in various phases of clinical trials (structures: Fig. 1).

Although there is close structural similarity in many of these ACE inhibitors, a wide range of analytical methods have been employed for their determination in biological fluids. Captopril, being the first of this series, has been the most intensively studied, and a wide variety of analytical methods have been described. These include spectrophotometry [1], radio-TLC [2], HPLC [3-8], GC [9, 10], GC-MS [11-15], and radioenzymatic [16-18] and RIA [19] methods. The analytical methods for other inhibitors appearing after captopril have, however, been studied with more selectivity. Here the different analytical methods (except for RIA, which J.D. Robinson considers in the following article) are reviewed with respect to the principles, the nature of the procedures, and their strengths and limitations.

Spectrophotometry.- Spectrofluorometric determination of ACE inhibitors in plasma has been reported for captopril by Ivashkiv [1]. The total captopril level was determined following reduction of the -S-S- linkage in the dimer (disulphide) and covalently bound proteins with tri-n-butylphosphine. After sample purification on an XAD-2 column, captopril was derivatized with 1-(7-dimethylamino)-4-methyl-2-oxo-2H-1-benzopyran-3-yl)-1H-pyrrole-2,5-dione to form a fluorescent derivative. After acidification, the derivative was extracted into toluene and further purified on a C-18 cartridge. With excitation at 380 nm, the fluorescence of the dimethylformamide eluate was measured at 440 nm. The assay was reported to have a sensitivity limit >200 ng/ml of captopril, and hence to be suitable for patients having total captopril levels in the μg/ml range. Although sample preparation is laborious, this method suits small laboratories which do not have access to chromatographic or MS instruments.

Radio-TLC.- In this method [2], TLC is followed by radioactivity determination. Because of the reactivity of the -SH group in captopril, the drug in blood samples is converted to a stable derivative by adding NEM® to freshly collected samples. This method is of limited use because of the need to administer the radiolabelled drug.

®*Abbreviations besides those on previous p.*- BPM, ***N***-(4-benzoylphenyl)-maleimide; DAPM, ***N***-4-(dimethylaminophenyl)maleimide; NEM, ***N***-ethyl-maleimide; SBD-F, 7-fluorobenzo-2-oxa-1,3-diazole-4-sulphonate; i.s., internal standard. See Fig. 1 for structure of NEM derivative.

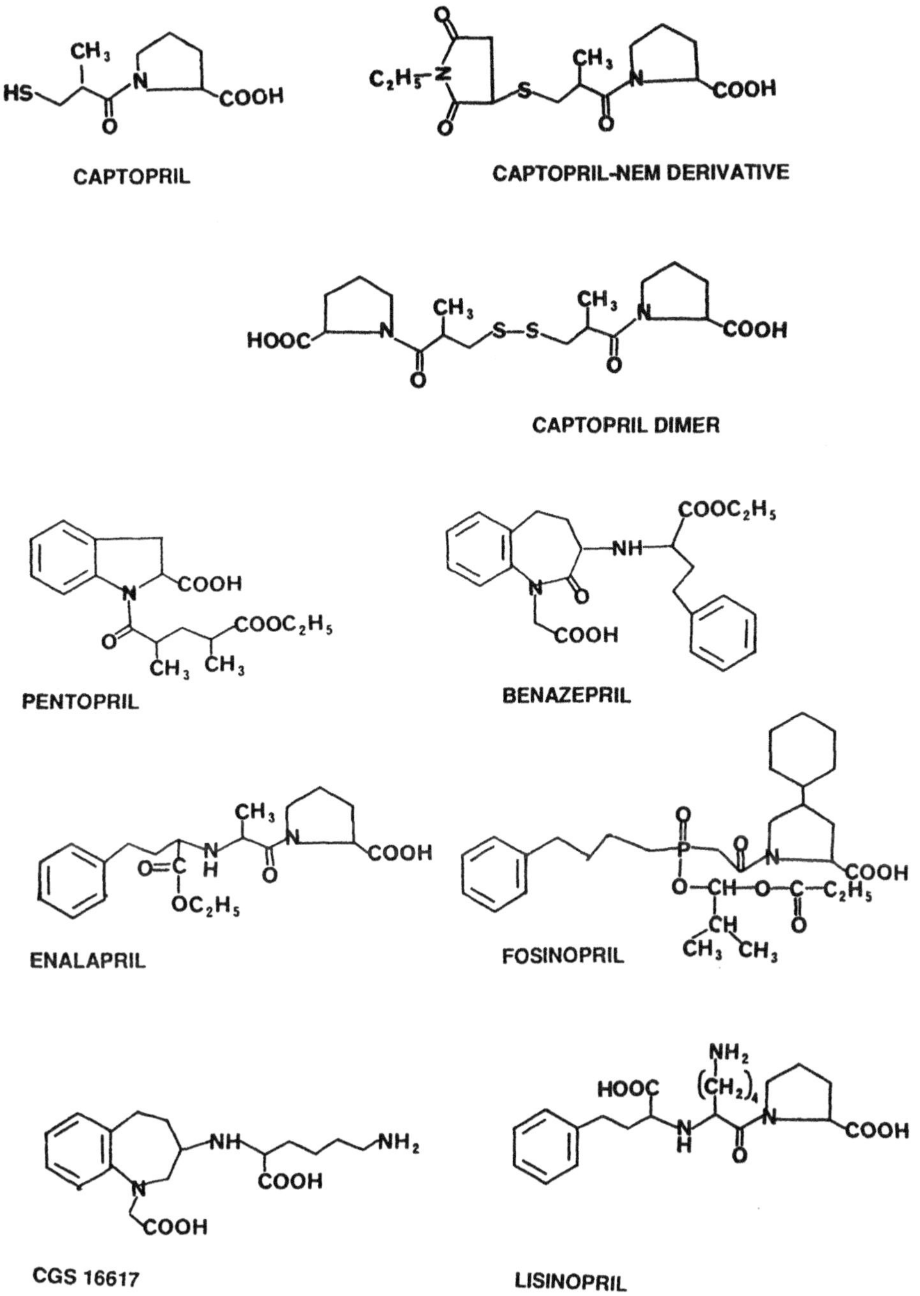

Fig. 1. Structures of compounds mentioned in the text, including prodrug esters.

HPLC (captopril: [3-8])

Captopril, rendered optically detectable.- Captopril lacks intrinsic fluorescence and high UV absorbance. Hayashi et al. [3] reported the formation of a UV-absorbing derivative by reacting the -SH group with BPM. The method was used to separately determine captopril and mixed disulphides in plasma and urine after oral administration of a captopril prodrug, DU-1219. Following derivatization of the unchanged captopril in plasma with BPM, the derivative was extracted with chloroform and run on a C-18 column using as mobile phase acetonitrile/methanol/1%(v/v) acetic acid (45:11:75) at 1 ml/min flow-rate. Captopril was measured with a UV detector set at 254 nm. Mixed disulphides of captopril in plasma were also determined by derivatization with BPM following their reduction with 0.8% tributylphosphine at 50° for 60 min.

Protein-conjugated captopril was determined by precipitating proteins in plasma with perchloric acid. The pellet was resuspended in water and neutralized with K_3PO_4. The protein-conjugated captopril was then reduced with tributylphosphine and analyzed for total captopril. This method is reported to have a sensitivity limit of 10 ng/ml for unchanged captopril and ~25 ng/ml for disulphide and protein-conjugated captopril.

Fluorescent derivatives have also been formed by reacting the -SH group of captopril with ***N***-(1-pyrene)-maleimide at pH 6.5, using the supernatant from deproteinized plasma to which i.s. had been added initially [4]. The resulting fluorescent adducts of both the captopril and the i.s. were extracted into ethyl acetate/benzene (1:1) mixture and chromatographed with detection at 390 nm (excitation at 340 nm). The reported sensitivity was 150 pmol/ml (32.5 ng/ml).

Toyo'Oka et al. [5] used a new fluorogenic reagent, SBD-F, which differs from more commonly used maleimide adducts of captopril. It is reported to have several favourable properties, e.g. high reactivity to thiols, lack of native fluorescence, good solubility in buffered matrices, and stability of both the reagent and the captopril derivative for >1 week at 4°. The method consisted of deproteinization of plasma with trichloroacetic acid, reduction of the oxidized captopril with tributylphosphine at pH 9.5, and then formation of the fluorophor adduct of captopril with SBD-F. The derivatives were then separated by RP-HPLC with detection at 515 nm (excitation at 385 nm). The sensitivity limit of the assay is ~10 ng/ml. This method appears to be relatively simple, selective and sensitive.

Captopril, with electrochemical detection.- Captopril has also been analyzed with electrochemical detection either as unchanged captopril [6, 7] or as an adduct of DAPM [8]. Perrett et al. [6] first reported an electrochemical oxidation of the free thiol at 0.07 V using a Au/Hg cell against Ag/AgCl as reference electrode.

Plasma was deproteinized with sulphosalicylic acid immediately after blood collection. The supernatant was frozen and stored at -20°. For analysis, the thawed sample was injected directly onto the column. Perrett et al. [7] later reported an improved method using a gold working electrode which was more selective and sensitive (2 pmol/ml). [Perrett discussed electrodes etc. earlier in this series - #C-1 in Vol. 14.-*Ed.*]

The electrochemical detection method (sensitivity ~10 ng/ml) reported by Shimada et al. [8] appears to have several advantages over those of Perrett et al. [6, 7]. It utilized DAPM to stabilize the reactive thiol in captopril besides forming a captopril-DAPM adduct which is electrochemically active - as applies likewise to captopril which, however, is unstable and hence needs -SH group protection immediately after blood collection. For stabilization, Shimada et al. used the adduct approach, preserving detectability; another innovation was their use of an i.s.

HPLC has also served to assay plasma for another ACE inhibitor, pentopril and its active metabolite CGS 13934 [20], using a 5 μm C-8 column and acetonitrile/tetrahydrofuran/pH 2.5 phosphoric acid (15:15:70) as mobile phase. Intrinsic UV absorbance at 254 nm allowed ~50 ng/ml to be detected.

GC

Bathala et al. [9] described GC-ECD determination of captopril following derivatization of the reactive thiol group with NEM and then of the carboxyl group with hexafluoroisopropanol in trifluoracetic anhydride. Jemal et al. [10] later demonstrated that the NEM itself is a good electrophoretic derivatizing agent and that the fluorinated ester obtained hardly surpassed the methyl ester in enhancing ECD responsiveness.

GC-NPD served to determine fosinopril and its active metabolite, SQ 27,519, in plasma [10]. Both carboxylic and phosphonic groups contained in the molecule were esterified, by methanolic HCl and hexafluoroisopropanol respectively. The method relied on N and P atoms in the analyte itself for detection; <10 ng/ml was detectable.

GC was also utilized for assaying pentopril and its active metabolite in urine. Methylation of the carboxyl group was achieved by reacting with methyl iodide and anhydrous caesium carbonate in dry acetone at 75° for 2 h. With NPD there was benefit to both sensitivity (~1 ng/ml using FID, as compared with 100 ng/ml with NPD) and selectivity.

GC-MS

Several analytical methods have been reported for captopril utilizing SIM-MS [11-15]. That of Cohen et al. [11] included stabilization of -SH by NEM at the outset and then solid-phase extraction using XAD-2 and carboxyl esterification with methanolic HCl. With electron impact (EI) MS, SIM detection was at m/z 230 with a fluoro analogue (m/z 248) as i.s. To determine total captopril [12], down to ~20 ng/ml, samples containing protein-bound captopril, disulphide dimer and mixed disulphides were reduced to captopril by treatment with tri-n-butylphosphine at 50° for 1 h in the dark. Positive chemical ionization (PCI) was the mode employed for the *S*-methyl metabolite of captopril [13]. Jemal et al. [14] later extended this approach to allow simultaneous determination of *S*-benzoylcaptopril and captopril utilizing a capillary GC column.

Drummer et al. [15] reported another GC-MS method which allowed simultaneous measurement of unchanged captopril and its three metabolites: captopril disulphide, *S*-methylcaptopril and *S*-methyl captopril sulphone. Following stabilization of captopril with NEM, the carboxyl group was derivatized with hexafluoroisopropanol and all derivatives were chromatographed on a packed column (3% OV-101, 2 m × 2 mm i.d.) in one temperature-programmed run and measured by SIM. Using urine or plasma, captopril, *S*-methylcaptopril and the dimer could be detected as low as 1, 10 and 25 ng/ml respectively. No *S*-methyl sulphone metabolite was detected in urine of either rat or man.

RADIOENZYMATIC INHIBITION (REI)

Monitoring plasma ACE activity has become an attractive way to follow circulating levels of ACE inhibitors [16-18, 22, 23]. Petty et al. [16] first suggested such an approach as a useful indicator of plasma levels of captopril and its active metabolites in the absence of a suitable direct analytical method. The basis is the 1:1 stoichiometric relationship in the binding of inhibitor to ACE. Three approaches have been used to determine the plasma concentration of the active inhibitor. The first is based on the specific binding of a radiolabelled ^{125}I-inhibitor [351A, *p*-hydroxy-benzamidine derivative of ***N***-(1-carboxy-3-phenylpropyl)-L-lysyl-L-proline] in a non-equilibrated system [17]. The radio-binding can then be specifically displaced by adding a non-labelled ACE inhibitor (drug analyte); the degree of displacement will correspond to the amount of inhibitor added and its relative ACE-binding potency. Free label is then separated by adsorption onto coated charcoal, and the radioactivity of the precipitate is counted in a gamma counter. Gronhagen-Riska et al. [18] used this method to measure several ACE inhibitors: captopril, enalapril, lisinopril, cilazapril, and benazeprilat (CGS 14831; the active inhibitor of benazepril). For captopril serum was diluted 1:100 immediately after collection to minimize *in vitro* oxidation/polymerization and reaction with proteins. The sample, however, still needs to be analyzed within 2 days. Sensitivity for captopril was 2 ng/ml and for other ACE inhibitors 0.25-5 ng/ml.

The second method is based on the determination of free ACE activity using a radiolabelled substrate, [^{3}H]hippuryl-glycyl-glycine [16]. Free enzyme acts on the substrate to release the radioactive product ([^{3}H]hippuric acid), which is selectively extracted (unlike substrate) into the scintillation cocktail for quantification. Swanson et al. [22] used this radioenzymatic method to quantify enalaprilat and the active metabolite (SQ 27,519) of the prodrug ester fosinopril. Biological samples containing active drug were treated with methanol to precipitate endogenous ACE. Supernatants after dilution with pH 8 HEPES buffer were incubated with the radioactive substrate and rabbit-lung ACE for 45 min. Following acidification, the [^{3}H]hippuric acid was extracted into a water-immiscible cocktail (mixture of toluene, ethyl acetate and standard scintillator; Ventrex Labs., Portland, ME). The ester prodrugs were analyzed by hydrolysis with NaOH.

In a variant method, Reydel-Bax et al. [23] determined the inhibitor concentration by measuring *in situ* the free ACE activity, this being inversely related to the amount of active inhibitor present. Free enzyme was reacted with a radiolabelled substrate and the radioactive product then selectively extracted into the scintillation cocktail for quantification. The concentration of circulating inhibitor is therefore measured in the patient's plasma by comparison with a standard curve made from results for addition of analyte to the same patient's blank plasma. Because of the high assay sensitivity, patients' plasma samples are diluted at least 5-10 fold with any blank plasma that is also used to derive the standard curve. This method, therefore, allows the use of one standard curve, from a blank plasma, that can be used with many patients' plasma samples. However, when inhibitor concentrations are very low and the plasma needs no dilution, the same subject's blank plasma is used to measure control activity. To analyze plasma samples containing prodrug ester, hydrolysis is performed with rat-plasma esterases rather than chemically. Altogether this radioenzymatic method appears to be the simplest and should be suitable for routine automated analysis of active ACE inhibitors in clinical laboratories.

CONCLUDING COMMENTS ON THE REI APPROACH

The above REI method [23] seems the most suitable one for monitoring plasma concentrations of active ACE inhibitors. This is because, once developed for one inhibitor, the REI can be universally adopted for others, in contrast with chromatographic methods. The REI procedure is relatively simple and rapid. It requires no chemical derivatization, nor any sophisticated analytical equipment, and is so sensitive that a minimal sample volume suffices (<0.1 ml). Although this method would not be expected to have absolute specificity, cross-reactivity is minimal, as expected because prodrug esters (pentopril, enalapril, benazepril, ramipril) hardly contribute. There is no metabolite problem for most of the non-sulphydryl active ACE inhibitors, other than captopril (they are mainly excreted unchanged, via the kidney and the bile). For captopril, plasma is diluted 1:100 immediately after collection to minimize polymerization and is analyzed within 2 days. The aptness of the REI approach for routine clinical drug monitoring is obvious.

References

1. Ivashkiv, E. (1984) *J. Pharm. Sci.* *73*, 1427-1430.
2. Migdalof, B.H., Singhvi, S.M. & Kripalini, K.J. (1980) *J. Liq. Chromatog.* *3*, 857-865.
3. Hayashi, K., Miyamoto, M. & Kripalini, K.J. (1980) *J. Chromatog.* *338*, 161-169.
4. Jarrott, B., Anderson, A., Hooper, R. & Louis, W.J. (1981) *J. Pharm. Sci.* *70*, 665-667.
5. Toyo'Oka, T., Imai, K. & Kawahara, Y. (1984) *J. Pharm. Biomed. Anal.* *2*, 473-479.
6. Perrett, D. & Drury, P.L. (1982) *J. Liq. Chromatog.* *5*, 97-110.
7. Perrett, D., Rudge, S.R. & Drury, P.L. (1984) *Biochem. Soc. Trans.* *12*, 1059-1060.
8. Shimada, K., Tanaka, M., Nambara, T., Imai, Y., Abe, K. & Yoshinaga, K. (1982) *J. Chromatog.* *227*, 445-451.
9. Bathala, M.S., Weinstein, S.H., Meeker, F.S., Jr., Singhvi, S.M. & Migdalof, B.H. (1984) *J. Pharm. Sci.* *73*, 340-344.
10. Jemal, M., Ivashkiv, E., Ribick, M. & Cohen, A.I. (1985) *J. Chromatog.* *345*, 299-307.
11. Cohen, A.I., Devlin, R.G., Ivashkiv, E., Funke, P.T. & McCormick, T. (1982) *J. Pharm. Sci.* *71*, 1251-1256.
12. Ivashkiv, E., McKinstry, D.N. & Cohen, A.I. (1984) *J. Pharm. Sci.* *73*, 1113-1117.
13. Cohen, A.I., Ivashkiv, E., McCormick, T. & McKinstry, D.N. (1984) *J. Pharm. Sci.* *73*, 1493-1495.
14. Jemal, M., Ivashkiv, E. & Cohen, A.I. (1985) *Biomed. Mass Spectrom.* *12*, 664-667.
15. Drummer, O.H., Jarrott, B. & Lousi, W.J. (1984) *J. Chromatog.* *305*, 83-93.
16. Petty, M.A., Reid, J.L. & Miller, S.H.K. (1980) *Life Sci.* *26*, 2045-2050.
17. Fyrquist, F., Tikkanen, I., Gronhagen-Riska, C., Hortling, L. & Hichens, M. (1984) *Clin. Chem.* *30*, 696-700.
18. Gronhagen-Riska, C., Tikkanen, I. & Fyhrquist, F. (1987) *Clin. Chem.* *162*, 53-60.
19. Duncan, F.M., Martin, V.I., Williams, B.C., Al-Dujaili, E.A.S. & Edwards, C.R.W. (1983) *Clin. Chim. Acta* *131*, 295-303.
20. Rakhit, A. & Tipnis, V. (1984) *Clin. Chem.* *30*, 1237-1239.
21. Tipnis, V. & Rakhit, A. (1985) *J. Chromatog.* *345*, 396-401.
22. Swanson, B.N., Stauber, K.L., Alpaugh, W.C. & Weinstein, S.H. (1985) *Anal. Biochem.* *148*, 401-407.
23. Reydel-Bax, P., Redalieu, E. & Rakhit, A. (1987) *Clin. Chem.* *33*, 549-553.

#C-2

THE ANALYSIS OF ACE INHIBITORS BY IMMUNOASSAY

J.D. Robinson and S. Lewis

Bioanalytical Department, Hoechst (UK) Ltd.
Walton Milton Keynes, Bucks. MK7 7AJ, U.K.

IA is one of the many techniques that have been used for the analysis of ACE inhibitors in biological fluids. The variety of IA procedures that can be employed are illustrated by the enzyme assays and RIA's for captopril, cilazapril, enalapril, ramipril and trandolapril. Nearly all of these compounds have been analyzed using different variations of the IA approach. In many cases the assays have been developed to measure concentrations of the active metabolite only, concentrations of the parent compound being determined following* in vitro *conversion to the metabolite. This article reviews the different IA approaches that have been adopted for analyzing ACE inhibitors and comments on their features of interest.*

The IA technique is just one of the many that have been described for the analysis of ACE inhibitors in plasma. The advantages of this technique over most of those described in the preceding article (A. Rakhit) lie in the sensitivity and specificity achievable using a small sample size. The technique also enables large numbers of samples to be processed in a relatively short time. The diversity of the IA approach is illustrated by the range of different assay procedures that have been developed for the analysis of ACE inhibitors. For example, EIA's have been developed for captopril and cilazapril while enalapril, ramipril and trandolapril have been determined using RIA methodology. Fig. 1 shows their structures.

ENZYME IMMUNOASSAYS (EIA's)

EIA's depend on the use of an enzyme-substrate reaction for the end-point detection system of the assay which, depending on the enzyme chosen, can be either photometric or fluorimetric and can itself

**Abbreviations*.- Ab, antibody; ACE, angiotensin converting enzyme; (E)IA, (enzyme) immunoassay; ELISA, enzyme-linked immunosorbent assay; BSA/HSA, bovine/human serum albumin; MCC, 4-(maleimidoethyl)cyclohexane carboxylic acid; RIA, radioimmunoassay.

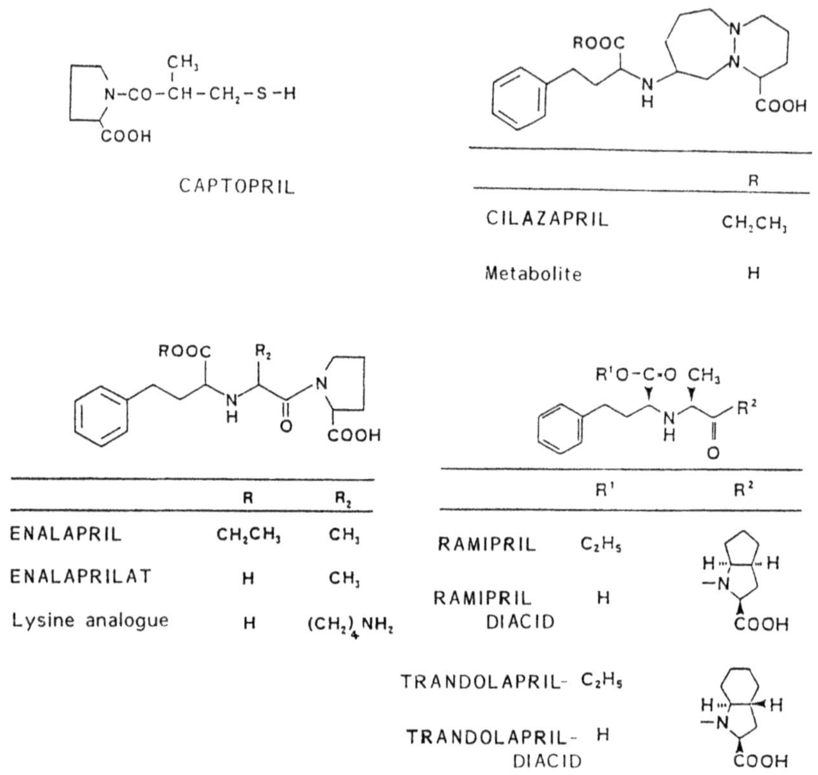

Fig. 1. Some ACE inhibitors: parent compounds and metabolites.

govern the sensitivity of the resulting assay. The assays for captopril and cilazapril exemplify variants in the EIA approach.

The captopril EIA was developed by Kinoshita et al. [1], utilizing Ab's raised in rabbits against a conjugate of captopril and bovine IgG with MCC as the coupling reagent. In developing the EIA they found that in order to determine the concentration of captopril in a sample they first had to react the sample with MCC to derivatize the captopril so that it was recognized by the Ab; this derivative coupled to an enzyme was used in the assay. The steps were as follows.-

(1) To the sample (100 µl) add normal plasma (100 µl) and MCC (50 µl).
(2) After incubating for 30 min at room temp., add captopril-MCC-β-D-galactosidase (100 µl) and antiserum (100 µl).

(3) After incubation for 16 h (overnight) at 4°, the Ab-bound and free fractions were separated using a solid-phase second Ab system (immunobead goat anti-rabbit), with incubation for 1 h at 37° and then centrifugation at 2500 rpm for 5 min.
(4) To the enzyme-containing pellet the substrate was added, and after 1 h at 37° the fluorescence of the product (7-hydroxy-4-methyl coumarin) was measured.

In this assay system the amount of fluorescence is inversely proportional to the concentration of captopril in the sample; with a 100 µl aliquot, 0.5 ng/ml is detectable. It is noteworthy that the underivatized drug did not react with the Ab, which would recognize the drug only after it had been derivatized with MCC. This illustrates a situation in which the Ab's were produced against the drug plus the bridging group and not the drug itself.

The EIA for cilazapril demonstrates a different approach, viz. ELISA. Cilazapril is one of the group of ACE inhibitors that are administered as prodrugs and are converted *in vivo* to an active diacid metabolite (Fig. 1). Tanaka et al. [2] raised Ab's to both the parent compound and the diacid by preparing lysine analogues and, using glutaraldehyde, coupling these to the carrier protein. Thus they were able to develop two separate ELISA's that were specific for either the parent compound or the metabolite; the enzyme that was coupled to one or the other was peroxidase. The assay steps were as follows.-

(1) The Ab was adsorbed onto a microtitre plate, and any non-specific binding sites were blocked by incubating with 0.01% HSA in pH 6.8 0.1 M phosphate buffer (100 µl).
(2) The sample (10 µl) and phosphate buffer (100 µl) containing 30% human serum were added.
(3) After 2 h at 4°, 'Cilazapril' peroxidase (50 µl) was added.
(4) After 20 h at 4°, then a wash with buffer, substrate (100 µl) was added.
(5) After 30 min at room temp., the reaction was stopped with 2 M H_2SO_4 and the absorbance at 490 nm was read.

The ELISA's for both compounds could detect as little at 30 pg/ml in a 10 µl plasma sample. The inclusion of 30% human serum was needed "to prevent any interference of serum (or plasma) components, which was sometimes observed to differ between individual sera" [2]. The assay was conducted at 4° since the cilazapril in the sample was hydrolyzed either spontaneously or by endogenous esterases to the active diacid metabolite.

RADIOIMMUNOASSAY (RIA)

Two different RIA's have been developed for the ACE inhibitor enalapril and its active metabolite enalaprilat [3, 4]. The same lysine analogue of enalapril (Fig. 1) was used in preparing the immunogens, but Hichens et al. [3] used albumin as the carrier protein while Worland & Jarrott used keyhole limpet haemocyanin. Another difference in the approaches of the two groups was in the nature of the tracer used: either a *p*-hydroxybenzimidate derivative of the lysine analogue was produced and then radioiodinated to give a ^{125}I-tracer [3], or the lysine analogue's free amino group was reacted with tritiated succinimidyl propionate to produce a tritiated tracer [4]. Both groups determined the concentration of the active enalaprilat in their RIA's and only measured enalapril itself after *in vitro* enzymatic conversion using rat liver homogenate or after base hydrolysis. The difference in the detectability (0.05 ng/ml [3] and 0.8 ng/ml [4]) may be a function merely of the differences between the two tracers, their specific activities and their efficiency of detection.

Worland & Jarrott found that addition of EDTA (to 25 mM) to their assay buffer, together with sheep serum (to 6%), helped to avoid the problems of endogenous ACE competing with the antiserum for both enalaprilat and their radiotracer. The EDTA has the effect of chelating the Zn atom of ACE; removal of this atom has been shown to decrease the affinity of ACE for enalaprilat by a factor of 25,000 [5].

The RIA's for ramipril and trandolapril have been developed using a common approach. The Ab's for both were raised in rabbits by Eckert et al. [6] by coupling the lysine derivative or ramipril diacid to BSA to produce the immunogen, and were from two consecutive bleeds of the same rabbit. A radioiodinated derivative of the same lysine analogue of ramipril diacid was used as the tracer for both RIA's. They are similar in procedure; only the diacid is measured, and any parent compound is determined after esterase hydrolysis. The steps in assaying ramipril or trandolapril were as follows.-

(1) The sample was incubated at 37° for 90 min, with porcine liver esterase added if total concentration (ester + diacid) was to be determined. Ab and tracer were then added.
(2) After 40 h (ramipril) or 16 h (trandolapril) at 4°, the second Ab (solid-phase) was added.
(3) After 1 h at room temp., the bound fraction was separated and counted.

The major difference between the two assays is in the incubation time, the time needed to reach equilibrium being longer for ramipril.

Temperature in the two RIA's.- In the development of the ramipril and trandolapril RIA's there was a noteworthy effect of temperature on the antigen-Ab reaction: the degree of binding of the tracer to the Ab was greater at 4° than at either room temperature or at 37° [6]. This conflicts with the normally accepted increase in reaction rate with increase in temperature, and may be related to the effects of endogenous material on the antigen-Ab interaction. More recently

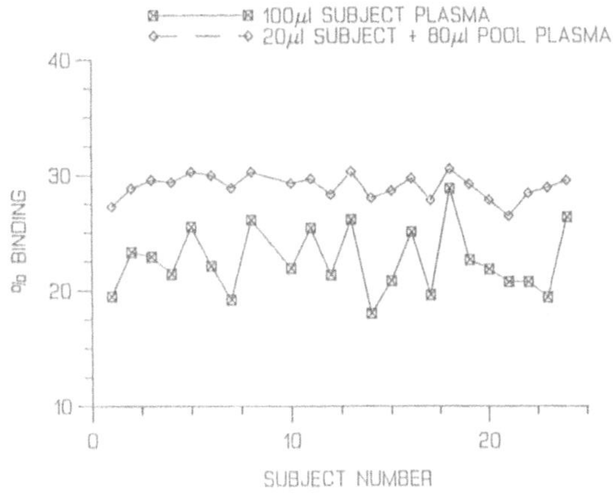

Fig. 2. Variation in the zero-binding values of pre-dose plasma samples, showing the effect of dilution with pooled normal plasma.

we have demonstrated (unpublished experiments) that plasma has an effect in both assays that differs between plasmas from different sources. This effect is seen as a variation between plasmas in the zero binding (i.e. the binding of the tracer to the Ab in the absence of standard material, expressed as % of the total amount of tracer added). This variation, illustrated in Fig. 2, can be reduced by diluting the plasma sample 1:5 in a pooled normal plasma. In Fig. 2 the C.V. for the zero binding with undiluted plasma was 12.5%, whereas after dilution it was reduced to 3.4%. Analysis of samples using this procedure has had the effect of reducing the spread of plasma concentrations for the drugs throughout a particular study, and has reduced the apparently large between-subject variations in blood drug levels.

CONCLUSIONS

IA's are widely applicable to the analysis of ACE inhibitors, both as EIAs and as RIA's. In all cases the assays that have been developed are both sensitive and specific, and can distinguish between the parent drug and its active metabolite. Many of the IA's described have demonstrated a matrix effect, the Ab binding to the antigen varying in the presence of different plasmas. A variety of techniques have been used to overcome this effect, or at least to reduce it to an acceptable level. The advantages of IA techniques for drug analysis have been demonstrated in respect of adapability and versatility. The same basic technique can be used in a variety of guises to produce sensitive and specific assays that can process large numbers of samples in a relatively short time.

References

1. Kinoshita, H., Nakamaru, R., Tanaka, S., Tohira, Y. & Sawada, M. (1986) *J. Pharm. Sci. 75*, 711-713.
2. Tanaka, H., Yoneyama, Y., Sugawara, M., Limeda, I. & Ohta, Y. (1987) *J. Pharm. Sci. 76*, 224-227.
3. Hichens, M., Hand, E.L. & Mulcahy, W.S. (1981) *Ligand Quarterly 4*, 43.
4. Worland, P.J. & Jarrott, B. (1986) *J. Pharm. Sci. 75*, 512-516.
5. Cordes, E.H., Bull, H.G. & Thornberry, H.G. (1984) in *Hypertension and the angiotensin system: Therapeutic approaches* (Doyle, A.E. & Bearn, A.G., eds.), Raven Press, New York, p. 167.
6. Eckert, H.G., Munscher, G., Oekonomopoulos, R., Urbach, H. & Wissmann, H. (1985) *Drug Res. 35*, 1251-1256.

#C-3

THE USE OF STABLE-ISOTOPE METHODOLOGY IN PHARMACOKINETIC STUDIES INVOLVING FLUNARIZINE

Robert Woestenborghs, Philip Timmerman, Achiel Van Peer and Jos Heykants

Department of Drug Metabolism and Pharmacokinetics
Janssen Research Foundation
B-2340 Beerse, Belgium

Co-administration of a drug and a suitable isotopomer can substantially reduce the number of study phases and subjects required in bioavailability studies. Flunarizine is particularly suited to stable-isotope methodology, because of its excellent GC properties and its long half-life in plasma. We therefore developed a sensitive GC-MS procedure for the simultaneous determination of flunarizine and its $\mathbf{d_4}$* *isotopomer. After analyte extraction from plasma, capillary GC-MS is performed in either the EI or the PCI mode using isobutane as reagent gas, resulting in sensitivity limits of 0.2 and 0.1 ng/ml respectively. With the EI mode, plasma levels were determined in 3 volunteers given a* d_0/d_4 *mixture (10 + 10 mg). The pharmacokinetics showed no isotope effect; thus the approach was well suited to bioequivalence studies.*

Flunarizine, (E)-1-[bis(4-fluorophenyl)methyl]-4-(3-phenyl-2-propenyl)piperazine (formula in Fig. 1), is a selective calcium entry blocker, widely used in cerebral and peripheral vascular disorders [1]. Its metabolism [2] and pharmacokinetics [3, 4] have been studied in experimental animals and man using several GC [5-7] and HPLC [8-10] methods. Flunarizine is extremely lipophilic (log P = 5.78) and strongly bound to plasma proteins (99.1%). Its distribution into body fat is extensive and, due to the slow redistribution from this depot, the plasma elimination half-life is ~3 weeks. Since inter-individual variation in plasma levels is considerable [3], the proper design of bioavailability studies is cumbersome, with respect both to study time and to the number of subjects required to demonstrate

Abbreviations:* **d, deuterium (4 atoms if $\mathbf{d_4}$); MS, mass spectrometry; EI, electron impact; (P)CI, (positive) chemical ionization; SIM, selected ion monitoring; i.s., internal standard.

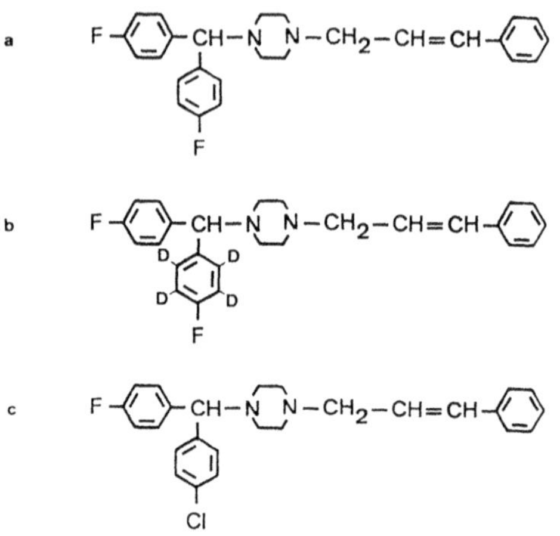

Fig. 1. Chemical structures of flunarizine (a), flunarizine-$\mathbf{d}_4$ (b) and the i.s. (c).

bioequivalence with sufficiently high statistical power. Stable-isotope methodology permits the simultaneous administration of a drug and its stable-isotope labelled analogue, and serves well to overcome these difficulties by reducing both intra- and inter-subject variability [11-13].

Here a method is described for the simultaneous determination of flunarizine and its $\mathbf{d}_4$ analogue in plasma. As the absence of any isotope effect is cardinal to pharmacokinetic use of this approach, the bioequivalence of the two drug species was studied in dosed volunteers.

MATERIALS

Flunarizine and the i.s. (R 70632) were synthesized in our research laboratories, and flunarizine-$\mathbf{d}_4$, with the 4 ^{2}H atoms in one *p*-fluorophenyl moiety, in this department (Fig. 1). The isotopic distribution as measured by 70 eV EI-MS was: d_0, 0.0%; d_1, 0.2%; d_2, 0.9%; d_3, 10.7%; d_4, 88.2%.

EXTRACTION PROCEDURE

Aliquots (2 ml) of plasma were spiked with 10 or 50 ng/ml of i.s., alkalinized with 2 ml 0.01 M NaOH and extracted twice with 4 ml 1.5% (v/v) isoamyl alcohol in n-heptane. The combined organic layers were back-extracted with 4 ml 0.05 M H_2SO_4, the acid layer was washed with 1 ml of the extraction solvent and then alkalinized with conc. ammonia. The analytes were then re-extracted twice with 3 ml of the heptane/isoamyl alcohol mixture and evaporated to dryness at 60° under N_2. The extraction residues were reconstituted in 50 µl of methanol and 5 µl aliquots were injected into the GC-MS.

GC-MS

Samples were injected through a moving-needle solventless injector (Alltech) mounted on a Hewlett-Packard 5985 B GC-MS system, equipped with a 10 m × 0.32 mm i.d. CP-Sil 19 CB fused silica capillary column (Chrompack) directly coupled to the ion source, which was operated in either the EI (70 eV) or PCI (isobutane) mode. Helium was used as carrier gas at a linear velocity, $\overrightarrow{u_0}$, of 60 cm/sec, and the GC temperatures were: injector, 295°; column, 280°; interface, 285°. For SIM in the EI mode, the instrument was focused to monitor the ion current intensities at m/z 287, 291 and 303 corresponding to characteristic ions for flunarizine-$\mathbf{d}_0$, flunarizine-$\mathbf{d}_4$ and the i.s. respectively. The ion source, with 1.2×10^{-5} torr pressure, was at 200° and the analyzer at 190°. In the PCI-SIM mode the corresponding settings were 0.6 Torr, 160° and 190°; the $[MH]^+$ ions of flunarizine-$\mathbf{d}_0$, flunarizine-$\mathbf{d}_4$ and the i.s. were monitored at m/z 405, 409 and 421 respectively.

CALIBRATION AND QUALITY CONTROL PROCEDURES

A series of calibration samples was prepared by spiking 2 ml aliquots of plasma with flunarizine-**d /d** at concentrations ranging from 0.5 & 0.05 to 50 & 50 ng/ml. After extraction and analysis as above, calibration curves for both analytes were constructed by plotting the peak area ratios A_{287}/A_{303} and A_{291}/A_{303} in the EI-SIM mode, and A_{405}/A_{421} and A_{409}/A_{421} in the PCI-SIM mode against the prepared flunarizine concentrations. Two series of quality control samples were independently prepared in blank human plasma to check the accuracy and precision of both methods. The first series comprised 7 (range 0.35-16.3 ng/ml), which were analyzed blind using the EI GC-MS method; the second series comprised 4 (range 0.19-1.89 ng/ml) and was used to control the PCI GC-MS method.

ASSAY VALIDATION PARAMETERS

The extraction procedure was comparable to our previous one [5], but included a washing step after the acid back-extraction. This narrows the solvent front in the GC-NPD method, results in pure GC-MS profiles in the sub-ng concentration range, and also permits less frequent instrument maintenance during the routine analysis of numerous samples. Nevertheless the extraction recoveries were still >90% for the flunarizine isotopomers and 70% for the i.s.

Although flunarizine has excellent GC properties on dimethyl-silicone phases [7], we selected the less thermostable CP-Sil 19 CB phase [a cyanopropyl/phenyl/methyl/vinyl (7:7:85:1) polysiloxane] because of the better separation of flunarizine from cholesterol. As shown in Fig. 2, interference-free chromatograms were obtained in either the EI or the PCI mode with retention times of 1.6 and 2.4 min for flunarizine and the i.s. Standard curves showed good

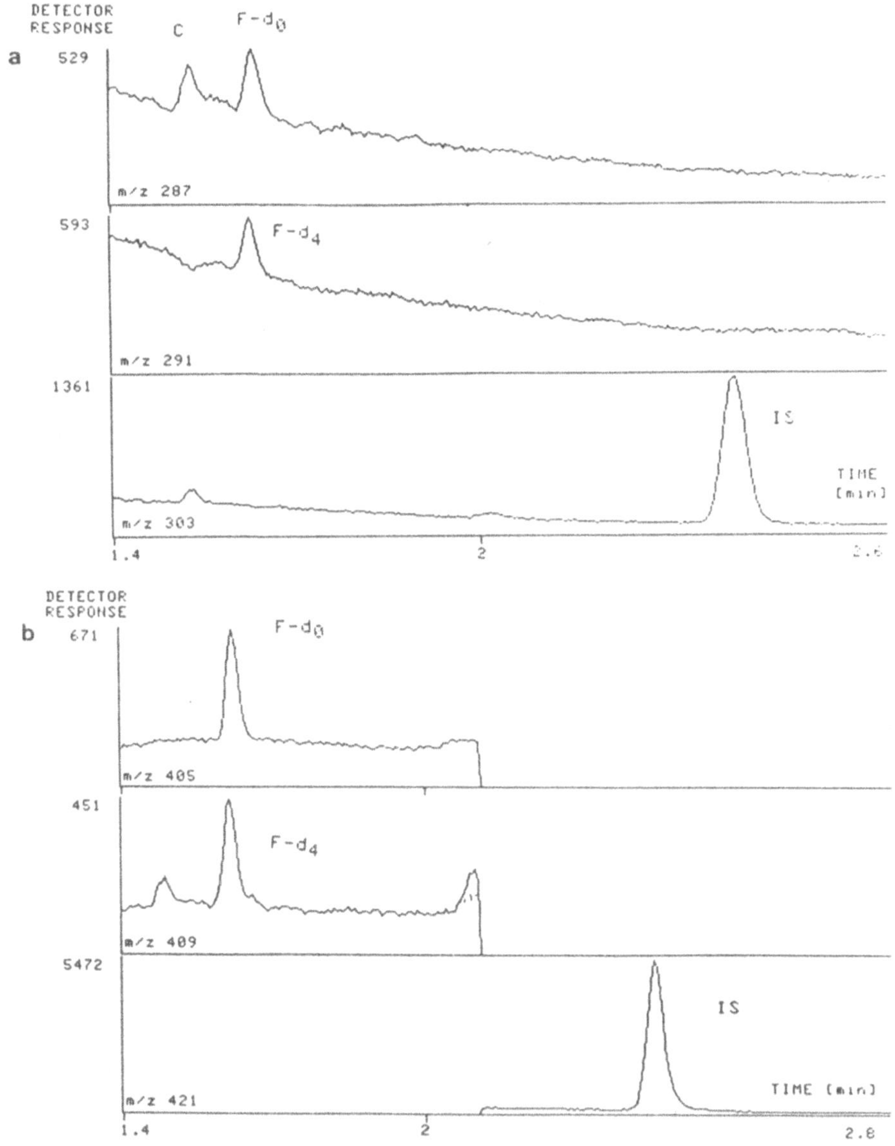

Fig. 2. Selected-ion recordings, (a) by 70 eV EI GC-MS, and (b) by isobutane-PCI GC-MS, of control plasma samples spiked as follows (F = flunarizine), ng/ml:

	F-d_0	F-d_4	i.s.
(a)	0.5; m/z 287	0.5; m/z 291	10; m/z 303
(b)	0.25; m/z 405	0.25; m/z 409	10; m/z 421

Table 1. Calibration curves for the assay of flunarizine-d_0/d_4 (x = concentration) in human plasma, with n calibration points: y = peak area ratio (flunarizine/i.s.).

Isotope & n		Range, ng/ml	i.s., ng/ml	Ionization mode	log y = *a*log x + *b*		Correlation coefficient
					a	*b*	
d_0:	8	0.25-50	50	EI	0.98880	-3.48035	0.9999
	7	0.5-50	50	EI	0.99403	-3.50641	0.9997
	7	0.1-20	10	EI	0.99231	-0.72961	0.9994
	5	0.05-2.5	10	PCI	0.92247	-0.72596	0.9997
	4	0.1-2.5	10	PCI	0.94033	-0.48291	0.9990
d_4:	8	0.25-50	50	EI	1.05100	-3.65329	0.9991
	7	0.50-50	50	EI	1.01221	-3.60968	0.9997
	6	0.50-20	10	EI	0.92674	-0.69266	0.9990
	5	0.05-2.5	10	PCI	0.97212	-0.79862	0.9993
	4	0.1-2.5	10	PCI	1.01140	-0.53370	0.9999

Table 2. Accuracy and between-assay precision of the assay with the two modes. RE = relative error (measure of accuracy).

Mode	Quality cont. samples	Isotope	RE	CV
EI	0.35-16.3 ng/ml range	d_0	4.6%	5.9%
		d_4	3.4%	5.1%
PCI	0.19-1.89	d_0	4.1%	3.6%
		d_4	2.1%	6.4%

linearity in either mode. Table 1 summarizes the regression parameters of some curves. Analysis of the quality control samples revealed a very satisfactory accuracy and between-assay precision in either ionization mode for both flunarizine isotopomers (Table 2).

CHOICE OF LABELLING POSITION AND SELECTION OF FINAL METHOD

As the metabolism of flunarizine occurs mainly in its cinnamyl moiety and also involves the formation of *N*-dealkylated metabolites [1, 2], the ^2H atoms were placed in one of the *p*-fluorophenyl groups to avoid metabolic isotope effects (see below); to achieve sufficient MS resolution from the unlabelled drug, 4 ^2H atoms were introduced. However, this labelling site has the limitation that the ion to be monitored in the EI mode is hardly abundant. In fact, the abundance of m/z 287 (291) is only 20% of that in the base ion peak (m/z 201), thus limiting the sensitivity in EI GC-MS. On the other hand, the use of a 'soft' ionization gas (isobutane) in the PCI mode allowed the monitoring of the more abundant molecular ions. Despite the better sensitivity of the PCI method (0.1 ng/ml) compared with EI

Fig. 3. Flunarizine plasma concentrations in a subject after intake of flunarizine-d_0/d_4 (10 + 10 mg) as an oral solution.

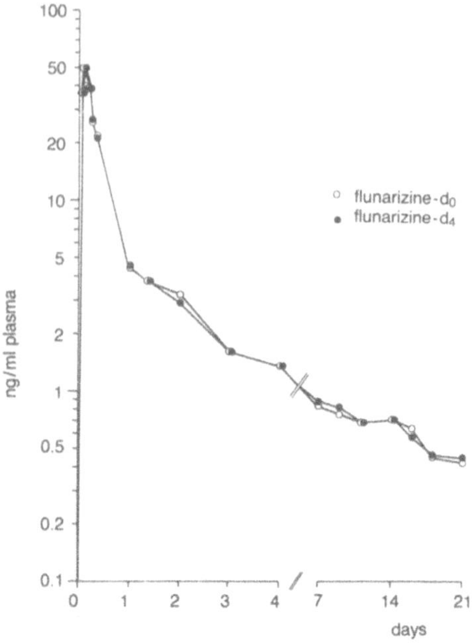

(0.2 ng/ml), we opted for EI because source contamination problems, caused by the isobutane-PCI, resulted in poor tuning reproducibility and/or more frequent maintenance of the GC-MS system. In contrast, the EI method was easier to perform and could routinely be applied to large numbers of samples without losing the excellent properties of the assay.

APPLICATIONS

The EI GC-MS method was accordingly chosen to evaluate possible metabolic isotope effects in 3 healthy male volunteers. The plasma concentration profiles (example in Fig. 3) and pharmacokinetic parameters showed that the isotopomers behaved identically. A bioequivalence study was then performed in 12 volunteers after simultaneous oral administration of a flunarizine-**d_0** capsule and a reference solution of flunarizine-**d_4** (5 mg of each). Identity in pharmacokinetic parameters was demonstrated, with a high degree of statistical confidence.

Acknowledgements

The authors are grateful to Mr. H. Lenoir for the synthesis of deuterated flunarizine and the i.s., to Mr. F. Van Rompaey for quality control, to Dr. Van de Velde for pharmacokinetic data analysis, and to Mrs. M. Sommen for manuscript preparation.

References

1. Holmes, B., Brogden, R.N., Heel, R.C., Speight, T.M. & Avery, G.S. (1984) *Drugs 27*, 6-44.
2. Meuldermans, W., Hendrickx, J., Hurkmans, R., Swysen, E., Woestenborghs, R., Lauwers, W. & Heykants, J. (1983) *Arzneim.-Forsch.* *33*, 1142-1151.
3. Heykants, J., De Cree, J. & Hörig, C. (1979) *Arzneim.-Forsch.* *29*, 1168-1171.
4. Michiels, M., Hendriks, R., Knaeps, F., Woestenborghs, R. & Heykants, J. (1983) *Arzneim.-Forsch.* *33*, 1135-1142.
5. Woestenborghs, R., Michielsen, L., Lorreyne, W. & Heykants, J. (1982) *J. Chromatog. 232*, 85-91.
6. Flor, S.C. (1983) *J. Chromatog. 272*, 315-323.
7. Kapetanovic, I.M., Torchin, C.D., Yonekawa, W.D. & Kupferberg, H.J. (1986) *J. Chromatog. 383*, 223-228.
8. Albani, F., Riva, R., Casucci, G., Contin, M. & Baruzzi, A. (1986) *J. Chromatog. 374*, 196-199.
9. Kobayashi, S., Taki, K. & Inoue, A. (1986) *Yakugaku Zasshi 106*, 217-220.
10. Nieder, M. & Jaeger, H. (1986) *J. Chromatog. 380*, 443-449.
11. Strong, J.M., Dutcher, J.S., Lee, W-K. & Atkinson, J. (1975) *Clin. Pharmacol. Ther. 18*, 613-622.
12. d'A. Heck, H., Buttrill, S.E., Flynn, N.W., Dryer, R.L., Anbar, M., Cairns, T., Dighe, S. & Cabana, B.E. (1979) *J. Pharmacokin. Biopharm. 7*, 233-248.
13. von Unruh, G.E., Eichelbaum, M. & Jengler, H.J. (1984) in *Drug Determination in Therapeutic and Forensic Contexts* [Vol. 14, this series] (Reid, E. & Wilson, I.D., eds.), Plenum, New York, pp. 27-37.

#NC(C)

NOTES and COMMENTS relating to

ACE INHIBITORS AND CALCIUM ANTAGONISTS

Comments relating to particular contributions:

#C-2 and #NC(C)-1 & -2: p. 165

#NC(C)-1

A Note on

ASSAY PROBLEMS WITH NIFEDIPINE

Peter S.B. Minty and ®Felicity A. Tucker

Department of Forensic Medicine and Toxicology
Charing Cross and Westminster Medical School
London W6 8RP, U.K.

®Servier Laboratories, Fulmer, Slough SL3 6HH, U.K.

Nifedipine is a calcium channel antagonist used in the treatment of angina and hypertension. It is sensitive, breaking down in daylight to its nitroso derivative and in UV light to its nitropyridine derivative [1]. Thermal decomposition during GC to the nitropyridine led Kondo et al. [2] to develop a method involving direct oxidation of serum with nitrous acid to the light-tolerant nitropyridine (Fig. 1).

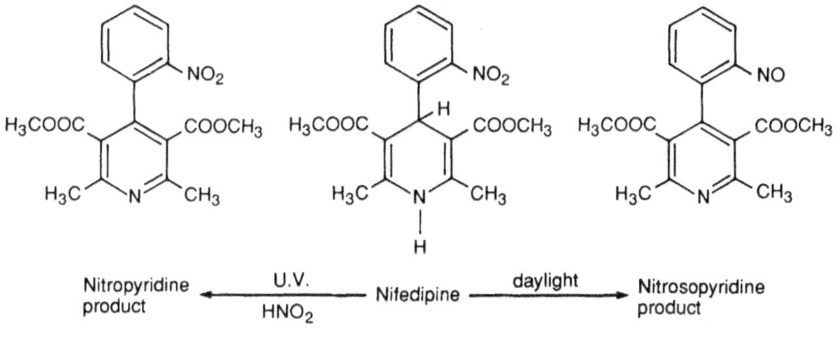

Fig. 1. Photo- and HNO_2-decomposition of nifedipine.

PACKED-COLUMN GC OF THE NITROPYRIDINE PRODUCT

The method [2] was modified in respect of the final extraction at alkaline pH (with 0.88 ammonia): diethyl ether was used instead of benzene for the sake of increased recovery, safety and higher volatility. Nitrendipine was used as the internal standard. For GC-ECD (Pye, with ^{63}Ni detector, at 275°) the 5 ft. × 0.125 ins. glass column, kept at 230° (injector at 230°), was packed with 2% OV7 on Chrom W-HP (100-120 mesh), and run with N_2 at 30 ml/min.

Within-day reproducibility was good (C.V. 6.3% for 6 assays of a patient's sample, mean 62.4 ng/ml). Calibration graphs were consistently linear with correlation coefficients >0.98. However, day-to-day reproducibility was poor (C.V. 16.4% on 4 samples) and occasionally varied by 100%. This was thought to be due to a combination of GC interferences (autosampler vial septa gave a peak which co-chromatographed with the nitropyridine) and possible variable oxidation. The latter could have been due to incomplete oxidation by the nitrous acid, followed by further oxidation due to peroxides in the ether. A suggestion by K-D. Rämsch (Bayer; pers. comm.) might have provided a remedy, viz. solvent extraction initially, then oxidation with NO_2 (bubbled through; a convenient procedure).

CAPILLARY GC OF NIFEDIPINE

A report [3] that the nitropyridine was a true metabolite led to a reappraisal of methodology. K-D. Rämsch (pers. comm.) found that nifedipine could be chromatographed intact by capillary GC. As success with packed-column GC has since been reported [4], the previously reported problem of thermal decomposition [2] may have been related to injector design, the inertness of fused silica columns not being pertinent.

The high sensitivity of the ^{63}Ni ECD on the Varian Vista 6000 was exploited to allow a single-step extraction [5]. Toluene (1 ml), serum (1 ml), 100 ng of nitrendipine as internal standard and 0.1 ml of 0.88 ammonia were mixed, and 1 µl of the organic phase was injected. The column was of vitreous silica, WCOT SE30 (12 m × 0.33 mm i.d.), with He at 2.5 ml/min and N_2 make-up at 30 ml/min. The solvent trapping effect was exploited by programming from 100° to 200° at 40°/min, then 200° to 260° at 10°/min, holding for 1 min. Splitless injection was used on a split/splitless injector, venting excess solvent after 0.7 min. Nifedipine eluted at 8.0 min, and i.s. at 9.2 min.

This assay method was accurate, and precise as shown by results for samples containing 5.0 or 100 ng/ml: within-day, 6.0 ng/ml (C.V. 8.2%, n = 5) and 102.0 ng/ml (6.1%, 5) respectively; between-day, 6.2 ng/ml (10%, 6) and 99.5 ng/ml (4.9%, 9). Subsequent studies on the photodecomposition of nifedipine in daylight showed that erythrocytes had a considerable protective effect: ~2% loss/h, compared with 7% lost in plasma over 15 min ($t_{½}$ 3.04 h; S.E. ±0.5 h). The $t_{½}$ in water was 1.64 (±0.5 h). Hence care in sample preparation is required once serum has been separated from whole blood.

Acknowledgements

F.A. Tucker was supported by a research grant from Bayer UK, and Dr. K-D. Rämsch of Bayer, Wuppertal, provided valuable technical advice.

References

1. Testa, R., Dolfini, E., Reschiotto, C., Secchi, C. & Biondi, P.A. (1979) *Il Farmaco 34*, 463-473.
2. Kondo, S., Kuchiki, A., Yamamoto, K., Akimoto, K., Takahashi, K., Awata, N. & Sugimoto, I. (1980) *Chem. Pharm. Bull. 28*, 1-7.
3. Dokladalova, J., Tykal, J., Coco, S.J., Durkee, P.E., Quercia, G.T. & Korst, J.J. (1982) *J. Chromatog. 231*, 451-458.
4. Tanner, R., Romagnoli, A. & Kramer, W.G. (1986) *J. Anal. Toxicol. 10*, 250-251.
5. Tucker, F.A., Minty, P.S.B. & MacGregor, G.A. (1985) *J. Chromatog. 342*, 193-198.

#NC(C)-2

A Note on

DETECTION OF THE METABOLITES OF SOME DIHYDROPYRIDINE-TYPE CALCIUM ANTAGONISTS IN URINE AND PLASMA

K-D. Rämsch and D. Scherling

Institutes of Clinical Pharmacology and Pharmacokinetics
Bayer AG, P.O. Box 101709, 5600 Wuppertal 1, F.R.G.

Nifedipine (NF), nitrendipine (NT) and nimodipine (NM) are potent calcium-entry antagonists that modify calcium-channel activity in different smooth muscles. As therapeutic agents they are widely used in the treatment of coronary heart diseases (NF), hypertension (NT) or cerebral disorders (NM). The calcium entry blockade is closely related to the intact dihydropyridine (DHP) ring system. Metabolites are usally pyridine analogues of the parent drugs without any effects on calcium channels; even though they are ineffective and non-toxic, their pharmacokinetic profiles have to be studied.

A common analytical procedure to detect neutral and free unconjugated acidic DHP metabolites requires a pre-extraction of plasma and urine samples under basic conditions with toluene, to remove the unchanged drugs and their primary uncharged metabolites. Hydroxy-carboxylic acids are then converted to corresponding γ-lactones by HCl. Lactones and carboxylic acids, extracted with ethyl acetate, are methylated with methyl iodide and tetrabutylammonium hydroxide in methanol. Extracts are purified using solutions of ascorbic acid and sulphuric acid before the final capillary-GC step.

Preferably quantitative analysis is performed on a cold-injection system with solvent split at 80-120°, rate 1.3°/sec. Samples are transferred to a 30 m DB-1 fused-silica capillary column at a rate of 12°/min, starting at 100° oven temperature. Chromatography follows at a rate of 20°/min up to 200° and 5°/min up to 280°. The nitro-substituted DHP-metabolites are detected using their electron-capturing properties (ECD). Detection limits of DHP-metabolites in plasma and urine range between 0.5 and 5 ng/ml depending on their retention times. Using this method it is possible to study plasma levels and the urinary excretion of metabolites of NF, NT and NM after administration of therapeutic doses to patients and healthy volunteers.

Ref. noted by Senior Editor, relevant to foregoing Forum Abstract:

Graefe, K.H., Ziegler, R., Wingender, W., Rämsch, K-D. & Schmitz, H. (1988) *Clin. Pharmacol. Ther. 43*, 16-22.

Comments on material in #C

Comments on **#C-2,** J.D. Robinson - IMMUNOASSAY OF ACE INHIBITORS
#NC(C)-1, P.S.B. Minty - NIFEDIPINE ASSAY
& **#NC(C)-2,** K-D. Rämsch - DIHYDROPYRIDINE DRUG METABOLITES

R. Heath asked whether the plasma interferences encountered in the immunoassay of ramipril and trandolapril warranted extraction procedures. **Robinson's reply:** we have not pursued this avenue because of possible problems such as possible blanks due to the requisite solvent and the need for tracer to follow drug recovery. **Comments by G.M. Purdy.**- Pre-assay heat treatment of samples warrants trial, being useful for diluted plasma although it does not work with undiluted plasma; in our laboratories (Pfizer) we used to perform an ACE bioassay on human plasma using a ^{14}C-labelled substrate for ACE: this worked well with dog plasma samples, but with monkey plasma there was assay interference due to an inherent ACE which varied between animals. The problem was overcome by using 100 µl of monkey plasma as the source of ACE and assaying all samples from one monkey by one calibration curve using its pre-dose plasma as the ACE source for the calibration curve furnished by spikes. Clinical samples also posed problems: to 100 µl assay samples were added 400 µl of a control batch of human plasma, thus making the volume of ACE sample up to the normal value of 500 µl. An extended-range bioassay was developed for human plasma by using 500 µl of pre-dose plasma from that patient as an ACE source for the calibration curve; this allowed 500 µl samples to be assayed, and also gave a measure of post-dose compared with pre-dose ACE activity. Placebo samples were included. No diurnal effects were seen.

Comment by A. Rakhit.- Your addition of pooled blank plasma seemed to have buffered the fluctuation in levels of endogenous interference with binding. This makes sense, since dilution of all samples with one source plasma merely adds interference of similar extent to all samples; hence pre-dose samples still show 30% binding. We have not found sample-to-sample variability.

P.S.B. Minty, answering G.P. Mould: it is by proper sampling that we avoid analyte decomposition in clinical samples. **Remark by T. Parton.**- Concerning your observation of varying dihydropyridine dehydrogenation in GC, I have experienced dehydrogenation of a pyrrolidine ring while using stainless steel needles; it was obviated by on-column injection with sheathed fused-silica needles. **K-D. Rämsch, replying to R. Heath.**- The plethora of metabolites of Ca^{2+} antagonists is attributable to the ester functionalities of these molecules, enabling many molecular interactions to occur. **R. Woestenborghs** wondered about the type of autosampler and vial that were used in extracting nimodipine from plasma into toluene, maybe with an adjustable sampling needle.

Supplementation by Senior Editor

Calcium antagonists feature in a thoughtful survey [1] which classifies them primarily in respect of molecular sites of action (and secondarily by 'profiles'). **'Group I'** includes nifedipine and dihydropyridines; **'II'** includes verapamil, diclofurine and diltiazem; **'III'** includes fendiline, cinnarizine, flunarizine, pimozide and trifluoperazine. (Diltiazem and cinnarizine are diphenylalkylamines.)

Assay procedures for ACE inhibitors and calcium antagonists, supplementing those in the foregoing articles, are listed in a Table on p. 219 (in 'NC' pages that follow Sect. **D**); assay problems are well considered in the diltiazem citation [26] as well as in a 1985 review recommended in the accompanying text (the first ref. in the list). For ACE inhibitors in rat serum, e.g. lisinopril, a radioinhibitor binding displacement assay has been described [2]: the results correlated well with RIA results.

The problem of light-sensitivity with nifedipine [see #NC(C)-1] and an analogue with a furoyl group instead of methoxy was minimized by using an internal standard of similar light-sensitivity, in a procedure where dog plasma was solvent-extracted prior to RP-HPLC; with electrochemical detection 1-2 ng/ml was measurable, and oxidized metabolites were excluded [3].

Supplementary references

1. Spedding, M. (1985) *Trends Pharmacol. Sci. 6*, 109-114.
2. Jackson, B., Cubela, R. & Johnston, C.I. (1987) *Biochem. Pharmacol. 36*, 1357-1360.
3. Huebert, N.D., Spedding, M. & Haegele, K.D. (1986) *J. Chromatog. 353*, 175-180.

Section #D

VARIOUS CARDIOVASCULAR DRUGS

#D-1

THE STUDY OF CHIRAL CARDIOVASCULAR DRUGS: ANALYTICAL APPROACHES AND SOME PHARMACOLOGICAL CONSEQUENCES

[1]Richard L. Lalonde, [1]Michael B. Bottorff and [1,2]Irving W. Wainer®

[1]Department of Clinical Pharmacy
University of Tennessee, Memphis
Memphis, TN 38163, U.S.A.

[2]Pharmaceutical Division
St. Jude Children's
Research Hospital
Memphis, TN 38101, U.S.A.

For chiral CV drugs, which are administered as racemates, differences between the stereoisomers in pharmacological action have long been recognized. The difficult task of ascertaining the fate of each isomer has been eased by the development of HPLC methodology with CSP's. This article deals with analytical approaches and, in outline+, pharmacological stereoselectivity aspects for propranolol, labetalol, disopyramide and verapamil as examples of chiral CV drugs for which information on plasma levels is clinically important.*

In the past few years it has become increasingly clear that a racemic mixture of an enantiomeric drug is actually a 50/50 mixture of two separate drug substances. The reason is that enantiomers often differ in potency, toxicity, pharmacological action, metabolism, plasma disposition and urine excretion kinetics. These differences are especially important in the clinical use of chiral CV drugs. For the individual stereoisomer of such drugs, the investigation of clinical pharmacology may appear to be a recent development; but pharmacological differences have long been recognized. For example, since 1914 quinidine has been preferred to its diastereomer, quinine, as an antiarrhythmic agent, being more active [1-3]. However, even though

® Addressee for any correspondence (at the hospital address).

* *Abbreviations.*- CV, cardiovascular; CSP, chiral stationary phase [OD: *see text*]; HDA, homochiral derivatizing agent; TPC, *N*-trifluoroacetyl-*l*-prolyl chloride.

+ *Note by Ed.*- The MS. has been abridged in respect of pharmacological content and citations (if two from one lab., earlier ref. may be omitted).

quinine has lower antimalarial activity, it has evolved as the drug of choice in malaria treatment because it also produces less cardio-depression.

The pharmacological differences between stereoisomers, both enantiomers and diastereomers, are often studied through the separate administration of each of the isomers. While this gives the investigator an idea of what will happen if the drug is marketed and administered as a single isomer, it may not reflect what happens when a racemic mixture is administered. This point is illustrated by the enantiomer-enantiomer interaction associated with the administration of the chiral CV drug propranolol. Its active (*S*)-form can decrease cardiac output in man and decreased liver blood flow in monkeys [4]; the (*R*)-form does not cause the same haemodynamic effects [4]. Since the systemic clearance of propranolol depends on liver blood flow, (*S*)-propranolol may decrease its own clearance and, when a racemic mixture is administered, may also reduce (*R*)-propranolol clearance. The administration of (*R*)-propranolol alone should result in a higher systemic clearance, since there will be no reduction in the liver blood flow.

The determination of the pharmacological fate of each enantiomer in a racemic mixture is not an easy task. The major difficulty is the identification and measurement of the isomers in biological fluids and tissues. This task has been eased by the development of HPLC CSP's which can separate the enantiomers from each other and from interfering peaks. These phases have been reviewed elsewhere [5-7] and will not be discussed here. The present topic is the application of chiral HPLC in the identification and understanding of differences between the enantiomers of chiral CV drugs.

β-BLOCKERS

Clinical pharmacology

All β-blockers have at least one asymmetric carbon atom in their side-chain and, except for timolol and penbutolol, are used clinically as racemic mixtures [8]. Fig. 1 shows representative molecular structures. The laevorotatory (*S*) isomers of both β-adrenergic agonists and antagonists are much more potent (× 100 in the case of propranolol) than the dextrorotatory (*R*) isomers [9, 10], due partly to stereoselective binding to adrenergic β-receptors. The equipotency of the propranolol isomers in membrane-stabilizing ('local anaesthetic') activity illustrates distinctions in pharmacological effects, and served to establish that effectiveness in suppressing essential tremor is due to the (*S*)-isomer's β-blocking effect and not to its membrane-stabilizing effect [11].

With propranolol, the prototype β-blocker (having both β_1 and β_2 activity), the (*R*)-form is preferentially metabolized in humans,

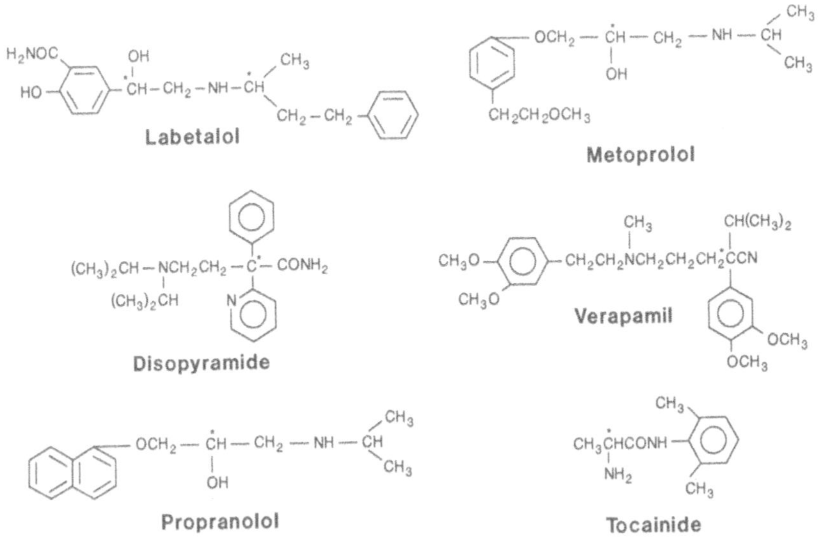

Fig. 1. Structures of some chiral CV drugs (* = asymmetric centre).

and the active (*S*)-form surpasses its plasma levels by 30-60% after oral administration [12, 13], reflecting lower intrinsic hepatic clearance due partly to higher plasma protein binding. The (*R*)-form is more prone to ring-oxidation, and shows lower excretions of unchanged drug, its glucuronide conjugate and side-chain oxidation products [12, 13]. A known decrease in propranolol clearance during chronic oral administration appeared to be stereoselective in the dog [14], but not in humans in preliminary studies by us (R.L. Lalonde et al.).

Stereochemical resolution on HPLC-CSP's

Diverse HPLC approaches have been used to resolve and quantify the enantiomers of β-blockers. The most common approach had been the conversion of the enantiomers into diastereomers using an enantiomerically pure derivatizing agent (HDA) followed by chromatography on an achiral HPLC system. For example, propranolol was resolved by Silber & Riegelman [15] and Hermansson & von Bahr [16] using (-)-TPC as the HDA, while Hermansson [17] used tert.-butoxycarbonyl-*l*-alanine and tert.-butoxy-*l*-leucine. In both methods, the resulting pair of diastereoisomers was resolved on an achiral octadecylsilane stationary phase, as used too by Walle et al. [18] to resolve the diastereomeric (+)-1-phenethyl isocyanate derivatives of (*R*)- and (*S*)-4'-hydroxypropranolol sulphate.

The determination of enantiomeric purity through the synthesis and separation of diastereomeric derivatives inherently entails the danger of inaccurate results due to possible enantiomeric contamination of the HDA. This danger was pointed out by Silber & Riegelman [15]

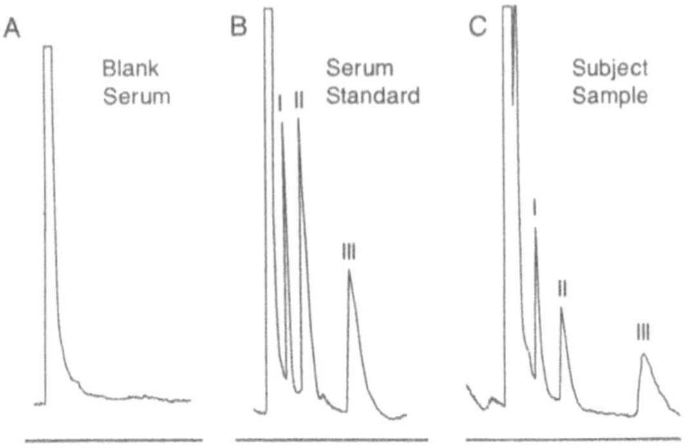

Fig. 2. Resolution of (*R*)- and (*S*)-propranolol (II & III resp.) on the OD-CSP. (C): serum from a subject 1 h after taking 160 mg racemic drug. (I) = verapamil (300 ng spiked into the serum); in (B) 150 ng racemic propranol was spiked in. Mobile phase: hexane/2-propanol/*N,N*-dimethyl-octylamine (92:8:0.01 by vol.). *From ref. [27], by permission.*

who found that commercial (-)-TPC was contaminated with 4-15% of the (+)-enantiomer and that the reagent rapidly racemized during storage. An additional complication is that enantiomers may have quite different rates and/or equilibrium constants when they react with the HDA, resulting in the generation of diastereomeric products differing in proportions from the starting enantiomeric composition [19]. These problems can be avoided by directly resolving the enantiomers using chiral chromatography.

The enantiomers of a number of β-blockers have been resolved using HPLC-CSP's. With one based on (*R*)-*N*-(3,5-dinitrobenzoyl)phenylglycine, Pirkle et al. [20] and Wainer et al. [21] resolved propranolol. It was first converted to an amide [20] or oxazolidone derivative [21] and then chromatographed using a hexane/2-propanol mobile phase. The latter method has been used in pharmacokinetic and pharmacological studies [21, 22]. Pronethalol and metoprolol can be stereochemically resolved without derivatization on a cyclodextrin-based CSP [23]. Many β-blockers have also been resolved on a CSP based upon α_1-acid glycoprotein [24, 25], as oxazolidone derivatives [24] or underivatized [25].

Okamoto et al. [26] recently reported the development of a new HPLC-CSP based upon cellulose-tris(3,5-dimethylphenylcarbamate) coated on macroporous silica (OD-CSP). The OD-CSP can resolve the underivatized enantiomers of a number of β-blockers with relatively high efficiency and short retention times, markedly surpassing previous HPLC-CSP resolution. Straka et al. [27] recently reported use of the OD-CSP to study the serum pharmacokinetics of propranolol enantiomers, with a simple extraction of the drug and no derivatization. Fig. 2 shows representative chromatograms. For propranolol the stereochemical selectivity (α) was 2.2 and the stereochemical resolution factor (Rs) was 3.7. Fig. 3 shows a concentration-time curve for the two enantiomers following ingestion of 160 ng of racemic propranolol.

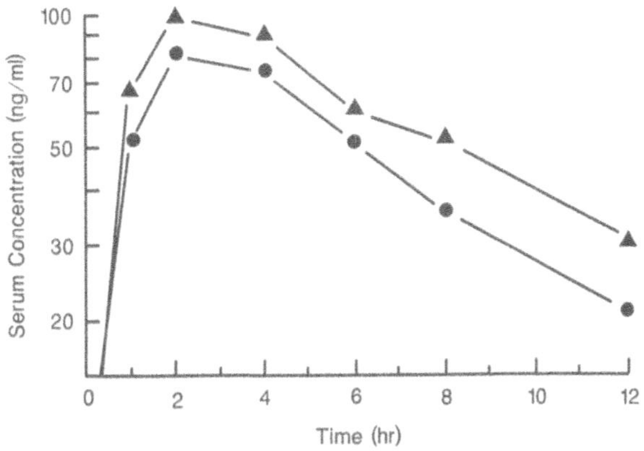

Fig. 3. Serum concentration-time curve for (*R*)-, ●, and (*S*)-propranolol, ▲, in a healthy subject after ingestion of a 160 mg dose of racemic propranolol. *From ref. [27], by permission.*

COMBINED α- AND β-BLOCKERS

The prototype of this class of drugs is labetalol, which has two asymmetric carbons in its side chain (Fig. 1). The marketed product is a mixture of two diastereomers, each being a racemic mixture of two enantiomers, the (*R*,*R*) and (*S*,*S*) enantiomeric pair and the (*R*,*S*) and (*S*,*R*) enantiomeric pair. As for propranolol, labetalol is a non-selective β-blocker, whereas its effects on post-synaptic α-receptors are similar to those of prazosin. The racemic mixture is typically 4-8 times more potent at β- than at α-receptors, depending on the model used [28]. However, the overall 'combined' effects on adrenergic receptors are actually caused by a complex interaction of the 4 stereoisomers, which differ significantly in their relative effects on the different receptors.

In respect of α- and β-blocker potency compared with racemic labetalol, the 4 stereoisomers rank as follows [28, 29].- (*R*,*R*) [undergoing clinical trials; 'dilevalol']: β-, × 3-4; α-, one-third. (*S*,*R*): α-, the most potent (slightly > racemic). (*R*,*S*): β-, 7-10 times less potent; α-, relatively weak. (*S*,*S*): β-, weak; α-, weak. Therefore labetalol's α-blocking activity is mostly attributable to (*S*,*R*) and to a lesser extent (*S*,*S*), and β-blocking activity to (*R*,*R*) and to a lesser extent (*R*,*S*). Little or no information is available on the individual pharmacokinetics of the 4 stereoisomers. Interestingly, it is reported that the ratio of β- to α-blocking effects is higher after i.v. than after oral administration [30].

Stereochemical resolution on HPLC-CSP's

With use of an α_1-acid glycoprotein HPLC-CSP, the (*R*,*R*)- and (*S*,*S*)-isomers of labetalol were resolved with an α of 1.36, and (*R*,*S*) and (*S*,*R*) with an α of 2.10. Unfortunately the peaks for (*S*,*S*) and (*R*,*S*) overlapped, and the separation was not applicable to pharmacokinetic studies following administration of a mixture of these. Applicability has come (Wainer et al., unpublished) from an approach,

Fig. 4. Resolution of labetalol stereoisomers - (*S*,*R*), A; (*S*,*S*), B; (*R*,*S*), C; (*R*,*R*), D - on an α_1-acid glycoprotein CSP. Mobile phase: see text.

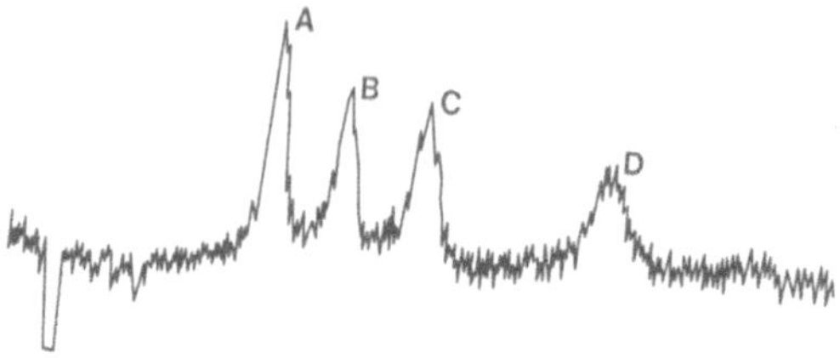

built upon this initial assay, with a different mobile phase: 0.02 M pH 7.5 phosphate buffer modified with 0.01 M tetrabutylammonium bromide and 0.5% 2-propanol, at 0.5 ml/min flow-rate and with the column at 30° throughout. Under these conditions the (*S*,*S*) and (*R*,*R*) enantiomers were resolved with α = 1.72 whilst α for (*S*,*R*) and (*R*,*S*) was 1.59 without a significant overlap of any of the peaks of interest. Fig. 4 shows a representative chromatogram. The 4 enantiomers are now undergoing pharmacokinetic study.

ANTIARRHYTHMIC AGENTS

Most chiral antiarrhythmic agents are administered as racemates. These drugs include disopyramide, mexiletine, tocainide, sotalol and flecainide. Since the interaction of such drugs with receptors, enzymes and transport proteins is often stereoselective, the individual enantiomers usually differ in antiarrhythmic efficacy, side-effects, protein binding and pharmacokinetics. An example is disopyramide (Fig. 1). In man the *d*- is more potent than the *l*-isomer in prolonging QTc intervals, but the isomers are equipotent as myocardial depressants, suggesting differing subcellular mechanisms for these two effects; the *d*-isomer also exerts stronger anticholinergic effects [31]. Given these differences in potencies, inter- and intra-patient variations in the pharmacokinetics of *d*- and *l*-disopyramide could produce a wide range of serum *d*-/*l*-disopyramide ratios, resulting in variability in pharmacological effects. Pharmacokinetic parameters are similar if the isomers are administered separately, but differ if a racemic mixture is administered, since the *d*- displaces the *l*-isomer from serum proteins [32-34]. The enantiomers also differ in plasma clearance, renal clearance and elimination half-life, with higher values for the *l*-isomer.

Stereochemical resolution using HPLC-CSP's

The resolution of the enantiomers of disopyramide was accomplished (α = 1.75; illustrated in Fig. 5) by Hermansson & Eriksson [35] with a CSP as for labetalol and, as mobile phase, pH 6.2 phosphate buffer containing 4.3% v/v 2-propanol and 1.95 mM dimethyloctylamine. With an achiral RP-2 pre-column linked to the main column, <15 ng/ml of each isomer could be detected in serum, allowing pharmacokinetic studies.

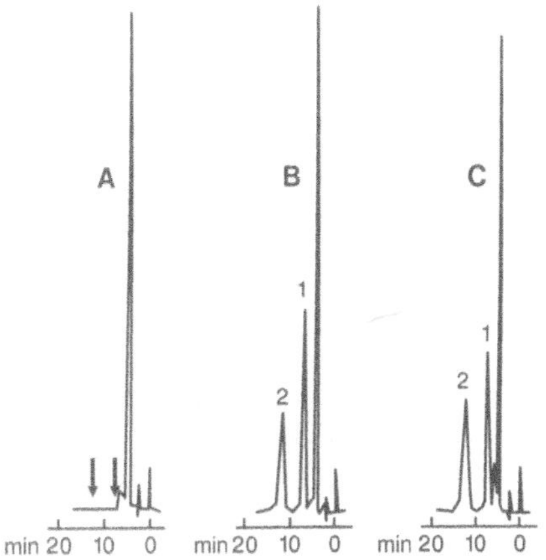

Fig. 5. Resolution of disopyramide stereoisomers - (*R*), 1; (*S*), 2 - on an α_1-acid glycoprotein CSP. A, blank plasma; B, blank plasma spiked with racemic disopyramide (1.5 µg/ml); C, plasma from a patient obtained 1 h after administration of racemic disopyramide. Mobile phase: see text. *From ref. [35], by permission.*

VERAPAMIL (CALCIUM CHANNEL ANTAGONIST)

Verapamil (Fig. 1) is the prototype calcium channel antagonist used in treatment of angina, supraventricular arrhythmias and hypertension. It is administered as a racemic mixture in which the *l*-isomer is 3-10 times more potent in producing negative inotropic and negative dromotropic effects in isolated myocardial tissue, intact animals and humans [36-38]. Pharmacokinetic stereoselectivity is observed. Serum shows a higher *d*/*l*-isomer ratio after oral (5:1) than after i.v. administration (2:1) of a racemic mixture, indicating preferential (first-pass) hepatic removal of the more active *l*-isomer, and higher bioavailability for *d*-verapamil [38-40]. Although the isomers differ in pharmacokinetic parameters when individually administered i.v., the elimination half-lives (~4-5 h) are similar since offsetting differences in distribution volume and clearance are observed [40]. Serum protein binding is higher for *d*- (~94%) than for *l*-verapamil (~88%), which at least partly explains the lower intrinsic hepatic clearance for the *d*-isomer following oral administration. It remains to be seen whether there is stereoselectivity in an effect already mentioned for propranolol - that with repeated dosing verapamil clearance falls and therefore bioavailability increases.

Stereochemical resolution using HPLC-CSP's

The enantiomers of verapamil have been resolved (α = 1.03) on a cyclodextrin-based HPLC-CSP using a gradient of acetonitrile and 1% aqueous triethylammonium acetate that changed from 10:90 to 20:80 over 10 min [23]. The chiral calcium channel blockers nisoldipine and nimodipine were also resolved (α = 1.04 and 1.05 respectively) on this sytem using a mobile phase composed of acetonitrile/1% aqueous triethylamine acetate (30:70). The enantiomers of verapamil have also been resolved (α = 1.75) on an α_1-acid glycoprotein HPLC-CSP [25]. The mobile phase was composed of 0.02 M pH 7.0 phosphate buffer containing 0.005 M decanoic acid and 0.05 M tetraethylammonium bromide.

CONCLUSION

The number and types of chiral CV drugs are rapidly increasing. The ability to separate the stereoisomers of these compounds is also growing, as illustrated by a recent review of the application of HPLC-CSP's to the resolution of biologically active molecules which cites >150 examples of drug substances which have been thus resolved [41]. When these two developments are coupled with the understanding of the pharmacological differences between enantiomers, it is likely that we shall see a decrease in the administration of racemic mixtures. In their place will be the administration of single enantiomers or of a controlled mixture of the isomers designed to produce a specific physiological response.

References

1. Frey, W. (1918) *Wien. Klin, Wochenschr. 55*, 849-853.
2. White, N.J., Looareesuman, S. & Warrell, D.A. (1983) *J. Cardiovasc. Pharmacol. 5*, 173-175.
3. Alexander, F., Gold, H. & Katz, L.N. (1947) *Pharmacol. Exp. Ther. 90*, 191-201.
4. Nies, A.S., Evans, G.H. & Shand, D.G. (1973) *Am. Heart J. 85*, 97-102.
5. Wainer, I.W. (1986) in *Bioactive Analytes, including CNS Drugs, Peptides and Enantiomers* [Vol. 16, this series] (Reid, E., Scales, B. & Wilson, I.D., eds.), Plenum, New York, pp. 243-257.
6. Lindner, W. & Petersson, K. (1985) in *Liquid Chromatography in Pharmaceutical Development: An Introduction* (Wainer, I.W., ed.), Aster Publg. Corp., Springfield, OR, pp. 63-131.
7. Wainer, I.W. (1987) *Trends Anal. Chem. 6*, 125-134.
8. Williams, K. & Lee, E. (1978) *Drugs 30*, 333-354.
9. Weiner, N. (1985) in *The Pharmacological Basis of Therapeutics* (Gilman, A.G., Goodman, L.S., Rall, T.W. & Murad, F., eds.), MacMillan, New York, pp. 176-210.
10. Barrett, A.M. & Cullum, V.A. (1984) *Br. J. Pharmacol. 34*, 43-55
11. Larsen, T.A. & Teravainen, H. (1982) *Acta Neurol. Scand. 66*, 289-294.
12. Silber, B., Holford, N.H.G. & Riegelman, S. (1982) *J. Pharm. Sci. 71*, 699-704.
13. Olanoff, L.S., Walle, T., Cowart, T.D., Walle, U.K., Oexmann, M.J. & Conradi, E.C. (1986) *Clin. Pharmacol. Ther. 40*, 408-414.
14. Bai, S.A., Wilson, M.J., Walle, U.K. & Walle, T. (1983) *J. Pharmacol. Exp. Ther. 227*, 360-364.
15. Silber, B. & Riegelman, S. (1980) *J. Pharmacol. Exp. Ther. 172*, 643-648.
16. Hermansson, J. & von Bahr, C. (1980) *J. Chromatog. 221*, 109-117.
17. Hermansson, J. (1982) *Suecica 19*, 11-20.
18. Walle, T., Christ, D.D., Walle, U.K. & Wilson, M.J. (1985) *J. Chromatog. 341*, 213-216.

19. Krull, I.S. (1977) in *Advances in Chromatography, Vol. 16* (Giddings, J. C., Grushka, E., Cazes, J. & Brown, P.R., eds.), Lippincott, Philadelphia, p. 146.
20. Pirkle, W.H., Finn, J.M., Schreiner, J.L. & Hamper, B.C. (1981) *J. Am. Chem. Soc. 103*, 3694-3696.
21. Wainer, I.W., Doyle, T.D., Donn, K.H. & Powell, J.R. (1984) *J. Chromatog. 306*, 405-411.
22. Donn, K.H., Powell, J.R. & Wainer, I.W. (1985) *Clin. Pharmacol. Ther*, *37*, 191.
23. Armstrong, D.W., Ward, T.J., Armstrong, R.D. & Bessley, T.E. (1986) *Science 232*, 1131-1135.
24. Hermansson, J. (1985) *J. Chromatog. 325*, 379-384.
25. Schill, G., Wainer, I.W. & Barkan, S.A. (1986) *J. Chromatog. 365*, 73-88.
26. Okamoto, Y. Kawashima, M. & Aburatani, R. (1986) *Chem. Lett.*, 1237-1240.
27. Straka, R.J., Lalonde, R.L. & Wainer, I.W. (1988) *Pharmaceut. Res.*, in press.
28. Sybertz, E.J., Sabin, C.S., Pula, K.K., Vliet, G.V., Glennon, J., Gold, E.H. & Baum, T. (1981) *J. Pharmacol. Exp. Ther. 218*, 435-443.
29. Brittain, R.T., Drew, G.M. & Levy, G.P. (1982) *Br. J. Pharmacol. 77*, 105-114.
30. Richards, D.A., Maconochie, J.G., Bland, P.E., Hopkins, R., Woodings, E.P. & Martin, L.E. (1977) *Eur. J. Clin. Pharmacol. 11*, 85-90.
31. Pollick, C., Giacomini, K.M., Blaschke, T.F., Nelson, W.L., Turner-Taniyasu, K., Briskin, V. & Popp, R.L. (1982) *Circulation 66*, 447-453.
32. Siddoway, L.A. & Woosley, R.L. (1986) *Clin. Pharmacokin. 11*, 214-222.
33. Lima, J.J., Boudoulas, H. & Shields, B.J. (1985) *Drug Metab. Dispos. 13*, 572-577.
34. Giacomini, K.M., Nelson, W.L., Pershe, R.A., Valdiviesco, L., Turner-Taniyasu, K. & Blaschke, T.F. (1986) *J. Pharmacokin. Biopharm. 14*, 335-356.
35. Hermansson, J. & Eriksson, M. (1984) *J. Chromatog. 336*, 321-328.
36. Ferry, D.R., Glossman, H. & Kaumann, A.J. (1985) *Br. J. Pharmacol. 84*, 811-824.
37. Giacomini, J.C., Nelson, W.L., Theodore, L., Wong, F.M., Rood, D. & Giacomini, K.M. (1985) *J. Cardiovasc. Pharmacol. 37*, 469-475.
38. Echizen, H., Vogelgesang, B. & Eichelbaum, M. (1985) *Clin. Pharmacol. Ther. 38*, 71-76.
39. Vogelgesang, B., Echizen, H., Schmidt, E. & Eichelbaum, M. (1984) *Br. J. Clin. Pharmacol. 18*, 733-740.
40. Eichelbaum, M., Mikus, G. & Vogelgesand, B. (1984) *Br. J. Clin. Pharmacol. 17*, 453-458.
41. Wainer, I.W. & Alembik, M.C. (1987) in *Chromatographic Chiral Separations* (Crane, L. & Zief, M., eds.), Dekker, New York, pp. 355-387.

#D-2

DEVELOPMENT OF A CHIRAL CAPILLARY GC METHOD FOR THE QUANTITATION OF THE ENANTIOMERS OF CROMAKALIM IN BIOLOGICAL FLUIDS

B.E. Davies

Department of Drug Metabolism & Pharmacokinetics
Beecham Pharmaceuticals Research Division
Harlow, Essex CM19 5AD, U.K.

Requirement	*A sensitive and specific method to allow cromakalim enantiomers to be determined in biological fluids down to 1 ng/ml.*
End-step	*GC on a Chirasil-VAL capillary column linked to a thermionic detector set up to detect nitrogen.*
Sample preparation	*Extract cromakalim and internal standard (i.s.) at alkaline pH using 2% isoamyl alcohol in toluene. Evaporate off the solvent and reconstitute in toluene for GC.*
Comments	*The enantiomers are well resolved; linearity is observed over the calibration range 1-40 ng/ml. The method is suitable for both pharmacokinetic and toxicokinetic studies.*

Cromakalim, BRL 34915 (structure overleaf), (±)6-cyano3,4-dihydro-2,-2-dimethyl-*trans*-4-(2-oxo-1-pyrrolidinyl)2H-benzo-[b]-pyran-3-ol, is a novel orally effective antihypertensive agent which relaxes blood vessels via K^+-channel activation. A method has been reported for quantitatively determining it in biological fluids using capillary GC linked to a nitrogen-specific detector [1]. A sensitive analytical technique was required that would enable BRL 38226 (3*R*,4*S*) and BRL 38227 (3*S*,4*R*), the (+) and (-) enantiomers of cromakalim respectively, to be resolved and quantitated in biological fluids. The following approaches were investigated:

- chiral derivatization with naphthylethyl isocyanate followed by non-chiral HPLC on a silica column;
- chiral HPLC using commercially available chiral stationary phases;
- chiral derivatization with menthyl chloroformate followed by non-chiral capillary GC;
- chiral capillary GC on a fused silica Chirasil-VAL capillary column.

The latter (Chirasil-VAL) approach gave good chromatography with adequate resolution of the enantiomers. The column was coupled to a thermionic detector, set up to specifically detect nitrogen, to provide adequate sensitivity.

MATERIALS

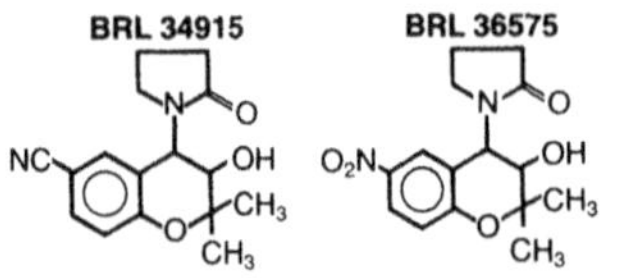

The panel shows the structures of cromakalim and the i.s., (±)*trans* BRL 36575. Solutions of each were made up in ultra-pure water (Elgastat Spectrum). Solvents were obtained from Rathburn. Other reagents were of AnalaR grade. BSTFA (bis-trimethylsilyl-trifluoroacetamide) containing 10% TMCS (trimethylchlorosilane) was obtained from Regis. Chirasil-VAL-L and -D columns were obtained from Chrompack U.K.

EQUIPMENT AND CHROMATOGRAPHY

The following were used: a Hewlett-Packard 5880 gas chromatograph equipped with a thermionic detector specifically set to detect nitrogen, a universal split/splitless capillary injector in the splitless mode, and an HP 7672 autosampler. Data were recorded and peak-height ratios calculated using a Multichrom Data System (VG Data Systems). The detector was at 300° and the injector at 250°, and the oven was programmed from 110° to 195° at 25°/min. The He head pressure was set at 30 psi with an N_2 make-up gas flow rate of 30 ml/min.

The column, Chirasil-VAL-D, was of fused silica, 25 m × 0.22 mm i.d. with a 0.06 µm film thickness. Peak-height ratios, relative to the i.s., were plotted against analyte concentrations, and standard curves obtained by least squares analysis.

SAMPLE PREPARATION

Extraction.- An appropriate aliquot of aqueous i.s. solution was added to plasma or urine (usually 1 ml), followed by 0.5 ml of 0.5 M NaOH. After mechanical extraction for 15 min with 3.5 ml of toluene containing 2% (v/v) isoamyl alcohol and centrifugation, the organic layer was removed and evaporated to dryness under N_2 at 50°.

Derivatization.- The effects of temperature, time, type of solvent and concentration of BSTFA on the derivatization reaction were studied. Fig. 1 shows a representative experiment. The final conditions reckoned to be optimal were as follows:- 50 µl of BSTFA (containing 10% v/v TMCS)/acetonitrile (1:1 by vol.) was added to the extraction residue and heated at 60° for 20 min, followed by evaporation to

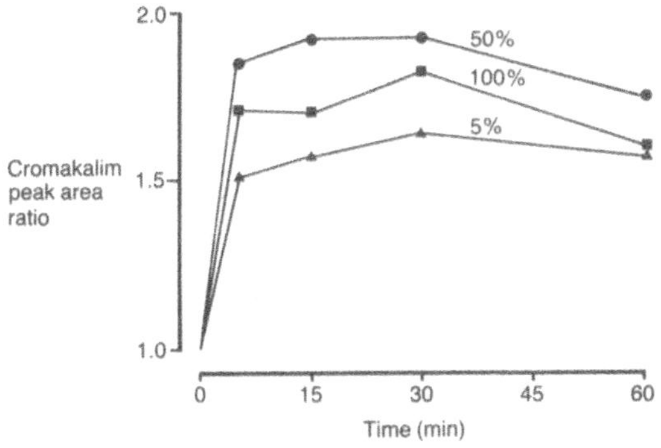

Fig. 1. The time course and yield of the derivatization procedure as affected by BSTFA concentration (% v/v in acetonitrile).

dryness under N_2. The derivatives were reconstituted in 30-50 µl toluene, and 2-4 µl was chromatographed after automated splitless injection.

ASSAY PARAMETERS

Chromatography.- Fig. 2 shows representative chromatograms for analysis of samples of diluted rat blood with and without added cromakalim. The BRL 38227 and 38226 peaks were well resolved with corrected retention times of 9.9 and 10.3 min respectively. These chromatograms were obtained using a Chirasil-VAL-D column from which BRL 38227 elutes first. If a Chirasil-VAL-L column is used, the elution order of the enantiomers is reversed. Capacity factors (k') were 16.5 for BRL 38227 and 17.1 for BRL 38226; the separation factor (α) was 1.04, and the resolution factor (R) was 2. The i.s., which is also a racemic mixture, does not resolve under these chromatographic conditons.

Linearity.- The peak height ratios of both BRL 38227 and 38226 varied linearly with concentration over the range studied, 1-40 ng/ml. Fig. 3 shows a typical calibration curve thus obtained for BRL 38227 in diluted rat blood; the curve for BRL 38227 was identical. The slope value, intercept, relative S.D. and limit of reliable determination were 0.0655, 0.004, 3.7% and 1 ng/ml respectively.

Reproducibility.- The following C.V. values were obtained for the assay of replicate human plasma samples containing BRL 38226 at three levels:

Fig. 2. Typical chromatograms obtained using this method for the analysis of cromakalim in diluted rat blood.

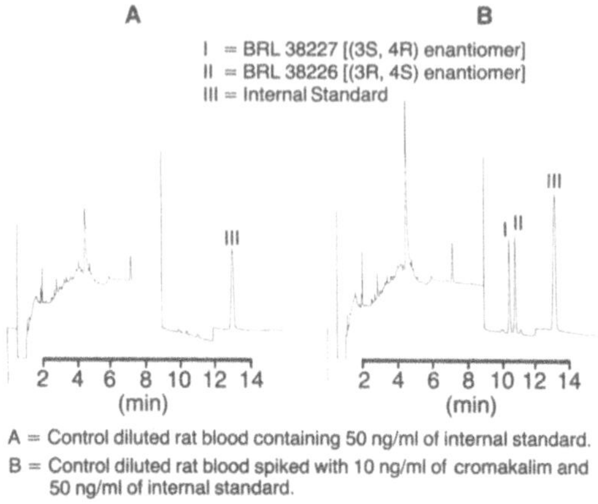

Fig. 3. Typical calibration curve for the analysis of BRL 38227 in diluted rat blood.

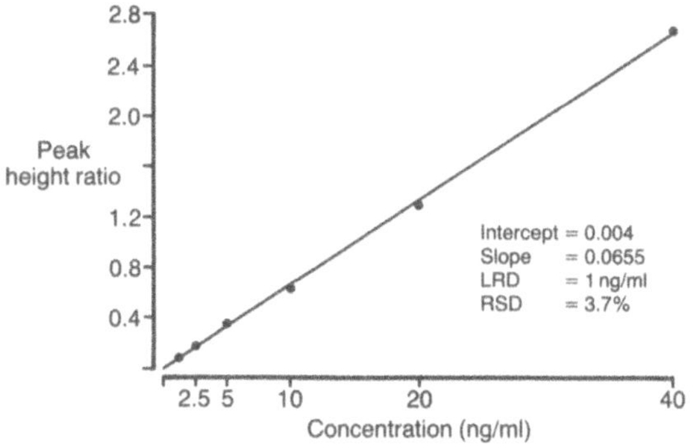

- within-day: 13%, 5% and 7% for 1, 5 and 40 ng/ml respectively;
- between days (consecutive days): 10%, 7% and 4% respectively.

Since assay precision markedly deteriorates at 1 ng/ml, this was set as the limit of reliable determination. BRL 38227 gave C.V.'s virtually identical with those for BRL 38226.

Accuracy.- To determine accuracy, spiked plasma samples were assayed blind alongside the test samples. Fig. 4 shows the results for a large number of such samples. The assayed values are mostly within 10% of the actual spike values; a few fall outside this range but are within 15%.

COMMENTS

The analytical technique developed to resolve the enantiomers of cromakalim is sensitive, reproducible and accurate and has been in routine use for >18 months. During this time biological samples from toxicology, metabolism and pharmacokinetic studies in animals and from human studies have been successfully analyzed. Column lifetime varies considerably and appears to depend partly on the number of injections made and partly on the temperature to which the column is programmed and then held. There is also a need to ensure that

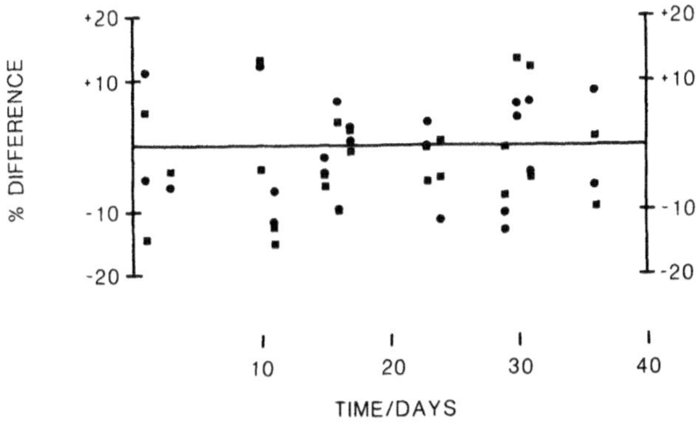

Fig. 4. Quality control samples assayed over a period of 6 weeks, plotted as % difference from nominal (spiked) value against duration of the study. BRL 38227, ●; 38226, ■.

the carrier gas is free of oxygen. Frozen storage is allowable; the drug is stable for at least 6 months when stored in biological fluids at -20°, as is now being verified with the individual enantiomers.

Acknowledgements

The author thanks Dr. J. Lough, Mr. V. de Biasi, Dr. T.S. Gill and Mr. J.E. Peppiatt for their valuable contributions to this work.

Reference

1. Davies, B.E. (1987) *Br. J. Clin. Pharmacol.* *24*, 273P.

#D-3

THE DETERMINATION OF PLASMA AMIODARONE AND DESETHYLAMIODARONE BY HPLC

B.J. Starkey, A.J.E. Green and G.P. Mould

Department of Clinical Biochemistry and Nutrition
St. Luke's Hospital, Guildford, Surrey GU1 3NT, U.K.

Since amiodarone may have serious side-effects, it is important to ascertain its levels in plasma. We have developed a method more robust than existing HPLC methods. The sample, with i.s.⊗ added, is solvent-extracted from 1 M pH 6.0 acetate buffer, these conditions being optimal for the drug and its desethyl metabolite. RP-HPLC with UV detection is then performed. This validated assay is rapid and needs only small samples of serum, which may be pre-stored at 4° or -20°; if plasma stored at -20° is used, it may need centrifugation after thawing.*

Amiodarone is a powerful anti-arrhythmic agent. Its measurement, together with that of its major metabolite, desethylamiodarone, is important to assess patient compliance and attainment of therapeutically effective levels, and to avoid undesirable extra-cardiac side-effects (which occasionally include hepatotoxicity [1]). G-I absorption of the drug is erratic and unpredictable (oral bioavailability ranging from 22% to 86%), and the rate of elimination (mainly by metabolism and biliary excretion) varies considerably: ½-life 3.2 to 79.7 h even after a single dose and >3 months after cessation of long-term therapy.

For measurement of amiodarone to be useful, the method must be reliable and an optimum therapeutic range must be defined to aid interpretation of clinical results; the range is still ill-defined (probably between 1.0 and 2.5 μg/ml for the drug [2]), and little guidance exists concerning the metabolite. Since the major metabolite, desethylamiodarone, is thought to be pharmacologically active, it too warrants measurement.

---

* In *in vivo* contexts, the term 'plasma' is preferred to 'serum'.- *Ed.*

⊗ i.s., internal standard; TMAC, tetramethylammonium chloride; NP, straight-phase; RP, reversed-phase.

Table 1. Recent methods for determination of amiodarone and desethylamiodarone. HPLC columns other than 'SiO_2' (NP) were RP. Phase compositions are by vol. Conventional chemical abbreviations are used: e.g. MeOH, methanol; MeCN, acetonitrile; also EA, ethylamine. The 'ng/ml' column is sensitivity [with sample vol., ml].

Column	Mobile phase	Initial step (& i.s.)	ng/ml	C.V. %	Ref.
C-18	$MeOH/NH_4OH$	MeCN pptn. (L8040)	25 [1]	2.2-5.6	3
C-18	MeCN/10 mM KH_2PO_4/2 mM TMAC, 90:9:1	Me-t-Bu ether, pH 6.0 (L8040)	?	?	4
C-18	MeOH/water/NH_4OH, 94:4:2	MeCN pptn. (L8040)	20 [0.1]	1.0-11.5	5
C-18	MeCN/water/H_3PO_4/diEA, 80:20:0.1:0.005	MeCN/$ZnSO_4$ pptn. (none)	50 [0.1]	6.5-6.7	6
SiO_2	MeOH/Et_2O, 80:20, +15 µl triEA/1	Hexane (triflupromazine)	15 [1]	5.4-8.8	7
C-8	MeOH/MeCN/10 mM NH_4 acetate, 40:40:10	Hexane, pH 5.5 (triflupromazine)	15 [1]	5.5-10.2	8
CN	MeCN/20 mM pH 3.0 phosphate, 60:40	BuCl, pH 8.0 (fenethiazine)	20 [0.5]	6.9-8.3	9
C-8	MeCN/10 mM pH 3.5 phosphate, 62:38	CN solid phase (L8040)	250 [0.1]	5.7-8.3	10

Although many of the older methods for amiodarone failed to assay this metabolite, most methods reported recently (surveyed in Table 1) do determine both the parent drug and the metabolite. Our objectives have been to develop a robust assay for amiodarone and its desethylated metabolite, to establish an optimum therapeutic range for amiodarone, and to investigate interrelationships between drug and metabolite concentrations in plasma. This article describes the development and optimization of an assay which incorporates the most desirable features of several previously published methods, and summarizes our initial experience of clinical use of the assay.

OPTIMIZED ASSAY

This is shown schematically in Scheme 1. Using this procedure with standards (spiked into drug-free serum) in the range 0-6 µg/ml, amiodarone and desethylamiodarone concentrations down to 0.1 µg/ml were readily detectable, allowing routine monitoring of patients' plasma drug concentrations. Procedural losses were accounted for by reference to the i.s., L 8040 [2-ethyl-3-(3,5-dibromo-4-dipropylaminopropoxylbenzoyl)benzothiophene].

Extraction efficiency is shown in Fig. 1 where sodium acetate buffers in the pH range 4-9 were added. Addition of 100 µl pH 6.0

SERUM (standard/control/test), 200 µl

Add 100 µl 1 M Na acetate pH 6.0, 40 µl i.s. (L 8040), 2 ml hexane; mix (Rollamix, 20 min) and centrifuge (2000 rpm, 5 min)

Organic layer

Evaporate down (N_2, room temp.)

Residue

Reconstitute in 100 µl methanol, and load 50 µl onto column:-
5 µm Spherisorb ODS-2 column, 250 × 4.5 mm;
mobile phase: 0.88 ammonia/methanol (3:97 by vol.), 1.2 ml/min;
detection at 242 nm; measure peak height ratios *vs.* i.s.

Scheme 1. Assay procedure for amiodarone and desethylamiodarone. Retention times 4.6 and 4.2 min respectively; 6.0 min for the i.s.

1 M acetate to 200 µl serum resulted in a final pH of 6.5, giving 92%, 78% and 93% extraction for amiodarone, desethylamiodarone and i.s. respectively.

Chromatography.- The column (C-18) and mobile phase were similar to those previously described [3]. The proportion of conc.(25% w/w) ammonia added to methanol as a competing base affected the relative retention volumes of the analytes (Fig. 2). At NH_3 concentrations below 0.5% (w/v) amiodarone eluted prior to desethylamiodarone, whereas at >1% this order was reversed. With 3% as chosen, the peaks were sharp and well resolved, and the run time was <8 min (Fig. 3).

Precision was ascertained with aliquots, pre-stored at -20°, of spiked normal serum. For amiodarone and desethylamiodarone ('des') at the stated µg/ml levels, the C.V.'s were as follows:
- **intra-assay**: drug at 0.81, 12%, and des at 0.76, 9.2%; drug at 1.04, 5.1%, and des at 1.18, 3.6%.
- **inter-assay**: drug at 0.59, 10.8%, and des at 0.76, 12.7%; drug at 1.18, 5.4%, and des at 1.16, 5.0%.

Interferences.- Possible interfering drugs were added to spiked normal serum containing amiodarone and desethylamiodarone at 5 µg/ml. Aliquots were extracted and assayed, to disclose any effect upon the peak-height ratios of analyte to i.s. or any appearance of additional peaks. The apparent concentrations of amiodarone and desethylamiodarone were unaffected by the addition, to 20 µg/ml, of propranolol, atenolol, pindolol, methyldopa, isosorbide dinitrate or spironolactone. Likewise, spiking of amiodarone (2 µg/ml), desethylamiodarone (2 ug/ml) and i.s. into serum from hospitalized patients taking therapeutic doses of digoxin, frusemide, amiloride, atenolol or nifedipine showed that the presence of these drugs or their metabolites had no effect on the values for amiodarone or desethylamiodarone.

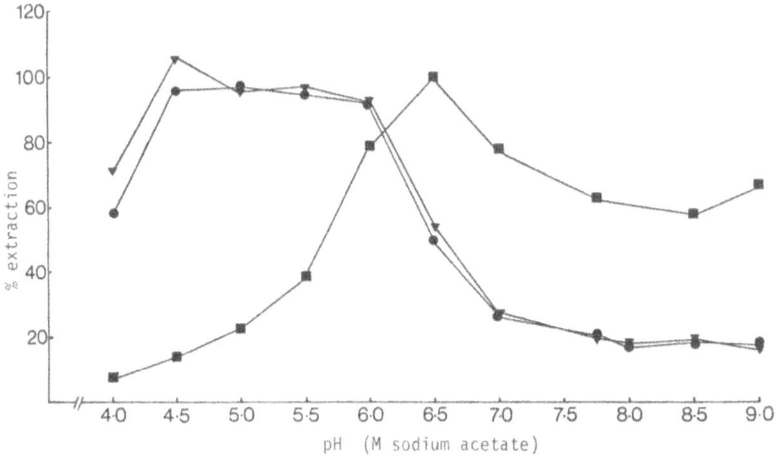

Fig. 1. Influence of pH on extractability for amiodarone (▼), desethylamiodarone (●) and i.s. (■), spiked into serum.

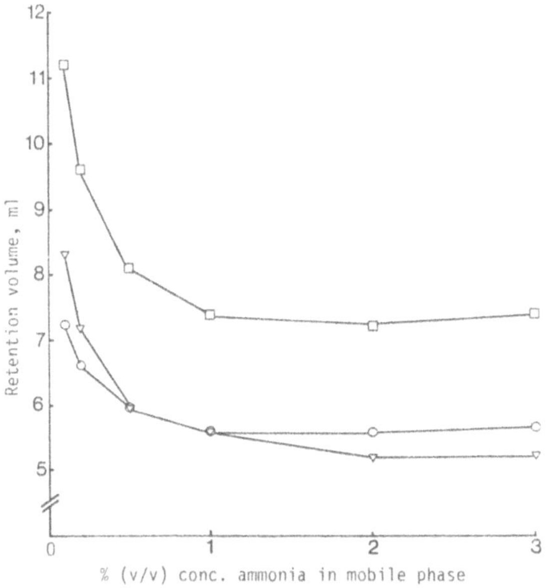

Fig. 2. Effect of ammonia concentration in mobile phase on retention of amiodarone (o), desethylamiodarone (▽) and i.s. (□), spiked into serum.

CLINICAL STUDIES

Preliminary measurements on patients receiving oral amiodarone (200-300 mg daily) gave plasma amiodarone concentrations between 0.20 and 1.36 µg/l. The time course is shown in Fig. 4. After a single i.v. dose (5 mg/kg over 2 h) of amiodarone the proportion of desethylamiodarone steadily increased whilst that of the parent drug decreased with an elimination half-life of ~96 h over the 5 days studied; after 4-5 days the ratio of drug to metabolite approached unity.

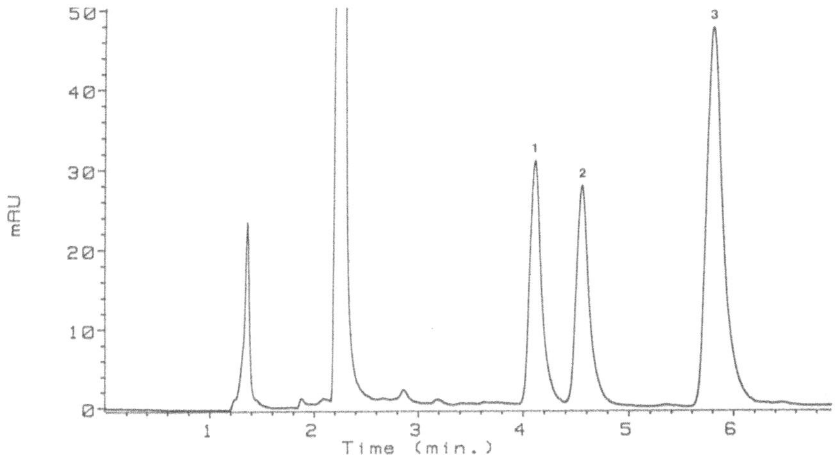

Fig. 3. Chromatogram of patient sample. **1**, desethylamiodarone (4.1 µg/ml); **2**, amiodarone (3.8 µg/ml); **3**, i.s.

Fig. 4. Time course of plasma amiodarone (●) and desethylamiodarone (■) levels following i.v. amiodarone administration.

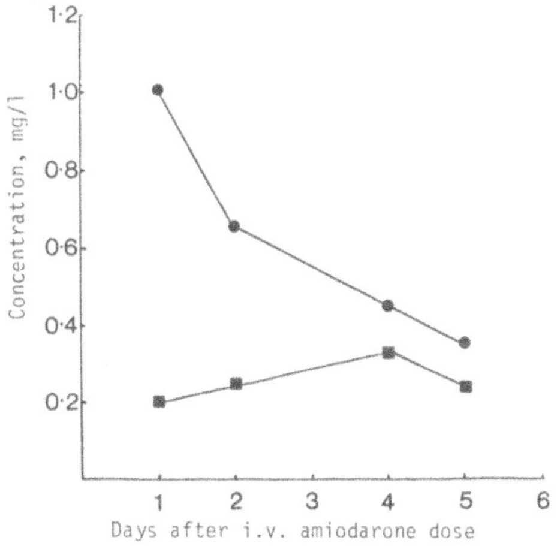

DISCUSSION

Many HPLC assays have been described for amiodarone and, in the more recent ones, for desethylamiodarone also. Approaches to specific preparation have been based on solvent extraction, acetonitrile deproteinization or solid-phase extraction (Table 1). Our initial experiments using acetonitrile precipitation yielded chromatograms which were not as clean as those with the present extraction method. This would be expected since, with the chosen pH conditions, many irrelevant compounds would not be extracted and so would not appear on the chromatogram. Furthermore, the precipitation methodology, by diluting the sample, resulted in a 3-fold decrease in drug concentration whereas our method results in a 2-fold increase. The method described also lends itself more easily to adaptation for greater sensitivity if required. Extraction of the analytes into alternative solvents was not investigated since hexane proved very effective, was easy to evaporate, and was readily available in a suitable grade of purity (HPLC grade).

The chosen mobile phase, methanol/water/ammonia, was similar to that employed by others [3, 5]. It had the advantage that the ammonia content may be adjusted to achieve the desired separation and retention time (Fig. 2). A disadvantage may be that the high mobile-phase pH may, in prolonged use, damage the column packing; but we have seen little deterioration in column performance provided that columns are washed thoroughly with a water/methanol mixture after use.

For the i.s. we have not investigated possible alternatives (Table 1) to L 8040 since this was readily available (Labaz-Snofi UK Ltd.). It closely resembles amiodarone in extractability (Fig. 1) and has a suitable retention time and absorption spectrum.

Interference studies demonstrated the high specificity of the assay. Many of the drugs tested would not be extracted from serum into hexane under the pH conditions used. The fact that no interference from digoxin could be demonstrated is important since in cases where digoxin and amiodarone are co-administered, close monitoring of amiodarone and desethylamiodarone concentrations may be necessary because of the well-documented drug interaction [11].

The assay precision, which benefits from using L 8040 as i.s., accords with that of previously reported methods and is more than adequate for clinical use, as is the sensitivity. Sensitivity can readily be increased, down to 10 ng/ml, by increasing the sample size and adjusting the extraction solvent volume, or by increasing the proportion of extract taken for chromatography. Already it appears that plasma amiodarone concentrations are variable, 0.2-1.4 µg/ml. In agreement with other workers [8] our initial observations suggest that at steady state the drug:metabolite ratio in plasma approximates to unity. Our time-course study showed for an i.v. dose a notably long elimination half-life and, in agreement with a previous report [12], near-equality of drug and metabolite concentrations at 4-5 days.

In summary, the assay is robust, simple, precise, substantially interference-free and, from preliminary studies, suitable for patient monitoring. Studies to assess the relationship between amiodarone concentration and therapeutic efficiency, and to establish a valid therapeutic concentration range, are currently in progress.

References

1. Latini, R., Tognoni, G. & Kates, R.E. (1984) *Clin. Pharmacokin.* *9*, 136-156. [This group's assay methodology is pertinent.- *Ed.*]
2. Rotmensch, H.H., Belhassen, B., Swanson, B.N., Shoshani, D., Spielman, S.R., Greenspan, A.J., Greenspan, A.M. & Vlasses, P.H. (1984) *Ann. Intern. Med.* *101*, 462-469.
3. Plomp, T.A., Engles, M.& Robles de Medina, E.O. (1983) *J. Chromatog.* *273*, 379-392.
4. Shipe, J.R. (1984) *Clin. Chem.* *30*, 1259.
5. Weir, S.J. & Ueda, C.T. (1985) *J. Pharm. Sci.* *74*, 460-465.
6. Lam, S. (1986) *J. Chromatog.* *381*, 175-178.
7. Bliss, M., Mayersohn, M. & Nolan, P. (1986) *J. Chromatog.* *381*, 179-184.
8. Muir, K.T., Kook, K.A., Stern,C. & Gardner, K.M. (1986) *J. Chromatog.* *374*, 394-399.
9. Hutchings, A., Spragg, B.P. & Routledge, P.A. (1986) *J. Chromatog.* *382*, 389-393.
10. Pollack, P.T., Carruthers, S.G. & Freeman, D.J. (1986) *Clin. Chem.* *32*, 890-893.
11. Vitale, P., Jacono, A., Reyero, E.G. & Zeuli, L. (1984) *Clin. Trials J.* *21*, 199-206.
12. Holt, D.W., Tucker, G.T., Jackson, P.R. & Storey, G.L.A. (1983) *Am. Heart J.* *106*, 840-846.

#D-4

THE ASSAY OF HYDRALAZINE: A REVIEW

P.H. Degen

Research and Development, Pharmaceuticals Division
Ciba-Geigy Ltd., CH-4002 Basle, Switzerland

It has long been difficult to determine hydralazine in biological fluids for biopharmaceutical and pharmacokinetic purposes. The drug can react with a number of endogenous compounds, e.g. pyruvic acid or α-ketoglutaric acid, to form acid-labile hydrazones. Being a base, hydralazine is extractable from biological fluids only at basic pH values where it is unstable, being degraded mainly to phthalazine. A derivatization procedure which can be carried out directly in biological fluids is therefore essential.

Reaction with nitrite under acidic conditions produced tetrazolophthalazine, which is both extractable and stable. However, acid-labile hydrazones of hydralazine are hydrolyzed and also react with nitrite. The method therefore detects the sum of free hydralazine and hydralazine liberated from acid-labile hydrazones. Although the method sufficed for studying the bioequivalence of dosage forms, it was unsuitable for investigating the pharmacokinetics of the unchanged drug. Various attempts to selectively measure free hydralazine have been published. The formation of a pyrazole, using 2,4-pentanedione at a neutral pH, with final GC-NPD determination, resulted in a specific and sensitive method for free hydralazine.

Hydralazine, 1-hydrazinophthalazine (Fig. 1), is a potent vasodilator, well established for the treatment of hypertension. It is unstable at pH >7, and not extractable at lower pH. Thus, for its quantitative determination, hydralazine must be derivatized in the biological material to render it extractable. This can be achieved in a number of different ways. Carbonyl compounds such as aldehydes (*p*-methoxybenzaldehyde) or ketones (cyclohexanone) form extractable hydrazones. Alternatively, reaction with nitrous acid forms an extractable tetrazolophthalazine, with pentanedione an extractable pyrazole, and with formaldehyde or acetaldehyde a ring-closed derivative (triazolophthalazine or methyltriazolophthalazine). Hydralazine has

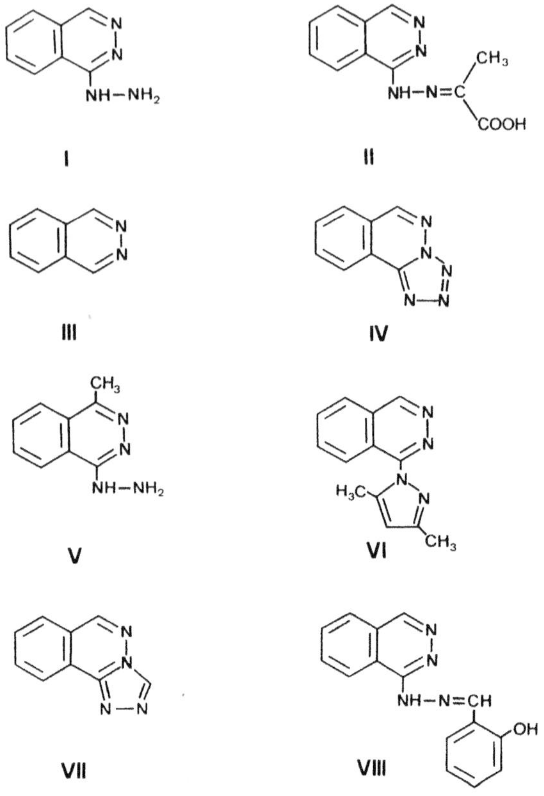

Fig. 1. Structures of:
I, hydralazine;
II, hydralazine pyruvic acid hydrazone;
III, phthalazine;
IV, tetrazolophthalazine;
V, 4-methylhydralazine;
VI, 1-(3,5-dimethyl-1-pyrazolyl)phthalazine;
VII, triazolophthalazine;
VIII, hydralazine salicylic acid hydrazone.

in fact itself been utilized as a derivatizing agent for determining nitrite or formaldehyde [1].

Hydralazine is extensively metabolized and is excreted almost exclusively in the form of metabolites [2, 3]. Plasma levels of unchanged hydralazine after a therapeutic dose are in the low nmol/ml range, and the biological half-life is very short. On the other hand, concentrations of 'apparent' hydralazine are considerably higher (by at least × 6) than those of the unchanged hydralazine [4, 5]. Table 1 lists the many published analytical procedures for 'apparent' hydralazine, free hydralazine and also for numerous metabolites. Methods employed range from spectrophotometry, GC-ECD, GC-NPD and GC-MS to HPLC. At the present time GC and HPLC assays are the most suitable measurement methods for hydralazine. However, the instability of hydralazine and its metabolites in solution does present analytical problems.

STABILITY

Most investigators involved with the assay of hydralazine have evaluated its stability, in aqueous solution and/or in plasma [7,

Table 1. Methods for the assay of hydralazine in biological fluids. **A** = 'apparent', **F** = free. P = plasma, S = serum, U = urine. LOQ = limit of quantification, μg (μ) or ng (n)/ml [for μmol/l, μ × 5]. HZ, hydrazine. **Z**, azol; **z**, azin (hence tetraZophthalze = tetrazolophthalazine). *Conventional abbreviations include* Me, methyl; EtOH, ethanol; ac., acid; NPD, nitrogen-phosphorus detection.

Assayed	Reaction/derivatization & extraction (ex.), and **end-step**	LOQ	Ref.
A; U	ninhydrin hydrzophthalze complex at pH 3; $CHCl_3$ ex.; 460 nm **photometry**	1-2μ	6
A; P	p-OHbenzaldehyde HZ at pH 6; $CHCl_3$ ex.; 365 nm **photometry**	0.1μ	7
A; P	improved version of above	50n	8
A; P, U	p-methoxybenzaldehyde HZ at pH 3; benzene/MeOH (93:7) ex.; 355 nm **photometry**	0.1μ	9
A; P, U	tetraZophthalze at pH 0.2; benzene ex.; **GC-ECD**	10 n	10
A+**M**; P,U	For **A**: as above. For **M**: silylation. **GC-MS**	?	11
F & **A**; P	tetraZophthalze, before & after MnO_2-oxidation; benzene ex.; **GC-ECD**	10 n	12
F; S	cyclohexanone HZ at neutral pH; **GC-MS**	?	#
pure*	pyrZe at neutral pH; Et acetate ex.; **GC-NPD**	?	13
pyruvic ac. HZ; P	$CHCl_3$ ex.; → 3-trifluoro-Me-triZophthalze; **GC-ECD**	10 n	14
F; P	pyrZe at neutral pH; Et acetate ex.; **GC-NPD**	10 n	15
A; P	triZophthalze at pH 5; **HPLC**	3 n	16
F; P	p-methoxybenzaldehyde HZ at neutral pH; **HPLC**	5 n	ø
A; S	triZophthalze at pH 1; toluene ex.; **GC-NPD**	20n	17
F & **A**; P	**F**: tetraZophthalze at pH 5.5, & **A**: at pH 1.0; $CHCl_3$ ex.; **HPLC**	1 n	4
pyruvic ac. HZ; P	deproteinized with MeOH; **HPLC**	10 n	18
F; U	pyrZe at neutral pH; CH_2Cl_2 ex.; **GC-NPD**	0.1 μ	19
F+pyruvic ac. HZ; P	**F**: p-methoxybenzaldehyde HZ at neutral pH; Et acetate ex., the pyruvic ac. HZ (**PH**) being co-extracted; **HPLC**	**F**: 2 n; **PH**: 25n	5
F; P	p-methoxybenzaldehyde HZ at neutral pH; hexane ex.; **HPLC**	1 n	20
F; S	triZphthalze; butanol ex.; 470 nm **photometry**	2 μ	21
F & **M**; U	direct injection (LOQ is amount injected); **HPLC**-electrochem. det.	(≤1 μ)	22
F & **A**: P	tetraZophthalze at neut. pH (**F**) or pH 1 (**A**); $CHCl_3$ ex.; **HPLC**	1 n	23
F; U	**adsorptive stripping voltametry**	2 μ	24
M; P	CH_2Cl_2 ex.; **HPLC**	0.5 μ	25
F; P	salicylaldehyde HZ at neutral pH; heptane/CH_2Cl_2/iso-pentanol ex.; **HPLC**-electrochem. det.	0.2 n	26

*assays on tablets   # A 1977 paper, ø a 1979 paper: same lab. as for ref. 5

Fig. 2. Concentrations of 'apparent' hydralazine in plasma of healthy volunteers treated with a single oral dose of 100 mg of hydralazine. Analysis according to Jack et al. [10] before and after 1 month's storage at -20°.

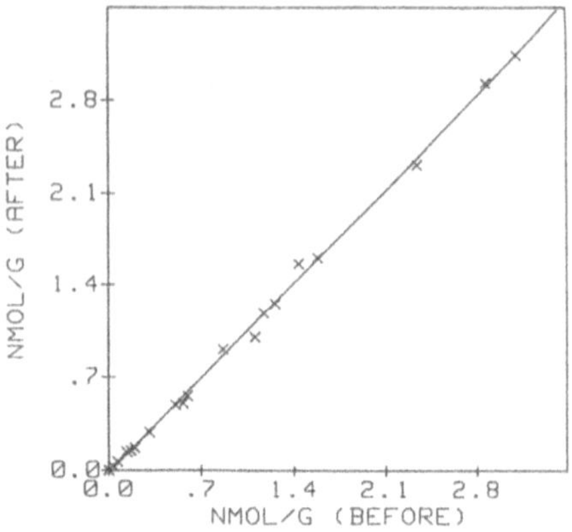

12, 14, 18, 27-29]. Stability is critically affected by oxygen, pH if >7, ion type and concentration, and temperature. The rate of oxidative degradation can be reduced by adding ascorbic acid or EDTA [5, 12-14] and by working at low pH. In pH 8 glycine buffer hydralazine is fairly stable (5% loss/h; room temp.), whereas in pH 8 phosphate buffer degradation is fast (80%/h; room temp.) [7]. The identified degradation products include phthalazine, phthalazinone and di-phthalazinylhydrazine [7, 27-29]. Reidenberg et al. [28] recovered ~65% as phthalazine (Fig. 1) after 69 h in pH 12 phosphate buffer at room temperature.

Hydralazine solutions in 0.1 M HCl are stable for ~4 days even at 20°, and for several weeks at 4°. However, in spiked plasma samples hydralazine is very unstable, and even storage at -20° does not ensure stability, since breakdown starts during the thawing of the sample. Hence spiked plasma samples should be used immediately after preparation. The instability observed in spiked biological samples refers to free hydralazine only. Free hydralazine (0.22 μg/ml) incubated with fresh human plasma was completely converted to the pyruvic acid hydrazone in ~2 h [4].

In contrast, plasma samples from pharmacokinetic studies are stable for at least one month at -20°. The differences between immediate measurement and measurement after one month at -20° did not exceed the methodological variation when apparent hydralazine was measured by the procedure of Jack et al.[10] (Fig. 2). One sample collected from a volunteer was thawed and re-frozen 6 times (over 10 days), an aliquot being removed each time for analysis of apparent hydralazine. The % deviation from the first analysis never exceeded the methodological variation. The explanation for this seeming discrepancy between findings with spiked plasma samples and samples obtained from treated volunteers must lie in the fact that according to Zak et al. the hydralazine (measured by a GC technique [10]) is present as metabolites whose hydrolysis is facile [12].

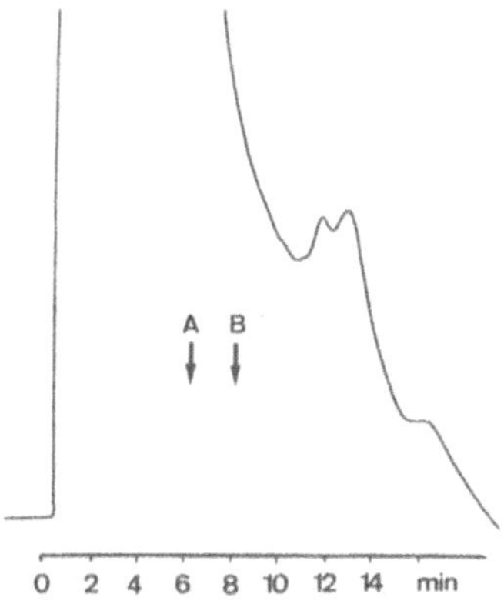

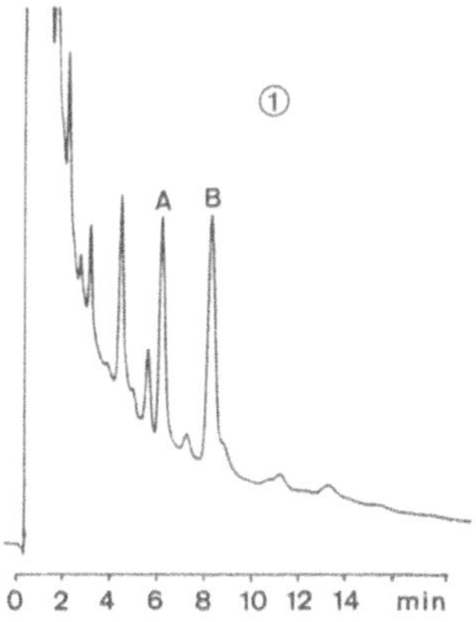

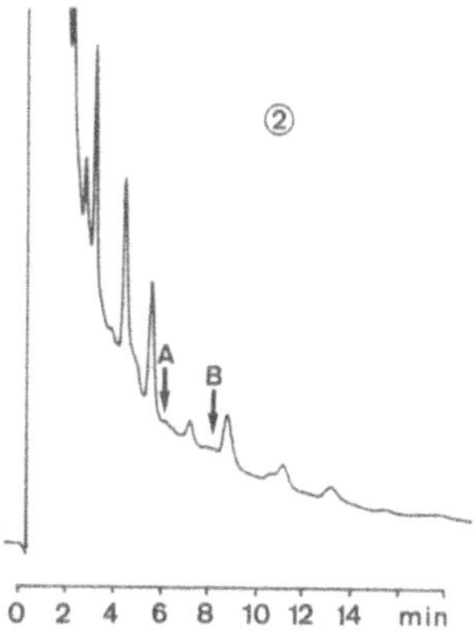

Figs. 3 *(above)* **& 4** *(right,* (1) & (2)*).* GC patterns for extracts of milk spiked with hydralazine (0.51 nmol; **A**) and internal standard (0.44 nmol; **B**) except for the blank, (2): only for Fig. 4 was a silica-gel purification step performed before GC.

METHODOLOGY

Amongst the different assay methods and their modifications (Table 1), the early spectrophotometric methods, original and improved, are non-specific and insufficiently sensitive for low doses of hydralazine. Higher sensitivity was obtained by Jack et al. [10] using the reaction of hydralazine with nitrous acid to form a tetrazolophthalazine (Fig. 1) with good GC-ECD properties and, with benefit to precision, using 4-methylhydralazine (Fig. 1) as an internal standard. This method proved reproducible in various laboratories and has been used extensively by Talseth [e.g. 30, 31] and Melander et al. [32] for bioavailability and pharmacokinetic studies in man. The method, being non-specific, records total circulating hydralazine: acid-labile hydrazones besides free drug. We have confirmed the notable efficiency of the method for bioavailability studies.

The determination of apparent hydralazine by the standard GC method as used for plasma or urine proved difficult for milk because of endogenous interferences (Fig. 3), obviated by an additional purification step. A toluene extract of the products from the reaction with nitrite was applied to a silica-gel column (silicagel 60, Merck; 5 cm × 5 mm i.d.), which was then washed with 4.5 ml of toluene/acetone (98:2 by vol.) and eluted with 20 ml ethanol. After drying down, the residue was re-dissolved in 0.5 ml toluene as the GC load sample. The resulting chromatograms (Fig. 4) are comparable with those obtained for plasma without the silica-gel column step.

Zak et al. [12] reported that part of the hydralazine measured in plasma by this GC technique consists of hydrolysis-prone hydralazine conjugates. Later, Reece et al. [14] demonstrated that almost all (~90%) the hydralazine detectable in plasma is present as pyruvic

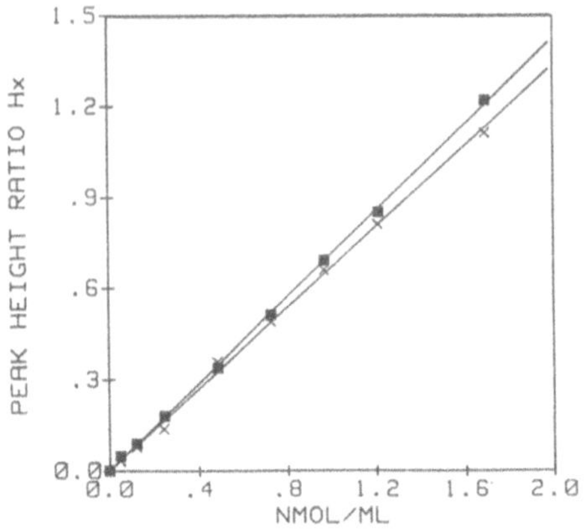

Fig. 5. Specificity test. Calibration curves without (×) and with (■) addition of pyruvic acid hydrazone.

acid hydrazone. This is due to the high normal concentration of pyruvic acid in plasma (~70 nmol/ml) which greatly exceeds that of hydralazine arising from therapeutic doses. Hence the drug determined via tetrazolophthalazine is referred to as 'apparent' hydralazine. Successive papers dealing with the selective determination of hydralazine then appeared [*inter alia:* 4, 5, 12, 13, 15, 19, 20, 22, 23, 26]. The most obvious solution to the problem was the derivatization of hydralazine at near-neutral or neutral pH. Only one derivatization procedure forms a stable derivative which also has excellent GC properties: it is based on the reaction of hydralazine with pentanedione to form a pyrazole (Fig. 1) [15]. The specificity was tested by adding pyruvate hydrazone to samples spiked with free hydralazine (Fig. 5).

After a therapeutic dose in man the levels of free hydralazine are very low, and the biological half-life is short [5]. Measurement of free hydralazine only is therefore inappropriate for routine determinations in bioavailability studies.

As most of the circulating hydralazine is in the form of pyruvic acid hydrazone, methods to selectively measure the hydrazone by HPLC have been developed [5, 14, 18, 22]. Other published methods are based on the formation of triazolophthalazine by reaction with formaldehyde (Fig. 1) [16, 17, 22]. But triazolophthalazine itself is a metabolite of hydralazine and should therefore not be chosen as a generated derivative. The most recent approach is based on the formation of a salicylaldehyde hydrazone at neutral pH (Fig. 1) followed by HPLC with electrochemical detection [26]. Thereby as little as 1 pmol/ml of free hydralazine is detectable in plasma.

CONCLUSIONS

The choice of method for the quantitative determination of hydralazine in plasma depends on the type of study to be performed. Specific methods for free hydralazine may be suitable for defining the pharmacokinetics of the unchanged drug but not for bioavailability studies. The determination of pyruvate hydrazone gives values which are very close to those described as 'apparent' hydralazine levels. However, hydrazones may be labile, and breakdown during an assay could give false results.

To evaluate the total amount of hydralazine in the circulation, formation of a stable derivative corresponding to the sum of free and 'bound' hydralazine serves to give representative and reproducible results.

References

1. Noda, H., Minemoto, M., Noda, A. & Ushio, T. (1986) *Chem. Pharm. Bull.* *34*, 3499-3501.
2. Schmid, K., Küng, W., Riess, W., Dollery, C.T. & Harland, S.J. (1987) *Arzneim.-Forsch.* *31*, 1143-1147.
3. Dubois, J.P., Schmid, K., Riess, W., Hanson, A., Henningsen, N.C.N. & Andersson, O.K. (1987) *Arzneim.-Forsch.* *37*, 189-193.
4. Reece, P.A., Cozamanis, I. & Zacest, R. (1980) *J. Chromatog.* *181*, 427-440.
5. Ludden, T.M., Ludden, L.K., McNay, J.L., Skrdlant, H.B., Swaggerty, P.J. & Shepherd, A.M.M. (1980) *Anal. Chim. Acta* *120*, 297-304.
6. Perry, H.M. (1953) *J. Lab. Clin. Med.* *41*, 566-573.
7. Schulert, A.R. (1961) *Arch. Int. Pharmacodyn.* *132*, 1-15.
8. Zacest, R. & Kock-Weser, J. (1972) *Clin. Pharm. Ther.* *13*, 420-425.
9. Zak, S.V., Bartlett, M.F., Wagner, W.E., Gilleran, T.G. & Lukas, G. (1974) *J. Pharm. Sci.* *63*, 225-229.
10. Jack, D.B., Brechbühler, S., Degen, P.H., Zbinden, P. & Riess, W. (1975) *J. Chromatog.* *115*, 87-92.
11. Haegele, K.D., Skrdlant, H.B., Robie, N.W., Lalka, D. & McNay, J.L. (1976) *J. Chromatog.* *126*, 517-534.
12. Zak, S.B., Lukas, G. & Gilleran, G.T. (1977) *Drug Metab. Dispos.* *5*, 116-121.
13. Smith, K.M., Johnson, R.N. & Kho, B.T. (1977) *J. Chromatog.* *137*, 431-437.
14. Reece, P.A., Stanley, P.E. & Zacest, R. (1978) *J. Pharm. Sci.* *67*, 1150-1153.
15. Degen, P.H. (1979) *J. Chromatog.* *176*, 375-380.
16. Proveaux, W.J., O'Donnell, J.P. & Ma, J.K.K. (1979) *J. Chromatog.* *176*, 480-484.
17. Angelo, H.R., Christensen, J.M., Kristensen, M. & McNair, A. (1980) *J. Chromatog.* *183*, 159-166.
18. Haegele, K.D., Skrdlant, H.B., Talseth, T., McNay, J.L., Shepherd, A.M.M. & Clementi, W.A. (1980) *J. Chromatog.* *187*, 171-179.

19. Facchini, V., Streeter, A.J. & Timbrell, J.A. (1980) *J. Chromatog. 187*, 218-223.
20. Ludden, T.M., Ludden, L.K., Wade, W.E. & Allerheiligen, S.R.B. (1983) *J. Pharm. Sci. 72*, 693-695.
21. Jendryczko, A., Drozdz, M. & Magner, K. (1984) *Rev. Roum. Biochim. 21*, 299-301.
22. Ravichandran, K. & Baldwin, R.P. (1985) *J. Chromatog. 343*, 99-108.
23. Rouan, M.C. & Campestrini, J. (1985) *J. Pharm. Sci. 74*, 1270-1273.
24. Wang, J., Tapia, T. & Bonakdar, M. (1986) *Analyst 111*, 1245-1248.
25. Lacagnin, L.B., Colby, H.D. & O'Donnell, J.P. (1986) *J. Chromatog. 377*, 319-327.
26. Wong, J.K., Joyce III, T.H. & Morrow, D.H. (1987) *J. Chromatog. 385*, 261-266.
27. McIsaac, W.M. & Kanda, M. (1964) *J. Pharmacol. Exp. Ther. 143*, 7-13.
28. Reidenberg, M.M., Drayer, D., DeMarco, A.L. & Bello, C.T. (1973) *Clin. Pharmacol. Ther. 14*, 970-977.
29. Lesser, J.M., Israili, Z.H., Davis, D.C. & Dayton, P.G. (1974) *Drug Metab. Dispos. 2*, 351-360.
30. Talseth, T. (1976) *Eur. J. Clin. Pharmacol. 10*, 183-187/311-317/395-401.
31. Talseth, T., Fauchald, P. & Pape, J.F. (1977) *Current Ther. Res. 21*, 157-168.
32. Melander, A., Danielson, K., Hanson, A., Rudell, B., Schersten, B., Thulin, T. & Wahlin, E. (1977) *Clin. Pharm. Ther. 21*, 104-107.

#D-5

APPROACHES TO THE ANALYSIS OF TWO CARDIOVASCULAR DRUGS IN PLASMA

R.D. McDowall®, J.C. Pearce, G.S. Murkitt, J.A. Jelly, W.J. Leavens, K.A. Fernandes and R.M. Lee

Department of Drug Analysis
Smith Kline and French Research Ltd.
The Frythe, Welwyn, Herts. AL6 9AR, U.K.

Methods for the analysis of two cardiovascular drugs in plasma involving the use of liquid-solid extraction and the AASP are described. In the first, AASP cassettes are eluted sequentially, with mobile phases of increasing acetonitrile content, to allow the quantification of the drug and its 4 metabolites. In the second a dedicated robotic autosampler (Gilson 222) is coupled with the AASP to produce a totally automated assay. All that is required is that centrifuged plasma samples be placed into the autosampler tray; sample preparation and HPLC analysis then follow without human intervention.*

At the 1985 Bioanalytical Forum, our group presented a review of liquid-solid sample preparation techniques with an application using the AASP for the analysis of a cardiovascular drug (SK&F 94120) [1]. Here we report the complete assay technique for SK&F 94120 and 4 metabolites, in a single plasma sample, using a novel approach of sequential elution from the solid phase. This is complemented by a method for fully automating the AASP which requires only that centrifuged plasma samples be presented to the instrument: all remaining sample preparation and the HPLC analysis and data collection are performed automatically.

ANALYSIS OF SK&F 94120 AND FOUR METABOLITES IN PLASMA

SK&F 94120, 5-(4-acetamidophenyl)pyrazin-2(1H)-one, when administered to animals is metabolized to a glucuronide and 3 other

* *Abbreviations.* - AASP, Advanced/Analytichem Automatic Sample Processor; t_r, retention time; i.s., internal standard. The prefix 'ortho' is implied in 'phosphoric acid'.

metabolites (P.M. Osborne et al., unpublished; and [1]). The assay reported previously was capable of analyzing the latter 3 metabolites but not the glucuronide which eluted immediately after the solvent front. In order to overcome this problem it was decided that sequential elution of the AASP cartridge with mobile phases of increasing strength was the best approach. The use of the AASP LC module's valve reset is an important factor for the success of this assay. During the first elution of the glucuronide, with a mobile phase containing 10% (v/v) acetonitrile, the glucuronide is eluted and measured by HPLC-UV. The cassette is then transferred to a second system whose mobile phase contains 20% acetonitrile, to quantify the parent drug and the 3 other metabolites.

Plasma samples for assay were first thawed at ambient temperature and then centrifuged at 2000 **g** for 10 min to remove fibrous material. An aliquot (100 µl) was transferred to a 1.5 ml polypropylene centrifuge tube. Water (100 µl) and i.s. solution (500 µl) were added to the sample and vortex-mixed. The i.s. (SK&F 94857, the propyl analogue of SK&F 94120) served to quantify the parent drug and SK&F 94120-MET II, III and IV. SK&F 94120 MET I (glucuronide) was measured using external standardization.

The AASP cassette was activated by passing 1 ml volumes of methanol and then water through each cartridge. Then 300 µl of diluted plasma was transferred to a newly prepared cartridge and air applied until the reservoir was empty. Water (1.5 ml) adjusted to pH 3 with phosphoric acid was then applied to wash off any highly polar material adsorbed to the sorbent. This was a modification of the existing method [1] for the parent drug, serving to prevent elution of the glucuronide from the extraction cartridge. The cassette was then removed from the manifold and transferred to the AASP HPLC systems for automated analysis.

The two systems were identical except for the mobile phases and the AASP instrument settings. The nature of the AASP assembly is shown in Fig. 1; details of the AASP can be found elsewhere [2, 3]. Each chromatographic assembly consisted of a Model 6000A (Waters Associates) pump, connected to an AASP LC module with the following settings:
- System 1: run time 6 min, cycle time 7 min, and valve-reset 0.3 min;
- System 2: run time 8 min, cycle time 7 min, and valve-reset 0.5 min.

Optimization of the valve-reset times was achieved by monitoring the recovery of each analyte from the cartridge following elution with varying volumes of the mobile phases.

The analytes were separated by a stainless steel column 300 mm × 3.9 mm i.d., packed with 10 µm C-18 µBondapak (Waters Assoc.) maintained at 35°. The column effluent was monitored by a Kratos 783 variable wavelength UV detector, set at 280 nm and 0.005 AUFSD. The detector signal was fed into an LDC Model 301 computing integrator.

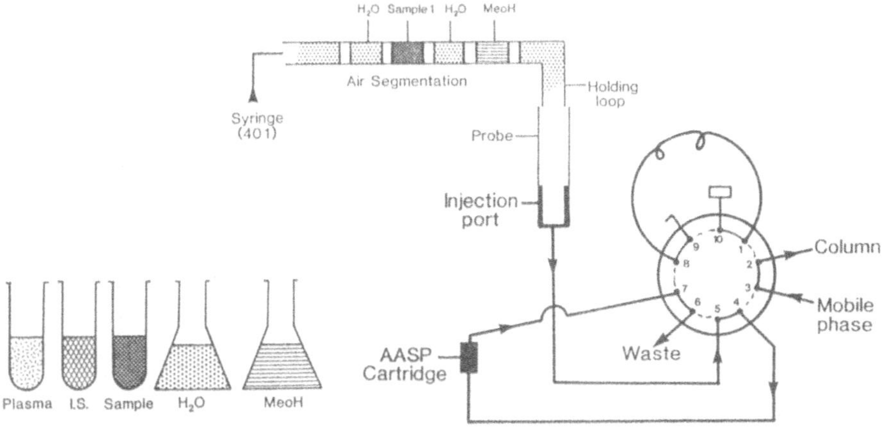

Fig. 1. Schematic diagram of the AASP-GILSON automated system (the Gilson module being relevant to SK&F 95654 assay; see text).

The mobile phase for the first chromatographic system (solvent system **1**) consisted of acetonitrile/10 mM ammonium acetate buffer, 1:9 by vol.; the pH of the mixture was adjusted to between 4.5 and 5.0 by adding phosphoric acid (300 µl/l). With 2.0 ml/min flow-rate, the t_r of SK&F 94120-MET I (glucuronide) was ~5 min. For solvent system **2** as used in the second assembly, the ratio was 2:8 rather than 1:9; at 2 ml/min the t_r's for SK&F 94120, i.s., MET IV and MET II were ~3, ~3.5, ~4.5 ~ and ~7 min respectively.

Results.- Fig. 2 (glucuronide) and Fig. 3 (the other analytes) show typical chromatograms, manifesting in each case good peak shape and adequacy of resolution for accurate quantification. Recoveries average 85% for parent drug and glucuronide and are quantitative for the other metabolites. The importance of the pH value in the initial water wash of the cartridge was clearly demonstrated: rapid loss of the glucuronide occurred if the pH were not reduced to 3. The precision and accuracy of the whole assay procedure were within 10% for all analytes over the concentration range 125-500 ng/ml.

FURTHER AUTOMATION OF THE AASP: ASSAY OF SK&F 95654 IN PLASMA

SK&F 95654, viz. (5-methyl-6[4-(4-oxo-1,4-dihydropyridin-1-yl)-phenyl]-4,5-dihydro-3(2H)-pyridazinone, exhibits potent positive inotropic and vasodilatory properties in animals, suggesting that it may be of value in the treatment of congestive heart failure in man. An assay to measure the drug in plasma was required; but a constraint in using the AASP is the off-line sample preparation stage.

Further automation of the AASP is achieved by connecting a Gilson 222 autosampler [4] to the AASP Valco 10-port valve via port 5 and

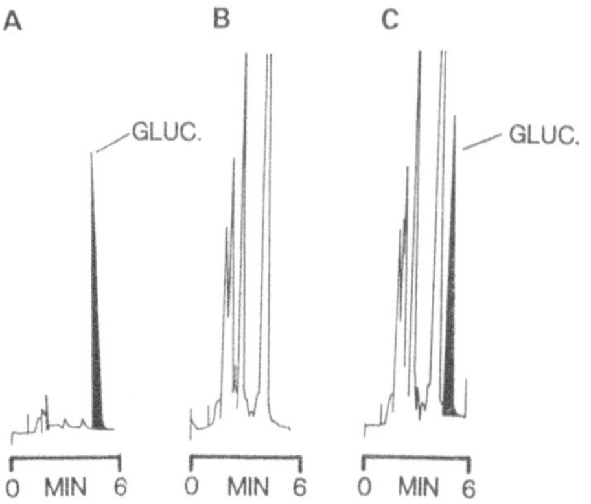

Fig. 2. Typical chromatograms (solvent system **1**) of (**A**) SK&F 94120-*O*-glucuronide, (**B**) extract of control dog plasma (= blank), and (**C**) as for (**B**) but spiked (0.25 ng/ml) with glucuronide.

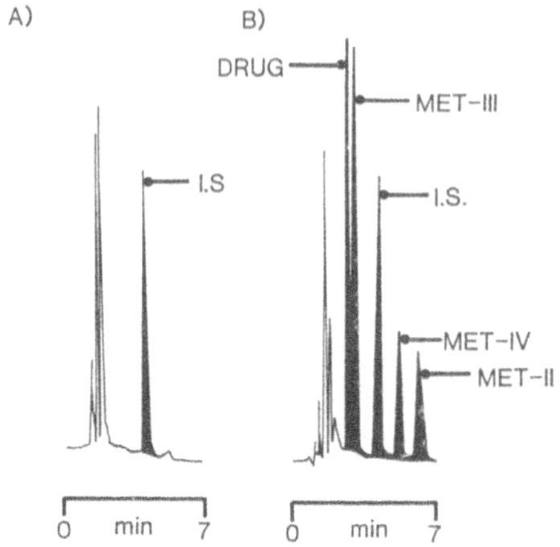

Fig. 3. Typical chromatograms (solvent system **2**) of control plasma spiked with (**A**) i.s., (**B**) i.s. + drug + 3 metabolites [see end of article].

disconnecting the purge pump [5]. This means that in the load position the AASP cartridge is on-line to the Gilson autosampler (Fig. 1) which picks up a train of solvents and samples, segmented by air, and pumps them through the cartridge. The order of solvents and samples is the reverse of that needed to activate the phase, apply the sample and wash the cartridge before elution. At the appropriate time, the HPLC mobile phase is switched through the cartridge, allowing elution of the compounds of interest directly onto the analytical column. The Gilson 222 assumes responsibility as the master controller of the system and communicates with the AASP via a 12 V interface relay box; thereby it can 'remote-start' the AASP, advance the cassette and switch the Valco valve at the pre-determined time.

In the only manual part of the procedure, plasma samples for the assay of SK&F 95654 were first thawed at room temperature and centrifuged (microfuge) at 10,000 **g** for 1 min to remove fibrous material that might otherwise block the extraction cartridges. Plasma (100 µl) was then transferred to a 1.5 ml polypropylene tube and placed in the Gilson 222 sample rack.

The chromatographic assembly consisted of a Model 114M pump (Beckman) set to deliver solvent at 1.5 ml/min, and an AASP LC module set up thus: mode, remote; first cartridge, 1; no. of samples, 99; run time, 0; cycle time, 0; valve-reset, 0. The analytes were separated on a stainless steel column, 300 × 3.9 mm i.d., packed with 10 µm µBondapak-phenyl (Waters Assoc.) and maintained at 35°. It was protected by a Rad-Pak CN (Waters) guard column. Detection was at 300 nm

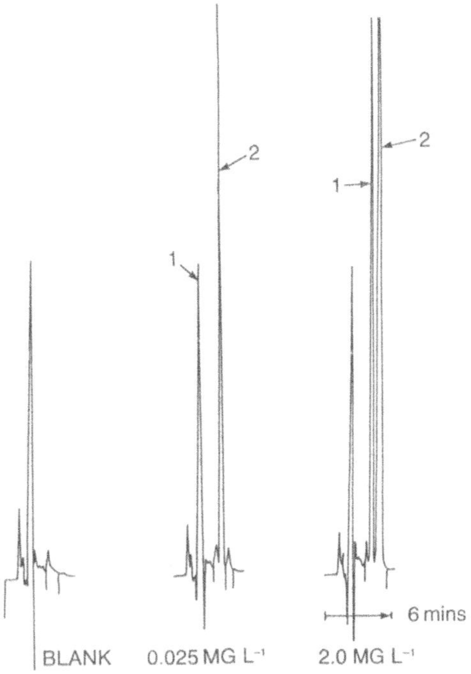

Fig. 4. Typical chromatograms of extracts of dog plasma spiked with SK&F 95654 (**2**) and i.s. (SK&F 94836; **1**).

with a Kratos 773 variable wavelength UV detector set at 0.001 AUFSD; the integrator was as for the SK&F 94120 assay.

The mobile phase was composed of acetonitrile and, pre-adjusted to pH 5.0 with phosphoric acid, 10 mM ammonium acetate; these were mixed (25:75) prior to use. A helium purge was maintained through the solvent reservoir throughout the chromatographic run.

Fig. 4 shows typical chromatograms, with t_r ~4.2 min for the i.s. (SK&F 94836) and ~4.9 min for SK&F 95654. No interfering endogenous compounds were observed in the extracts, and therefore the assay was deemed to be specific. The precision of the method was in the range 0.3-5.1%. Generally the accuracy was better than 2%, and in no case was there >5% inaccuracy.

CONCLUSIONS

The two assays presented here show the notable benefit from liquid-solid sample preparation and its capacity for automation when coupled with dedicated robotic units. The first method shows how, from a single extraction, a drug and 4 metabolites can be measured, even when one is a glucuronide. The second method was totally automated, easy to use, and involved minimal human input. It was robust, yet sensitive, accurate and precise, with a high throughput, making it ideal for a busy laboratory in a pharmaceutical company.

Furthermore, the strategy of linking liquid-solid extraction with HPLC has wide applicability, e.g. to on-line derivatization for which a reaction can be conducted either in the holding loop or on the solid phase. Time delays and washes could be introduced to allow such a reaction to proceed, and to remove any excess reagent prior to chromatographic analysis. Such approaches are under active consideration.

References

1.[⊗] McDowall, R.D., Pearce, J.C., Murkitt, G.S. & Lee, R.M. (1986) in *Bioactive Analytes, including CNS Drugs, Peptides and Enantiomers* (Reid, E., B. Scales & I.D. Wilson, eds.), Plenum, N. York, 235-242.
2. McDowall, R.D., Pearce, J.C. & Murkitt, G.S. (1986) *J. Pharm. Biomed. Anal. 4*, 3-21.
3. Pearce, J.C., Jelly, J.A., Fernandes, K.A., Leavens, W.J. & McDowall, R.D. (1986) *J. Chromatog. 353*, 371-378.
4. Lindsay, M. & Hawkes, D. (1985) *Lab. Pract.*, September, 45-51.
5. McDowall, R.D. & Pearce, J.C. (1988) *Analytical Developments in the Biomedical, Forensic and Pharmaceutical Sciences*, Plenum, New York, pp. 217-225.

[⊗] Fig. 3 in this ref. shows metabolic pathways, one giving I - the glucuronide, and a minor pathway giving, by gut microfloral action on the parent drug, metabolite II:- 2-(4-acetamidophenyl)pyrazine. The latter, after absorption, is further metabolized to its *N*-oxide (metabolite III) or is deacetylated to metabolite IV.

#NC(D)

NOTES and COMMENTS relating to VARIOUS CARDIOVASCULAR DRUGS

Comments relating to particular contributions:

#D-2 & -5, and #NC(D)-1 to -3: p. 217

Table with supplementary assay refs., for Sects. #B & #C also: p. 218

#NC(D)-1

A Note on

QUANTITATIVE ANALYSIS OF FLESINOXAN IN PLASMA AND URINE AT THE pg/ml LEVEL USING GC-ECD

M.P. van Berkel, H. de Bree and K. Sierat

Duphar Research Laboratories, P.O. Box 2
1380 AA Weesp, The Netherlands

Ceelan et al. [1] developed an HPLC method for the determination in plasma of flesinoxan (given as the hydrochloride; DU 23973), a relatively large and polar molecule of low volatility, shown on left in the diagram below. However, on the basis of the results of the first safety and tolerance study the dose was lowered to 0.25 mg/person, resulting in plasma levels in the pg/ml range. For the pharmacokinetic monitoring of the drug at this low dose, the HPLC method was unsuitable. A new analytical procedure had to be developed using a completely different strategy aiming at utmost sensitivity. GC-ECD[⊗] is a highly sensitive technique which could be suitable after hydrolysis of the drug to the FBz and then esterification with PFBB:

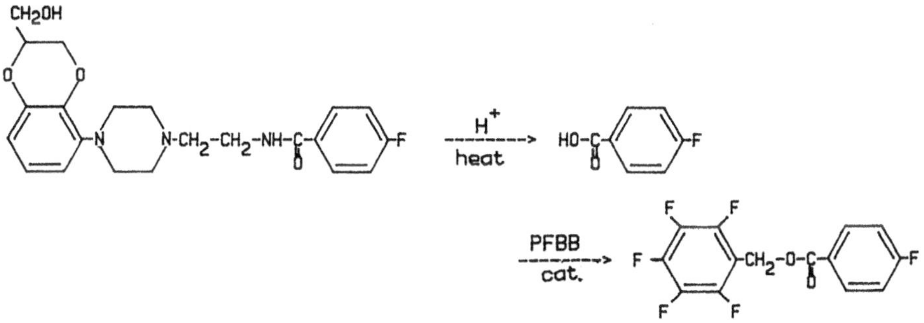

Materials.- Flesinoxan and the i.s. (Fig. 1) were obtained from our own laboratories; both were of >98% purity. From 3- and 4-FBz (Merck, FRG; 821795 and 818537) the PFB esters were prepared according

[⊗] ECD denotes electron-capture detector; FBz, fluorobenzoic acid or benzoate; MIBK, methyl isobutyl ketone; PFB(B), pentafluorobenzyl (bromide); d.f., degrees of freedom; i.s., internal standard. 'Ether' is diethyl ether. Drug and i.s. concentrations are for hydrochloride.

Fig. 1. The internal standard (i.s.), DU 122049.

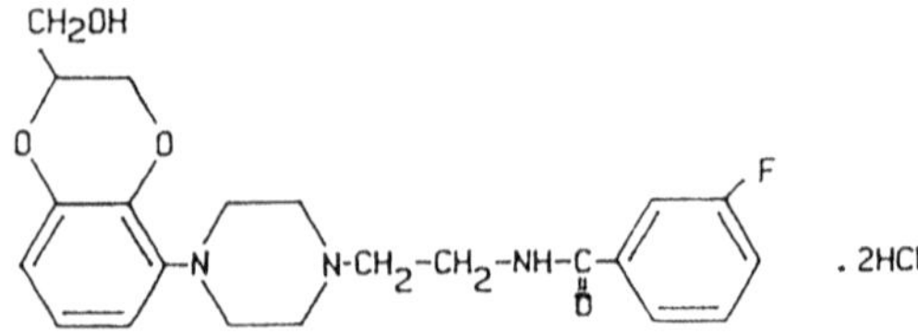

to Knapp [2]; after additional purification each was shown by GC-ECD to be >99% pure. All organic solvents were distilled twice using a 2-m vigreux in an all-glass apparatus. All chemicals were of analytical grade. Tripropylamine (Fulka, Switzerland; 93240) was purified by filtration through activated silica. Aqueous solutions were made in double-distilled water and washed 3 times with iso-octane.

Stock solutions containing ~10 mg flesinoxan or i.s. in 100 ml methanol were prepared every 3 months, and used as freshly prepared ×1000 dilutions in appropriate solvents. For 3-FBz and 4-FBz as used for optimization of some of the steps, a 10 mg/100 ml stock solution in diethyl ether was freshly diluted ×1000 with methanol prior to use. The esters as also used to check the derivatization, clean-up and GC procedure were dissolved (10 mg/ml) in and diluted ×100 with iso-octane. All stock solutions were stored at 4°.

Sample preparation.- To 5 ml plasma or urine, 10 ng of i.s. and 1 ml 25% (w/w) ammonia were added, and the mixture extracted with 5 ml MIBK. The organic phase was washed with 2.5 ml 1 M NaOH. If the expected concentration was <500 pg/ml, ~10 ml plasma or urine was taken, and 2 ml ammonia and 10 ml MIBK were used.

Hydrolysis and clean-up.- The MIBK phase was evaporated to dryness, 200 µl 15% (w/v) HCl was added, and the vials were closed tightly and kept at 110° for 16 h. The samples, after cooling and adding 300 µl 5 M NaOH, were washed with 2 ml ether/n-pentane (4:1 by vol.). The aqueous phase was acidified with 400 µl 15% HCl and extracted with 2 ml ether/n-pentane. The resulting organic phase was washed with 200 µl water.

Preparation of derivatives.- The organic phase was evaporated down to ~400 µl at room temperature with a gentle stream of N_2, transferred to a 500 µl conical vial, and taken gently to dryness with N_2. The residue was dissolved in 50 µl of a freshly prepared derivatization reagent: 400 µl tripropylamine, 200 µl PFBB (Pierce, U.S.A.; 58220) and 2 ml acetone. After 15 min (room temp.) 10 µl 100% ethylamine (Fluka; 02940) was added to react with the excess of PFBB (room temp., 60 min). The mixture was acidified with 50 µl 35% (w/v, aq.) perchloric acid and extracted with 50 µl iso-octane/ether.

HPLC clean-up.- The organic phase was injected completely, using a valve (Rheodyne, U.S.A.; 7125) with a 50 µl sample loop, onto a

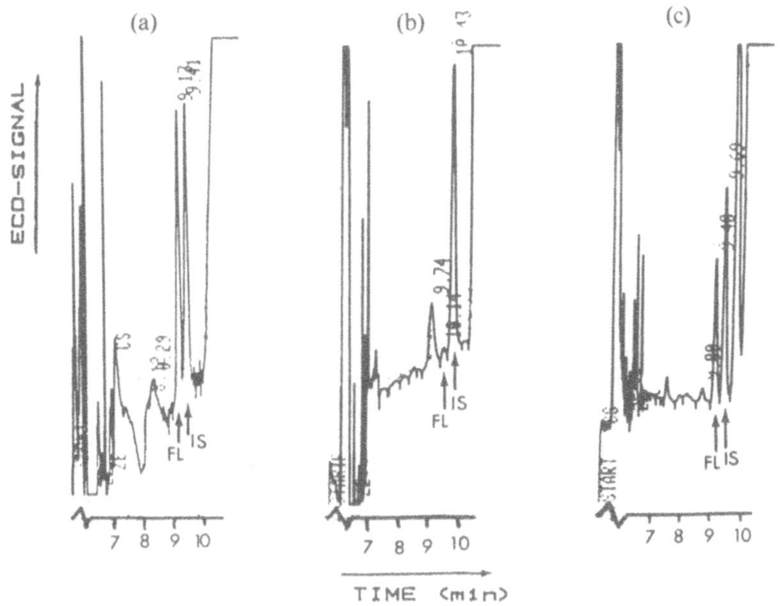

Fig. 2. GC-ECD of the PFB derivatives of 4-FBz (FL) and 3-FBz (IS), with 1 µl injections: a standard solution corresponding to 0.5 pg each of drug and i.s. (a), and plasma with 10 ng i.s. in the original 10 ml: (b) blank, (c) with 7 ng drug in the original 10 ml.

Zorbax 5 µm Silicagel column (200 × 4.6 mm). Iso-octane/ether/acetonitrile (940:50:10 by vol.) was passed through at 1.5 ml/min with a Waters M6000 solvent delivery system. The fraction (~600 µl) containing both 3- and 4-FBz was collected.

GC-ECD (Fig. 2).- The ECD in the Pye Unicam GVC instrument was of the ^{63}Ni type, run at 320°. The bonded-phase column, operated at 165°, was of fused silica, FSOT 007 CPS-2 (25 m × 0.23 mm; Quadrex Scientific, U.K.). The N_2 carrier gas flow-rate was 2 ml/min. An on-column injection system (Pye Unicam PU 4700 auto-injector) at 250° was used to inject 1 µl of the collected HPLC fraction. To avoid any interference from late-eluting peaks in the next run, the column was raised to 230° at 10°/min, and kept 4 min at 230°, after the derivates of flesinoxan and i.s. had been eluted. Their peak-height ratios were used for calibration.

ASSAY DEVELOPMENT AND PERFORMANCE

Choice of i.s.- From a number of isomers, homologues and analogues, the 3-fluoro isomer of flesinoxan was selected. No step in the procedure significantly affected the peak-height ratio of a 1:1 mixture of analyte and i.s. The PFB derivatives co-eluted in the HPLC clean-up step, but were finally separated by GC.

Table 1. Flesinoxan/i.s. peak height ratios for plasma samples spiked in the low detection range. The s values are S.D.'s.; t-values (Student; 6 d.f.) are calculated with respect to the blank.

pg/ml (& no. of determns.)	Mean ratio	s	t
0 (5)	0.039	0.028	-
53 (3)	0.053	0.018	0.75
105 (3)	0.102	0.005	3.68
211 (3)	0.191	0.019	8.10

Optimization of extraction etc.- Recoveries (checked with ^{14}C-flesinoxan and -metabolites) in the first step were <50% with solvents less polar than MIBK, and were ~90% only if ammonia were added. With the washing step to remove any acidic metabolites that could contribute to the total 4-FBz after hydrolysis (phenols, carboxylic and sulphonic acids), overall recoveries were ~80% for plasma and ~60% for urine. For hydrolysis to the FBz, trial of several mineral acids, temperatures and times led to the conditions chosen. The described 2-stage extraction before derivatization (~70% rec.) was necessary to remove hydrolyzed endogenous materials co-extracted from plasma and urine. Only acetone gave near-quantitative derivatization under mild conditions (15 min, room temp.). With tripropylamine present the yield was >95%, still ~90% overall after the perchloric acid extraction which readily removed both the product of reaction of excess PFBB with ethylamine (***N***-Et-PFbenzylamine) and the two amines.

Suitability of samples for GC.- To obviate GC interferences the clean-up on a silica column was vital; the derivatives were so non-polar that their capacity factor was small and precluded mere use of a silica cartridge. Their t_r's were checked in every fifth HPLC run by 265 nm detection, to ensure quantitative collection of the drug and i.s. derivatives. Neither showed any fall in concentration after 3 months' storage (10 ng/ml) in iso-octane at 4°.

Performance of the assay.- Good linearity was found with fully processed plasma samples, viz. blanks spiked with drug hydrochloride in the range (0.2-5 ng/ml) relevant to initial pharmacokinetic and clinical studies with 0.25 mg dosage. In this range the overall drug recovery through all assay steps is ~30% with plasma, or ~20% with urine for which linearity was likewise demonstrated. The **determination limit** was ~100 pg/ml of plasma (see 't' values in Table 1), but somewhat higher (~500 pg/ml) for urine because of the more variable sample composition. **C.V.** values were 9% within-day (13 d.f.) and 10% between-day (16 d.f.). Selectivity and sensitivity needs are met.

References

1. Ceelen, P.R.J., et al. (1986) in V. 16, this series (Reid, E., et al., eds.), 297.
2. Knapp, D.R. (1979) *Handbook of Analytical Derivatization Reactions*, Wiley-Interscience, New York, p. 47.

#NC(D)-2

*A Note** on

ASSAY OF NITROGLYCERIN AND METABOLITES BY CAPILLARY GC

Stephen H. Curry and Hae-Ryun Kwon

College of Pharmacy, University of Florida
J. Hillis Miller Health Center
Gainesville, FL 32610, U.S.A.

Satisfactory procedures are needed for the assay in plasma of nitroglycerin (glyceryl trinitrate, GTN) and denitrated metabolites, two isomeric dinitroglycerols (1,2- and 1,3-GDN) and mononitroglycerol (GMN; 2 isomers). Besides glycerol itself, metabolism also furnishes inorganic nitrite and, therefrom by reaction with GSH (forming GSSG), nitrate; the desirability of being able to measure these 4 analytes was pointed out at the 1985 Bioanalytical Forum ([1], wherein p. 111 shows metabolic pathways) but is not pursued here.

Early methods for GTN, now obsolete, entailed colorimetry through coupling reactions after breaking down the GTN to nitrate and nitrite, obviously precluding ester-type and even organic *vs.* inorganic distinctions. For GTN we described, with important sample-handling precautions, an effective HPLC (CN-silica) method with detection by the thermal energy analyzer (TEA) [2]. We have thereby been able to investigate the metabolic processes including that involving GSH. The method is highly specific and is free from problems inherent in GC (e.g. thermal decomposition of GTN), but has barely adequate sensitivity and involves extremely costly equipment. The latter holds also for reported GC-MS methods.

The prevalent approach is GC with electron capture detection (ECD). GC-ECD with packed columns enabled us to measure unmetabolized GTN; coupled with selective solvent extraction, it gave readable GC peaks amongst a forest of contaminating peaks. However, the GTN concentrations being extremely low, some analysts have obtained very poor chromatograms and, it seems, a high incidence of false positives and negatives. Thus, reports of the presence of unmetabolized GTN in plasma after oral dosing now appear to have been a false-positive

* *Editor's compilation from notes (lacking refs.) and a laboratory protocol furnished by S.H.C., who can answer queries on refs. etc.*

observation. The GC-ECD packed-column system has never been refined to include metabolites, in which interest is increasing.

GC-ECD methods for GTN with capillary columns have been described (e.g. by Z. Penton as cited by Ed. in [2], p. 373). The method of P.K. Noonan et al., as used to study oral bioavailability [3], is designed for both GTN and GDN's. We find it necessary to conduct two assays, one for GTN with pentane for extraction and DB-1 column assay, and the other involving hexane/ethyl acetate (1:1 by vol.) extraction for the two GDN's: no one column will separate the drug and metabolites adequately, and the concentration ranges for drug and metabolites are very different.

The DB-1 column and essentially the reported conditions [3] (but not using pentane/ether as extractant) serve us well (using 2 ml plasma) by virtue of the selective solvent extraction which removes the low-concentration parent drug (50 pg/ml is detectable) from the metabolites. Practical points, with either method, include pre-silanization of all glassware, and drying down of the solvent extract under N_2, then redissolving in ethyl acetate (50 µl; 0.5 µl taken for GC). For the metabolites (1 ml plasma; 0.2 ng/ml of 1,2- or 1,3-GDN detectable; t_r's 5.0 and 5.5 min), a GB-60 column (Analabs, CT; 25 m × 0.25 mm) is effective, following silylation of the GDN's (with BSTFA; Pierce Chemical Co.). The split injector and detector are set at 200°, and the column at 130° with He carrier and N_2 make-up gas (1.2 and 30 ml/min). The internal standard is 1-bromonaphthalene. The extraction solvent is comparable in polarity to pentane/ether and is less explosive.

Using the GC-ECD system with capillary columns, we showed [4] that, after swallowing GTN, plasma contains the GDN's but no GTN, i.e. the metabolites are crucial. Since the metabolic denitration is probably involved in the mechanism of action of GTN, we need to focus attention on conversion of GDN's to GMN's, which HPLC-TEA has shown to be present in plasma. At present the GC-ECD capillary column system is inapplicable to the GMN's.

References

1. Curry, S.H. (1986) in *Bioactive Analytes, including CNS Drugs, Peptides and Enantiomers* [Vol. 16, this series] (Reid, E., Scales, B. & Wilson, I.D., eds.), Plenum, New York, pp. 109-112.
2. Curry, S.H., Algozzini, G. & Yu, W. (1984) in *Drug Determination in Therapeutic and Forensic Contexts* [Vol. 14, this series] (Reid, E. & Wilson, I.D., eds.), Plenum, New York, pp. 367-368.
3. Noonan, P.K., Patrick, K. & Benet, L.Z. (1986) *J. Pharm. Sci. 75*, 241-243 [method: Noonan et al. (1984) *ibid. 73*, 923-927].
4. Curry, S.H. & Aburawi, S. (1985) *Biopharmaceutics & Drug Disp. 6.* 235-280.

#NC(D)-3

A Note on

HPLC-FLUORESCENCE METHOD FOR THE DETERMINATION OF THE NEW β_1-ADRENORECEPTOR BLOCKING AGENT NEBIVOLOL IN HUMAN PLASMA

Robert Woestenborghs, Luc Embrechts and Jos Heykants

Department of Drug Metabolism and Pharmacokinetics
Janssen Research Foundation
B-2340 Beerse, Belgium

Requirement	An assay method specific and sensitive enough to study the pharmacokinetics of nebivolol in man and experimental animals.
End-step	RP-HPLC with fluorescence detection (Fig. 1; ex: λ_{288}; em: λ_{310}).
Sample preparation	Alkalinized plasma extracted with heptane-isoamyl alcohol (95:5 by vol.), back-extracted with dil. sulphuric acid and re-extracted after alkalinization.
Comments	Care needed in respect of (1) the purity of the chromatographic solvents because of the rather critical detector setting, and (2) the dissolution of the extraction residues because of the poor solubility of the compound and the internal standard (i.s.). Sensitivity limit 0.1 ng/ml.

The drug analyte, α,α'-[iminobis(methylene)]bis[6-fluoro-3,4-dihydro-2*H*-1-benzopyran-2-methanol], is a new potent and selective β_1-adrenoreceptor blocking agent [1, 2].

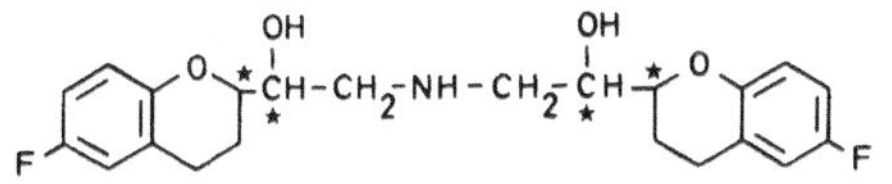

(mixture of two isomers, *SRRR* and *RSSS*; * denotes a chiral centre)

Sample preparation.- Plasma (2 ml) is spiked with 50 ng of i.s. (R 60750, the bis-7-methyl analogue of the drug), acidified with 1 M HCl (0.5 ml) and washed with 4 ml heptane. The plasma, after addition of 1 ml 1 M NaOH, is extracted twice with 3 ml 5% v/v isoamyl alcohol in heptane. Analytes are back-extracted with 3 ml 0.05 M H_2SO_4. The aqueous layer is again alkalinized (conc. ammonia, 150 µl) and re-extracted twice with 2.5 ml of the extraction solvent. The combined organic layers are dried down at 60° (N_2) and the residues reconstituted with 100 µl of the HPLC mobile phase.

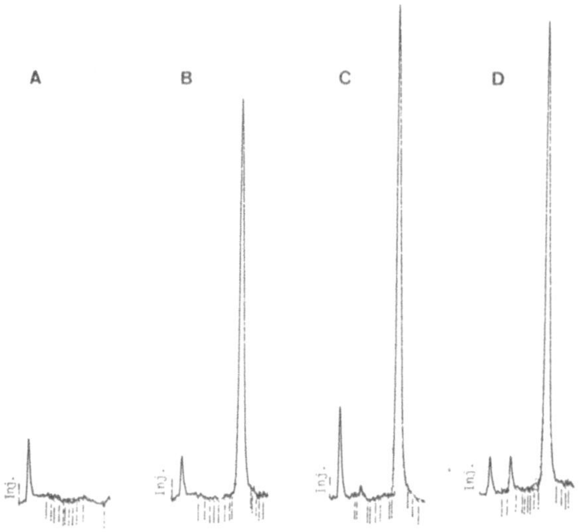

Fig. 1. HPLC (fluor.) of control plasma: blank (**A**), or spiked: 25 ng/ml of i.s. [t_r 254 sec] (**B**); 0.1 ng/ml of nebivolol [t_r 118 sec] & 25 ng/ml of i.s. [t_r 254 sec] (**C**); 0.5 ng/ml of nebivolol [t_r 117 sec] & 25 ng/ml of i.s. [t_r 250 sec] (**D**).

HPLC.- Samples (50 µl) are injected onto a 15 cm × 2.1 mm i.d. ODS-Hypersil (5 µm, Shandon) column, and eluted with 0.03% diethylamine in a 7:3 (by vol.) mixture of 0.01 M tetrabutylammonium hydrogen sulphate (TBAHS) and acetonitrile, with 0.9 ml/min flow rate. A Perkin-Elmer LS-4 fluorescence spectrophotometer was used for the monitoring.

COMMENTS

The method was found to perform satisfactorily (0.1 ng/ml measurable; accuracy and precision within 10%) with the following provisos.-

(1) The plasma should be pre-washed to remove the bulk of the fatty acids; this 'normalizes' different plasma types with respect to extraction recovery, especially for the i.s.

(2) Care has to be taken in re-dissolving the extraction residues because of the poor solubility of the analytes.

(3) The chromatographic solvents should be pure because the wavelength settings (not far apart) are critical for detection.

References

1. De Crée, J., Geukens, H., Leempoels, J. & Verhaegen, H. (1986) *Drug Dev. Res. 8*, 109-116.
2. De Crée, J., Geukens, H., Cobo, C. & Verhaegen, H. (1987) *Angiology 6*, 440-448.

Comments on material in #D

Comments on **#D-2**, B.E. Davies - CROMKALIN ENANTIOMERS
& **#NC(D)-3**, R. Woestenborghs - NEBIVOLOL ASSAY, *with debate on recoveries and internal standard types*

B.E. Davies, answering D. Dell: for Chirasil-Val-D capillary columns from Chrompak we found a variable and unpredictable lifespan, ranging from a few weeks to several months. **Answer to P.S.B. Minty:** it was because of the manufacturer's recommendation that we included NaCl - with good results - in the mobile phase for α_1-AGP columns.

Comment by R. Whelpton to R. Woestenborghs, who had encountered apparent recoveries of >100%: recoveries of ~100% may give false confidence and do need investigation! **P.H. Degen added** that even with extraction recoveries of only a few % a method can be valid, provided that an optimal i.s. is used (**S.H. Curry** agreed, assuming acceptable variance). **H. de Bree** remarked that a low overall recovery might be unacceptable to the FDA; he advocated 'extracted calibration graphs' that obviated concern about overall recovery and matrix effects - a policy which **Woestenborghs** felt should not be universally applied. **Curry** favoured extracted standards, as there seemed little merit in merely checking for signal quenching or possibly enhancement by the biological matrix by de Bree's proposed post-extraction calibration.

Comments on **#D-5**, P.H. Degen - HYDRALAZINE ASSAY
#NC(D)-1, H. de Bree - FLESINOXAN ASSAY
& **#NC(D)-2**, S.H. Curry - NITROGLYCERIN & METABOLITES ASSAY

P.H. Degen, answering P.S.B. Minty: hydralazine stabilization in the fresh blood sample is ensured by carrying out the derivatization reaction beside the patient! **A. Rakhit** wondered whether hydrazone is metabolized to hydralazine in the liver and is itself active; also, concerning flesinoxan (**de Bree**), whether any products of amide hydrolysis *in vivo* were separated from those obtained *in vitro*.

Remarks to S.H. Curry mainly concerned the HPLC-TEA method for GTN and its dinitrate metabolites - which was unacceptable to the FDA (reason not known). **A.J. Woodward** (of Simbec Research) commented on its routine use, approved by the FDA for pharmacokinetic studies: 50 or for the GDN's, 100 pg/ml is measurable, and all compounds can be assayed in one run - not the case with capillary GC-ECD, which does however give better sensitivity for TNG as needed in some transdermal patch studies (for which **Curry** favoured GC-MS; 50 pg/ml detectable). **Woodward** emphasized the need to prevent losses initially: the blood sample, collected into heparin + $AgNO_3$, is put on crushed ice and centrifuged within 2 min, and the plasma is snap-frozen with dry ice/methanol. *(See p. 222 for two recent reports.- Ed.)*

Table 1. Assay literature (cf. **#B, #C** & **#D**). ANTI-INFLAMMATORY DRUGS. *Non-obvious abbreviations*.- **en** if enantiomers distinguished (→ **dia**, diastereoisomer formation). Metabolites: des if *N*-desalkyl or -desacyl, otherwise met. (or, if 'active', **met.**). **Pl** = plasma (or serum), **Bl** = blood, **U** = urine (hyd. = conjugates hydrolyzed). WORK-UP: extraction: LL, liquid-liquid, LS, liquid/solid (cartridge; type indicated if not C-18); deprot. = deproteinized, deriv = derivatization (if not **dia**). ESTIMATION: cpGC = capillary GC, ECD = electron-capture detector, NPD = N/P detector; C-18 = HPLC with ODS-bonded phase, RP = other reversed-phase, NP = normal-phase (silica), grad. = gradient; detection mode stated (flu = fluorescence). Detection limit (Det.) is wt./ml of starting sample, where μ = μg, n = ng, p = pg.

Analyte(s) & sample type	Processing & measurement	Det.	Ref.
Arylpropionic acids (**P** *denotes* profen)			
Benoxa**P**/Flunoxa**P**, **Pl** or hyd.**U**:	LL; C-18, UV	?0.2 μ	3
en: Etodolic acid, **Pl** & hyd. **U**:	LL; →**dia**; cpGC-NPD	?	4
Flunoxa**P**, **Pl** & **U**:	LL; C-18, UV	50 n	5
Flunoxa**P**, **Pl** & **U**:	deprot., LL; C-18, UV	100 n	6
Ibu**P**, **Pl**:	deprot. (CH_3CN); C-18, UV	1 μ	7
Ibu**P** & mets., **Pl** & **U**:	deprot. (CH_3CN); C-18, UV	**Pl** 0.2/**U** 1 n	8
en: Ibu**P**, adipose tiss.:	LL &c, →**dia**; GC-MS(CI)	⊗	9
en: Ibu**P** & mets., **Pl** & **U**:	deprot. (CH_3CN); →**dia**; cpGC-MS	10 n	10
Keto**P**, **Pl** & hyd. **U**:	deprot., LL; C-18, UV	25 n	11
Keto**P** (topical), **Pl** & synov. fluid:	LL; C-18, UV(array)	~50 n	12
Procetofenic acid (Procetofen **met**), **Pl** & **U**:	LS; C-8, UV	0.5 μ	13
en: Tiaprofenic acid: **Pl, U, synov. fluid**:	LL; →**dia**; cpGC-NPD		14
Various anti-inflammatories			
Acemetacin, **Pl**:	LL; C-18, UV	10 n	15
Apyramide & Indomethacin (**met.**), **Pl**:	deprot./salt; C-18, UV	30 n	16
Chloroquine & des, **Pl** & **U**:	LL; C-18, UV (or flu)	3 (0.2) n	17
Diflusinal, **Pl** & **U**:	deprot.(CH_3CN); C-18, flu	2 μ	18
Fluocortin butyl (unstable!), **Pl**:	LL (fast!); C-18, peroxyoxal. lumin.	100 p	19
Indomethacin, **Pl** & **U**:	deprot.; C-18; +NaOH →flu	**Pl** 10 n	20*
Proquazone & met., **Pl** & **U**:	deprot., LL; C-18, UV	~1 n	21
Tenoxicam, **Pl**:	LL; C-18, UV	20 n	22
Tenoxicam & met., **U**:	LS; C-18, UV	50 n	23
Zidometacin, **Pl** & **U**:	LL; C-18, UV	250 n	24

⊗ A study of stereoselective uptake. *Also assay of sulindac, **Pl** & **U**

ANALYTICAL APPROACHES IN RELATION TO THERAPEUTIC CLASS: some refs. - *Senior Editor's compilation, especially relevant to pharmacokinetics*

Anti-inflammatories: Table 1 (& Sect. **B**). **'CV' drugs**: Table 1 opposite (& **C, D**); see a 1985 survey [1] of approaches and pitfalls. **Various**: see a 1980 survey of HPLC methods [2]. In this Series (list facing title p.) see V. 14 for anti-cancer drugs and V. 16 for CNS-acting drugs. EFFECTIVE REINFORCEMENT: *Citation Index* follow-up of refs.

Table 1, continued: CARDIOVASCULAR DRUGS. *References overleaf.*

Analyte(s) & sample type	Processing & measurement	Det.	Ref.
ACE inhibitors and calcium antagonists (& see p. 166)			
Diltiazem & des-*O*Ac, **Pl**:	LL; deriv, cpGC-ECD	3 (des 1) n	25
Diltiazem & des, **Pl**:	LL; C-18 +acid, UV	2.5 n	26
Nilvadipine, **Pl**:	LL: cpGC-MS(NCI)	10 p	27
Nimodipine, **Pl**:	LL; cpGC-ECD	2.5 n	28
Verapamil, **Pl**:	LL (simple; cf. lit.); cpGC-NPD	50 n	29
Verapamil, **Pl**:	LS; cpGC-NPD	1 n	30
β-blockers			
en: Pindolol etc., **Pl** & **U**:	LL; →**dia**; C-18, flu	15 n (**U**: 0.2 μ)	31
en: Betaxolol, **Pl**:	LL; →**dia**; C-18 +amine, flu	0.5 n	32
Atenolol or Metoprolol, **Pl**:	RP(CN), flu with no emit filter	10 n	33
Betaxolol, **Bl** & **U**:	LL; RP(CN), flu	10 n	34
Betaxolol, **Pl**:	LL; C-18, flu	4 n	35
Carvedilol & des, **Pl** & **U**:	LL (2-step): C-18, flu	0.4n (**U**: 0.8 n)	36
Levobunolol & **met.**(dihydro), **Bl** or **Pl**:	Process sample fast (enzymic reduction) deprot. & LL; C-18, flu	1 n	37
Nadolol, **Pl**:	Deprot.; C-18, flu	<50 n	38
Oxprenolol, **Pl**:	LL; C-8 +acid, UV	20 n	39
Pindolol, **Pl** & **U**:	LL; RP(CN), UV (non-specific)	3 n	40
Timolol, **Pl** & **U**:	LS; C-18 +amine; UV	2 μ	41
Antiarrhythmics			
Amiodarone & des, **Pl**:	LS(CN); C-8 +amine, UV	~1 n	42
Amiodarone & des, **Pl** & **tissues**:	LL; NP +acid, UV	25 n	43
en: Mixiletine, **Pl**:	LL; chiral LC (Pirkle 1A), flu	5 μ	44
Penticainide & des, **Pl** & **U**:	LL; NP +acid, UV	<1 μ	45
Propisomide, **Pl** & **U**:	LL (complex); pcGC-NPD	0.2 (**U**: 5) μ	46
Antihypertensives and vasodilators			
Cilostazol, **Pl**:	LS; C-18, UV	25 n	47
Doxazosin, **Pl**:	LL; C-18 +acid, flu	0.5 n	48
Fenoldopam + mets., **Pl** & hyd. **U**:	LL; C-18 +acid, electrochem.	0.05 (**U**: 5) n	49
Indoramin, **Pl**:	LL; RP(CN), electrochem.		50
Isosorbide-5-nitrate, **Pl**:	LL & deriv; cpGC-ECD	5 n	51
Naftidrofuryl, **Pl**:	LL; C-18, flu	20 n	52
Pinacidil & met., **Pl**:	LL; C-8, UV	5 n	53
Prazosin, **Pl**:	LL (selective); RP(phenyl), flu	15 p	54
'UK 33274' (Pfizer; similar to prazosin), **Bl** or **Pl**:	LL (complex); C-18, flu	1 n	55
Terazosin, **Pl** & **U**:	LL; C-18, flu	50 n	56
Tiodazosin, **Pl**:	LL; C-18, flu	1 n	57
Vincamine, **Pl**:	LL (not simple); cpGC-NPD	3 n	58
Various			
Digoxin + mets., **Pl**:	LS; C-8, RIA on fractions	15 p	59
Digoxin, **Pl**:	LL (complex); C-18, deriv & flu	0.5 n	60
Milrinone, **Pl**:	LS; C-18, UV	5 n	61
16-Acetylgitoxin (**met.** of pengitoxin), **Pl**:	RIA (pengitoxin cross-reacts 38%)	4 n	62
Sulmazole & mets., **Pl** (rat):	Deprot. & LS; C-18, grad., flu	100 n	63

References *for preceding two pages*

1. Ahnoff, M., Ervik, M., Lagerström, P-E., Persson, B-A. & Vessman, J. (1985) *J. Chromatog. 340*, 73-138.
2. Meffin, P.J. & Miners, J.O. (1980) in *Progr. Drug Metab. 4* [Bridges, J.W. & Chasseaud, L.F., eds.; Wiley], Chap. 5 (pp. 261-307).
3. Furlanut, M., Montanari, G., Colonna, V. & Perosa, A. (1983) *Int. J. Clin. Pharmacol. Res. 3*, 163-166.
4. Coutts, R.T., Pasutto, F.M., Jamali, F. & Singh, N. (1986) in *Development of Drugs and Modern Medicines* (Gorrod, J.W. et al., eds.), Horwood/VCH, 232.
5. Pedrazzini, S., Zanoboni-Mucciaccia, W. & Forgione, A. (1987) *J. Chromatog. 413*, 338-341.
6. Segre, G., Bianchi, E., Mascaretti, L., Quaglia, G. & Forgione, A. (1987) *Int. J. Clin. Pharmacol. Res. 7*, 243-250.
7. Lalande, M., Wilson, D.L. & McGilveray, I.J. (1986) *J. Chromatog. 377*, 410-414.
8. Shah, A. & Jung, D. (1986) *J. Chromatog. 378*, 232-236.
9. Williams, K., Day, R., Knihinicki, R. & Duffield, A. (1986) *Biochem. Pharmacol. 35*, 3403-3405.
10. Young, M.A., Aarons, L., Davidson, E.M. & Toon, S. (1986) *J. Pharm. Pharmacol. 38*, 60P.
11. Lempiäinen, M. & Mäkelä, A-L. (1985) *Int. J. Clin. Pharmacol. Res. 7*, 265-271.
12. Ballerini, R., Casini, A., Chinol, M., Mannucci, C., Giaccai, L. & Salvi, M. (1986) *Int. J. Clin. Pharmacol. Res. 6*, 69-72.
13. Ramusino, A.C. & Carozzi, A. (1986) *J. Chromatog. 383*, 419-424.
14. Singh, N.N., Jamali, F., Pasutto, F., Russell, A.S., Coutts, R.T. & Drader, K.S. (1986) *J. Pharm. Sci. 75*, 439-442.
15. Notarianni, L.J. & Collins, A.J. (1987) *J. Chromatog. 413*, 305-308.
16. Sauvaire, D., Cociglio, M. & Alric, R. (1986) *J. Chromatog. 375*, 101-110.
17. Bergqvist, Y. (1980) *J. Chromatog. 221*, 119-127.
18. Schwartz, M., Chiou, R., Stubbs, R.J. & Bayne, W.F. (1986) *J. Chromatog. 380*, 420-424.
19. Koziol, T., Grayeski, M.L. & Weiberger, R. (1984) *J. Chromatog. 317*, 355-366.
20. Stubbs, R.J., Schwartz, M.S., Chiou, R., Entwhistle, L.A. & Bayne, W.F. (1986) *J. Chromatog. 383*, 432-437; also (1987) *413*, 171-180⊗.
21. Lempiäinen, M. & Mäkelä, A-L. (1985) *J. Chromatog. 341*, 105-113.
22. Hermann, P., Körner, J. & Zinapold, K. (1986) *J. Chromatog. 374*, 95-102.
23. Dell, D., Joly, R., Meister, W., Arnold, W., Partos, C. & Guldimann, B. (1984) *J. Chromatog. 317*, 483-492.
24. Bonardi, G., Colombo, F., Guenzati, G., Lepore, A.M. & Sega, R. (1984) *Int. J. Clin. Pharmacol. Res. 4*, 419-423.
25. Grech-Belanger, B., Leboeuf, E & Langlois, S. (1987) *J. Chromatog. 417*, 89-98.
26. Abernethy, D.R., Schwartz, J.B. & Todd, E.L. (1985) *J. Chromatog. 342*, 216-220. *[Cf.* (1986) *382*, 377-381.*]*

⊗assay of sulindac & mets.

27. Tokuma, Y., Fujiwara, T. & Noguchi, H. (1985) *J. Chromatog. 345*, 51-58.
28. Jakobsen, P., Mikkelsen, E.D., Laursen, J. & Jensen, F. (1986) *J. Chromatog. 374*, 383-387.
29. Shukla, U.A., Stetson, P.L. & Ensminger, W.D. (1985) *J. Chromatog. 342*, 406-410.
30. Hoffman, D.J. & Higgins, J. (1986) *J. Chromatog. 374*, 170-176.
31. Hsyu, P-H. & Giacomini, K.M. (1986) *J. Pharm. Sci. 75*, 601-605.
32. Darmon, A. & Thenot, J.P. (1986) *J. Chromatog. 374*, 321-328.
33. Harrison, P.M., Tonkin, A.M. & McClean, A.J. (1985) *J. Chromatog. 339*, 429-433. [*Also* (1986) *381*, 168-174 and *382*, 215-224.]
34. Caqueret, H. & Bianchetti, G. (1984) *J. Chromatog. 311*, 199-205.
35. Canal, M. & Flouvat, B.J. (1985) *J. Chromatog. 342*, 212-215.
36. Reiff, K. (1987) *J. Chromatog. 413*, 355-362.
37. Hengy, H. & Kolle, E-U. (1985) *J. Chromatog. 338*, 444-449.
38. Moncrieff, J. (1985) *J. Chromatog. 342*, 206-211.
39. Godbillon, J., Duval, M. & Gosset, G. (1985) *J. Chromatog. 345*, 365-371.
40. Shields, B.J., Lima, J.J., Binkley, P.F., Leier, C.V. & MacKichan, J.J. (1986) *J. Chromatog. 378*, 163-167.
41. Lennard, M.S. & Parkin, S. (1985) *J. Chromatog. 338*, 249-252.
42. Pollak, P.T., Carruthers, S.G. & Freedman, D.J. (1986) *Clin. Chem. 32*, 890-893.
43. Kannan, R., Miller, S., Perez, V. & Singh, B.N. (1987) *J. Chromatog. 385*, 225-232.
44. McErlane, K.M., Igwemezie, L. & Kerr, C.R. (1987) *J. Chromatog. 415*, 335-346.
45. Walker, S., Flanagan, R.J. & Holt, D.W. (1987) *Biomed. Chromatog. 2*, 35-38.
46. Necciari, J., Mery, D., Sales, Y., Berthet, D. & Cautreels, W. (1985) *J. Chromatog. 341*, 202-207.
47. Akiyama, H., Kudo, S., Odomi, M. & Shimizu, T. (1985) *J. Chromatog. 338*, 456-459.
48. Cowlishaw, M.G. & Sharman, J.R. (1985) *J. Chromatog. 344*,403-407.
49. Boppana, V.K., Heineman, F.C., Lynn, R.K., Randolph, W.C. & Ziemniak, J.A. (1984) *J. Chromatog. 317*, 163-174.
50. Leelavathi, D.E., Seffer, E.F., Dressler, D.E. & Knowles, J. (1986) *J. Pharm. Sci. 75*, 421-423.
51. Marzo, A. & Treffner, E. (1985) *J. Chromatog. 345*, 90-95.
52. Walmsley, L.M., Wilkinson, P.A., Brodie, R.R. & Chasseaud, L.F. (1985) *J. Chromatog. 338*, 433-437.
53. Hamilton, M. Farid, K.Z. & Henry, D.P. (1986) *J. Chromatog. 375*, 359-367.
54. Reece, P.A. (1980) *J. Chromatog. 221*, 188-192.
55. Rubin, P.C., Brunton, J. & Meredith, P. (1980) *J. Chromatog. 221*, 193-195.
56. Patterson, S.E. (1984) *J. Chromatog. 311*, 206-212.
57. Forgue, S.T., van Harken, D.R. & Smyth, R.D. (1985) *J. Chromatog. 342*, 221-227.
58. Michotte, Y. & Massart, D.L. (1985) *J. Chromatog. 344*, 367-371.

59. Gault, M.H., Longerich, L., Dawe, M. & Vasdev, S. (1985) *Clin. Chem. 31*, 906-907.
60. Kwong, E. & McErlane, K.M. (1986) *J. Chromatog. 381*, 357-363.
61. Oddie, C.J., Jackman, G.P. & Bobik, A. (1986) *J. Chromatog. 374*, 209-214.
62. Weiler, E.W. & Lach, H-J. (1980) *Clin. Chim. Acta 104*, 337-343.
63. Bernstein, J.R. & Franklin, R.B. (1985) *J. Chromatog. 342*, 228-233.

NITROGLYCERIN METABOLITES [cf. #NC(D)-2 by S.H. Curry]
- *citations by Senior Editor*

Two recent reports concern assay of plasma for the metabolites. GC-ECD with a packed column of SP-2401 was used for the two mononitrates after diethyl ether extraction and redissolution in tetrahydrofuran containing phenylboronic acid; the detection limit was 10 ng/ml plasma [64]. Capillary GC-ECD was used to assay the two dinitrates (on OV-1, with continuous air-cooling of the on-column injector), after solid-phase extraction [65]. Each was detectable down to 250 pg/ml.

64. Scharpf, F., Yeates, R.A., Laufen, H. & Eibel, G. (1987) *J. Chromatog. 413*, 91-99.
65. Sioufi, A., Pommier, F. & Dubois, J.P. (1987) *J. Chromatog. 413*, 101-108.

Enoxime (and its sulphoxide metabolite) - an inotropic drug - is mentioned at the end of the 'NC' pages that follow Sect. **E:** its assay was by 'automated sequential trace enrichment' with initial dialysis.

Section #E

VARIOUS DRUGS AND SEPARATORY APPROACHES

#E-1

THE ROUTINE USE, AND COLUMN-STABILITY IMPLICATIONS, OF SEVERAL COLUMN-SWITCHING HPLC METHODS FOR DETERMINING DRUGS IN PLASMA AND URINE

J.B. Lecaillon®, C. Souppart, J.P. Dubois and *A. Delacroix

Laboratoires CIBA-GEIGY
Biopharmaceutical Research
Center, B.P. 308
92506 Reuil-Malmaison, France

*Laboratoire de Chimie
Industrielle, Conservatoire
National des Arts et Métiers
75003 Paris, France

Several possible configurations can be used for column-switching systems:
- trace enrichment on a pre-column and automatic back-flush or forward-flush injection of the purified sample on an analytical column;
- injection of the sample onto one column and automatic selection of a portion of the eluate (heart-cutting) for chromatography on a second column;
- combination of these two systems: trace enrichment and heart-cutting. These approaches are briefly described for metoprolol, CGP 6140 and oxiracetam.

The above methods for column-switching have been used for the determination of several thousands of biological samples, and gave satisfactory results for accuracy and precision: for spiked samples, the mean recoveries ranged from 97% to 107%, and the C.V.'s ranged from 5% to 8%.

For these methods and their clinical application, estimates of technician work-time for one determination ranged from 16 to 44 min; for a given method the longer work-times were mostly attributable to defective day-to-day conditions of the different elements involved in the system, including the columns. The use of automatic devices with cartridges for the sample preparation seemed a worth-while alternative for drug determination.

Several column-switching methods have been described in recent years [1-5]. This article aims to give some information on the routine use of these methods in our laboratory for quantitatively determining

® addressee for any correspondence

drugs in biological fluids, viz. metoprolol (β-blocking agent), CGP 6140 (anti-parasitic compound) and oxiracetam (nootropic compound).

Fig. 1 shows several possible systems for column-switching.
- Trace enrichment on a pre-column (or a cartridge) and automatic *back-flush* injection of the purified sample onto an analytical HPLC column. This is the system most frequently described.
- Trace enrichment on a pre-column (or a cartridge) and automatic *forward-flush* injection of the purified sample onto an analyticaL HPLC column. Compared with the back-flush approach the advantage is better protection of the analytical column against small particles contained in the sample; the drawbacks are the need to dilute the solute before injection into the analytical column and possible memory effects due to strongly retained compounds. With short pre-columns filled with large-diameter particles, these drawbacks can be minimized.
- Injection of the sample onto one column and automatic selection of a portion of the eluate (heart-cutting) for chromatography on a second column. In this system the retention times of the components on the first column should be stable. Memory effects, due to strongly retained compounds, can be eliminated by rinsing the first column with a strongly eluting solvent.
- Combination of the latter two systems: trace-enrichment + heart-cutting. This system is not frequently described, probably because three columns are needed; but it could give a high specificity for the determination of compounds of similar polarity.

As previously shown [6], the choice of the columns in the system should be guided by the polarity of the compounds to be determined. Reversed-phase (RP) chromatography is best adapted to the injection of biological aqueous samples. Nevertheless, for drugs of high polarity normal-phase (NP) chromatography can also be used.

The initial column should contain a packing material of high loading capacity, totally porous and of medium particle diameter (10-40 μm). The second should be more retentive, to reconcentrate the selected portion of the eluate from the first column, and should have good chromatographic efficiency (5 μm diam. particles). For bonded-phase columns the approximate order in respect of polarity is: diamino > amino > cyano > dimethylamino > nitro > diol > C_2 > C_8 > C_{18}. This gives an indication of the order in which the columns are to be used.

These considerations have been put into practice for the establishment of the column-switching HPLC methods used in this work.

EXPERIMENTAL

Equipment.- The automatic sample injector that enabled as much as several ml to be injected was Model 231 from Gilson (Villiers-le-Bel,

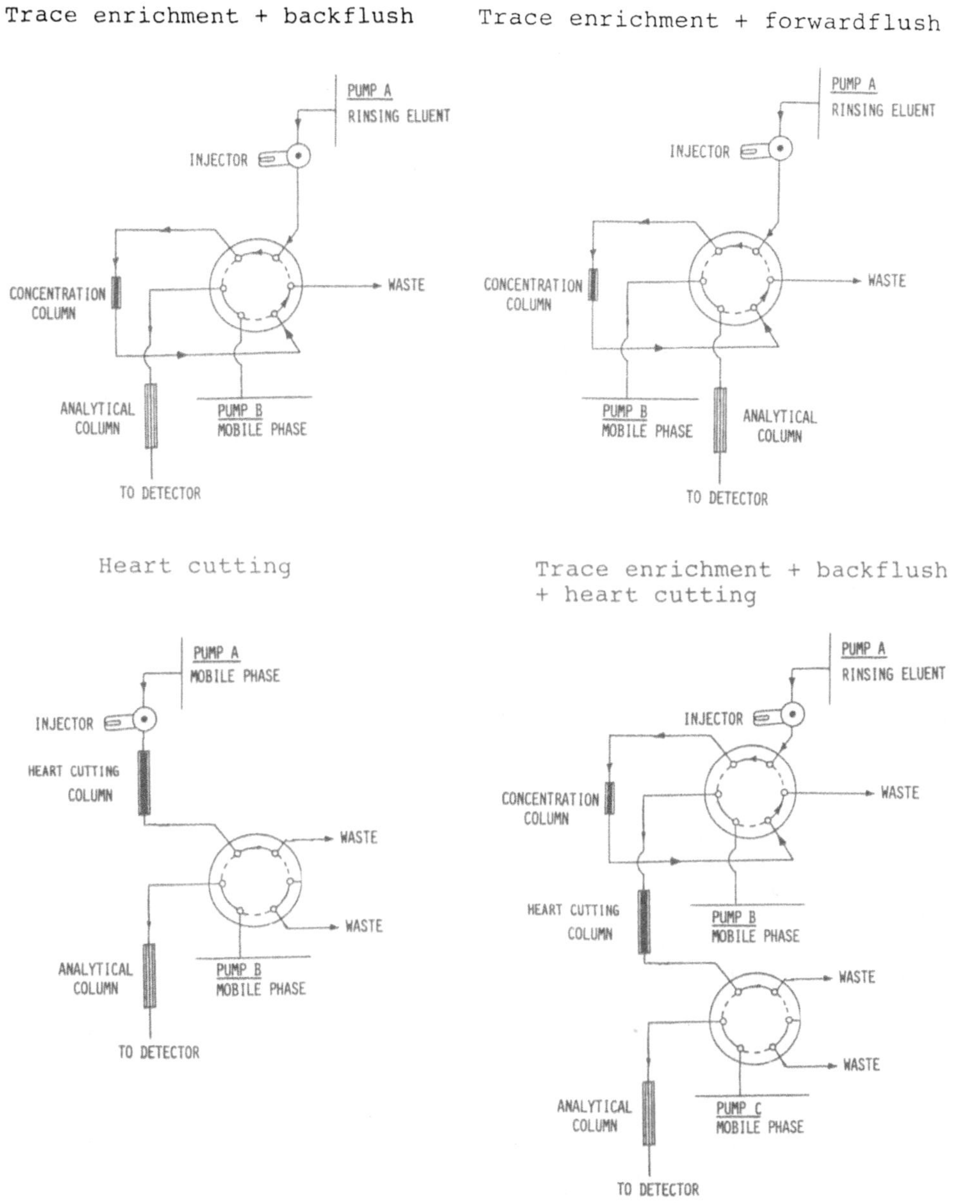

Fig. 1. Several possible column-switching arrangements.

France). Automatic injection of components loaded on cartridges was performed with an Advanced Automatic Sample Processor (AASP; Varian). The following HPLC pumps were used: Gilson Models 302 and 303, Constametric Model III (LDC) and, when an elution gradient was needed, Spectra-Physics Model SP 8700.

The air-actuated valves were Valco Model AH-CV-UHPa-N60 and Rheodyne Model 7000. For actuating the valves and sampler and controlling the SP 8700 pump a Spectra-Physics Model 4270 Computing Integrator was used. Monitoring was done with a Kontron Model SFM 23/B fluorescence detector or a Kratos Model Spectroflow 773 UV detector.

Solvents and reagents.- The synthetic reference compounds, provided by Ciba-Geigy Ltd. (Basle), were as follows: CGP 6140 - 4-nitro-4'-(*N*-methyl-piperazinyl-thiocarbonylamido)-diphenylamine; CGP 13,231 (*N*-oxide metabolite of CGP 6140); CGP 10,631, as internal standard (i.s.) for CGP 6140; oxiracetam - 4-hydroxy-2-oxo-1-pyrrolidineacetamide; CGP 14,998, as i.s. for oxiracetam; metoprolol tartrate - di-{ (±)-1-(isopropylamino)-3-[*p*-(2-methoxyethyl)-phenoxy]-2-propanol} L(+)-tartrate; and alprenolol, as i.s. for metoprolol. All solvents and reagents were of analytical grade (Merck, Darmstadt), acetonitrile was of HPLC grade (#412412; Carlo Erba, Milan), and methanol was of RPE-ACS quality (#414816; Carlo Erba). The sources of the column packings were Merck (LiChrosorb and LiChroprep), Macherey-Nagel (Nucleosil) and Phase Separations (Queensferry, U.K.; Spherisorb).

METHODS

In using the methods described below, the calibration curves were performed every 2 or 3 weeks, and spiked samples with known concentrations of compounds for validation of the methods were determined daily.

CGP 6140 and its *N*-oxide metabolite.- Two methods were used [7], based on (1) trace-enrichment and *back-flush*, and (2) trace-enrichment onto a cartridge and subsequent *forward-flush* (Fig. 1). For (1), plasma (450 µl) is added to the i.s. solution (45 µl) and to water/methanol (70:30 by vol.; 1.8 ml); or urine (120 µl) is added to the i.s. solution (60 µl) and to the water/methanol (1.1 ml). From the diluted sample, 2 ml (plasma) or 1 ml (urine) is injected onto a 25-40 µm LiChroprep RP-2 column (250 × 4.7 mm i.d.) and rinsed with water/methanol (90:10) at 5 ml/min for 5 min. The sample is then back-flushed and eluted onto the analytical column with the mobile phase.

For (2), using the AASP, each cartridge is conditioned with methanol and water before adding the diluted plasma or urine sample. After rinsing with 10 mM K_2HPO_4 (1.5 ml) to eliminate polar components, the lipophilic compounds are forward-flushed from the cartridge with the mobile phase and loaded onto the HPLC column.

For both methods, the HPLC column is 5 µm Spherisorb ODS-1 (250 × 4.6 mm i.d.), with acetonitrile/methanol/4 mM ammonia solution (54.5: 5 : 40.5 by vol.) as mobile phase at 1 ml/min. CGP 6140 and its *N*-oxide were detected at 405 nm. The quantitation limits were 20 ng/ml (= 50 nmol/l) in plasma and 100 ng/ml in urine.

Oxiracetam.- The method was based on heart-cutting and injection of the selected eluate onto the analytical column (Fig. 1). The term 'mixture' (by vol.) below signifies water/acetonitrile. To the solution of i.s. (CGP 14,998) are added plasma (250 µl), 4:1000 mixture (4.2 ml) and dichloromethane (0.8 ml). The diluted sample is centrifuged, and 1 ml is injected onto a 10 µm LiChrosorb-NH_2 column (250 × 4.7 mm i.d.) and eluted with 5:95 mixture at 2 ml/min. The selected heart-cut is reconcentrated on the analytical column, 5 µm Nucleosil -NH_2 (300 × 4.7 mm), and eluted with 10:90 mixture at 1 ml/min. During the analytical step the former column is flushed with a stongly eluting solvent, 50:50 mixture. This eliminates the strongly retained plasma components which, in running the next sample, may interfere with oxiracetam due to memory effects. Drug detection is at 200 nm. The limit of quantitation was ~150 ng/ml (~1 µmol/l) of plasma.

Metoprolol.- The method was based on trace-enrichment + back-flush + heart-cutting [8] (Fig. 1). To the solution of i.s. (alprenolol) and water (1.2 ml), plasma (1 ml) is added. The diluted sample is centrifuged, injected onto a 25-40 µm LiChroprep column (250 × 4.7 mm i.d.) and rinsed with water at 6 ml/min for 5 min. The sample is then back-flushed and eluted on the heart-cutting column with the mobile phase, acetonitrile/2 mM Na acetate in 20 mM acetic acid (75:25), at 1 ml/min. The selected heart-cut is reconcentrated on the analytical column, 5 µm LiChrosorb RP-8 (250 × 4.7 mm), and eluted with the mobile phase, acetonitrile/8 mM Na acetate in 80 mM acetic acid (75:25), at 1 ml/min. The metoprolol was determined by fluorescence detection with 280 nm excitation and 320 nm emission. The limit of quantitation was ~10 ng/ml (~30 nmol/l) of plasma.

RESULTS

CGP 6140 and its *N*-oxide metabolite

Method (1), trace-enrichment and back-flush (column-switching), was used for 640 samples from clinical studies, and method (2) using cartridges and the AASP (cartridge-switching) was applied to 500 samples. Spiked samples, comprising about half the total, were used to validate the methods. The calibration curves (with 4-6 concentration values) gave a mean correlation coefficient (**r** ±S.D.) of 0.999 ±0.001 (n = 22) with method (1) and 0.999 ±0.001 (n = 6) with method (2).

Column life.- The RP-2 column was replaced every 40 injections to ensure protection from endogenous plasma and urine components. The analytical column could be used for the determination of 150-200 samples but gave peak shoulders after such a series of determinations. In method (2) the analytical column appeared to survive longer than with (1), and >200 sample determinations could be performed; also, the solvent consumption was lower.

Table 1. Reproducibility and accuracy of the methods as found with spiked samples: mean recoveries [observed concn. as % of spiked-in concn.], ±S.D. (& no. of observations).

Drug	At the quantitation limit, %	General mean %
CGP 6140	98.8 ±9.9 (n = 23)	101.8 ±8.2 (n = 98)
CGP 6140 with AASP	97.2 ±8.0 (n = 5)	97.0 ±8.2 (n = 17)
Oxiracetam	102.0 ±8.0 (n = 35)	101.0 ±6.0 (n = 201)
Metoprolol	107.0 ±5.0 (n = 8)	103.4 ±5.5 (n = 25)

The cost of the casettes with the AASP system has to be considered with method (2); but it represents a minor element in the total cost of one sample analysis.

Oxiracetam

Of the total number of samples - 2200 in all, from clinical studies - about half were spiked samples used for method validation. The calibration curves gave a mean **r** ±S.D. of 0.999 ±0.001 (n = 43).

Column life.- The first column (LiChrosorb-NH_2) was replaced every 200 injections to ensure a good separation, on the analytical column, of the drug, i.s. and endogenous plasma and urine components. The analytical column could be used to determine ~400 samples, but thereafter gave poor separation of the drug and a plasma component.

Metoprolol

Of the 170 samples that were determined, from one clinical study, about one-tenth were spiked samples used to validate the method.

Column life.- The first (LiChroprep RP-2) column was replaced every 50 injections to ensure good protection of the heart-cutting column. Each column could be used for the determination of >200 samples.

DISCUSSION

The switching methods described here gave satisfactory results for precision and accuracy, as shown by the recoveries and the C.V.'s for the spiked samples (Table 1). For CGP 6140 and oxiracetam, which appear difficult to extract from plasma, these methods seemed well adapted. For metoprolol there is the option of determining the drug with a conventional extraction procedure prior to HPLC [9]. This alternative was preferred in most cases because the switching method needed 3 pumps, valves and columns and was found to be a burden when the sample number in the clinical study was relatively small.

Table 2. Minimum and maximum mean work-time (min) per determination for a technician, throughout the clinical studies: ratio of total work-time for a study to no. of determinations in the study (n).

Drug	Minimum time (& n)	Maximum time (& n)
CGP 6140	32 (522)	44 (975)
CGP 6140 with AASP	22 (228)	30 (348)
Oxiracetam	16 (648)	32 (120)
Metoprolol	23 (504)	-

Work-time per determination

Mean work-times were calculated as in the heading to Table 2. The total work-time included the time taken for the analytical work, the time to replace or repair any malfunctioning equipment, and the time for inspecting the chromatograms, for calculating concentrations and for transcribing the results into tabular form. The minimim and maximum work-time values for one determimination ranged from 16 to 44 min as shown in Table 2, which gives a good indication of how long one determination took in our laboratory with these systems.

Surprisingly, the oxiracetam method showed the lowest work-time despite its complexity. For CGP 6140, the work-time with the switching method was slightly longer than with the other method, due to the limited stability of the drug on the sample tray (~10 h only). The use of the AASP resulted in a decrease in work-time when compared to the switching method. Manufacturers may be offering some interesting variants of the cartridge approach.

The differences between minimum and maximum work-times within a method could mostly be attributed to defective column or equipment operation, and show the importance of the reliability of all elements of the systems.

Operational points

Some of the problems met during the establishment or use of the methods, and improvements found to be achievable, may be summarized as follows. For CGP 6140, the main problem-area was adsorption of the drug and its *N*-oxide onto glassware and tubing. This was overcome by including a small proportion of methanol in the rinsing solvent in the switching method. As is evident from the S.D.'s in Table 1, the reproducibility was poorer than for the other two drugs, and may reflect these adsorption problems.

Due to the need for high sensitivity in determining CGP 6140, a large plasma volume (~400 μl) was injected onto the column, the

Table 3. Influence of sample pre-treatment on column life in the determination of CGP 6140 in plasma (for which the amount loaded is tabulated). Following injection (injn.) onto the trace-enrichment column, the analyte was back-flushed onto the analytical column, to which the stability remarks refer. TCA = aq. trichloroacetic acid, 10% w/v; '10:90' denotes 10 parts (by vol.) added to 90 parts plasma.

Plasma	Pre-treatment	Observations & remarks
200 µl	None; direct injn.	Bad peak shape after 20 injns.
180 µl	TCA (10:90) → plasma	Bad peak shape after 20 injns.
130 µl	TCA (30:70) → plasma	Cpds. partly retained in the protein ppt.
290 µl	Water (70:30) → plasma	Bad peak shape after 90 injns.
390 µl	Water (80:20) → plasma	Change col. after 200 or (pre-col.) 50 injns.

stability of which in relation to sample pre-treatment is shown in Table 3. The method with the AASP was slightly easier to use routinely, mainly because the analytical column life appeared better than with the column-switching method.

For oxiracetam, a large interaction was needed between the computing integrator and the pumping system since during each determination an elution gradient was set up and the flow-rate was modified. Nevertheless, this system appeared to work with good reproducibility.

Due to the low limit of quantitation needed for determining metoprolol in clinical samples, large plasma volumes (~1 ml) were injected. The increase in pressure in the first two columns resulting from the injection of plasma could ba limiting factor. On the first column, using stainless steel frits of pore diameter 2 µm on both sides of the column, the pressure ranged from 20 to 100 bar after 100 sample injections (whereas it was <10 bar initially). On the second column, with stainless steel frits of pore diameter 2 µm on both sides of the column, the pressure increased to ~100 bar after injection of 50-100 samples. With teflon frits of pore diameter 15 µm at the top of the column, the pressure rose only to ~40 bar.

CONCLUSIONS

The switching HPLC systems have been routinely used for determing drugs in biological fluids, and have given satisfactory precision and accuracy. The following appeared to be of benefit in the routine use of such methods:
- dilution and centrifugation of the plasma;
- injection of the smallest volume of plasma required for the sensitivity needed;

- use of the largest possible diameters for particles of filling material in the columns;
- use of frits with pores of as large diameter as possible.

The column life and the reliability in the day-to-day operation of all devices comprised in the switching systems are important factors in increasing the number of determinations attainable per working day.

References

1. Apffel, J.A., Alfredson, T.V. & Majors, R.E. (1981) *J. Chromatog. 206*, 43-57.
2. Erni, F., Keller, H.P., Morin, C. & Schmitt, M. (1981) *J. Chromatog. 204*, 65-76.
3. Roth, W., Beschke, K., Jauch, R., Zimmer, R. & Koss, F.W. (1981) *J. Chromatog. 222*, 13-22.
4. Roth, W. (1983) *J. Chromatog. 278*, 347-357.
5. Nazareth, A., Jaramillo, L., Karger, B.L., Giese, R.W. & Snyder, L.R. (1984) *J. Chromatog. 309*, 357-368.
6. Lecaillon, J.B., Febvre, N. & Souppart, C. (1984) *J. Chromatog. 317*, 493-506.
7. Lecaillon, J.B., Sioufi, A., Souppart, C. & Dubois, J.P. (1987) *Chromatographia 24*, 876-880.
8. Lecaillon, J.B., Souppart, C. & Abadie, F. (1982) *Chromatographia 16*, 158-161.
9. Lecaillon, J.B., Godbillon, J., Abadie, F. & Gosset, G. (1984) *J. Chromatog. 305*, 411-417.

#E-2

MICELLAR LIQUID CHROMATOGRAPHY

John G. Dorsey

Department of Chemistry, University of Florida
Gainesville, FL 32611, U.S.A.

Micelles are notable for their ability to solubilize hydrophobic compounds in aqueous solution. Micellar solutions can help in improving analytical methodologies and in developing entirely new concepts in analytical chemistry; they have been studied especially as HPLC mobile phases. The micelles provide a hydrophobic site for interaction with solutes in the mobile phase and can replace traditional modifiers such as methanol or acetonitrile in RP-HPLC. We have studied the chromatographic capabilities of these unique mobile phases - capabilities that are unrealizable with traditional mobile phases. One such advantage is in gradient elution, never popular in repetitive analyses by RP-HPLC because of the slow re-equilibration of the stationary phase after a change in mobile phase composition. We have shown that micelle concentration gradients speed the elution of strongly retained compounds without altering stationary phase ocmposition. This allows a step-gradient back to initial conditions and obviates column re-equilibration. Furthermore, if the conductivity of the two solutions of different micelle concentrations is balanced, micellar gradients are compatible with EC detection - again in contrast with hydro-organic mobile phases. Serum or urine can be directly injected onto a RP column for therapeutic drug monitoring. Proteins in the sample are solubilized by the micelles and elute with the void volume, while drugs and metabolites are released and partition into the stationary phase. This eliminates the tedious clean-up steps traditionally necessary. Micellar mobile phases will never replace traditional hydro-organic mobile phases. Their chromatographic capabilities, however, point to a role in the laboratory of every chromatographer.*

The feasibility of replacing organic modifiers in the RP-HPLC mobile phase by an aqueous micelle solution was first demonstrated in 1980 by Armstrong & Henry [1]. Micellar chromatography offers

**Abbreviations:* RP, reversed phase; EC, electrochemical; CMC, critical micelle concentration; QSAR, quantitative structure-activity relationship; EMIT, enzyme-multiplied immunoassay.

analytical capabilities that are unavailable with traditional mobile phases. In fact, since that first publication, an entire field of separations using 'organized' media has emerged, as featured in a recent volume [2].

Micellar chromatography should not be considered a 'new' form of chromatography but rather a variant of traditional RP-HPLC. RP-HPLC is most rigorously described as a separation in which retention is decreased with a decrease in solvent polarity - which affects *all* solutes; it is here that micellar differs from traditional hydro-organic RP-HPLC. Only those solutes that interact with the micelle, either through electrostatic interactions or hydrophobic effects, are affected by a change in micelle concentration of the mobile phase. Micellar chromatography is, then, truly a secondary equilibrium method. The primary equilibrium, which affects all solutes, is the partitioning of the solute between the (bulk) mobile phase and the stationary phase, and the *secondary* equilibrium is the solute-micelle partitioning, which is now the controlling parameter for separation. Theoretical aspects of secondary equilibria, especially optimization, have been described recently by Foley & May [3, 4] and, as we are investigating, should prove useful for rapid optimization of separations with micellar mobile phases.

Like other secondary equilibria such as acid-base equilibria and ion-pairing methods, micellar mobile phases can be used to provide unique chromatographic selectivities for 'difficult' separations. However, unlike other secondary equilibria mthods that are applicable only to narrow ranges of compounds, micellar chromatography applies to a very wide range of compounds, the only requirement being that the solute partition to the micelle. This means that *all* hydrophobic compounds, and many hydrophilic compounds which are electrostatically attracted to the micelle structure, are candidates for separation by micellar chromatography. As RP-HPLC is generally the method of choice for hydrophobic compounds, the great majority of RP separations may be amenable to micellar chromatography.

In many forms of secondary equilibria separations, the concentration of the equilibrant, or the mobile phase component which participates in the secondary equilibrium, controls at least partially the strength and selectivity of the mobile phase. In micellar chromatography the concentration of micelles plays this role, which means that for all separations carried out with micellar mobile phases the *strength of the mobile phase can be changed while maintaining an unchanged bulk solvent composition.* This key feature of micellar mobile phases can overcome inherent problems and limitations of hydro-organic mobile phases, and expands capabilities in separation, gradient elution techniques, detection and sample separation. This is a prime reason, more compelling than low cost, low toxicity and chromatographic selectivity, for adoption of a technique that requires a

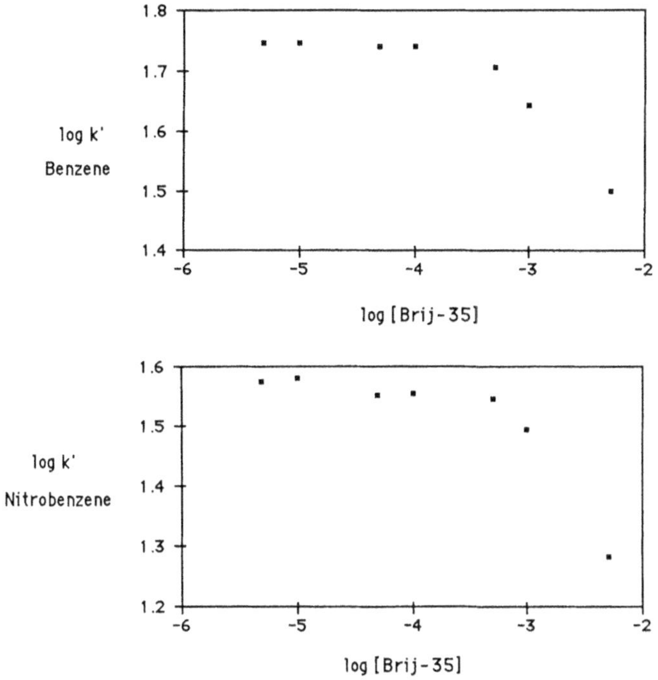

Fig. 1. Retentions as a function of Brij-35 concentration. *Upper graph:* benzene; *lower graph:* nitrobenzene.

new learning curve.- Routine use of micellar mobile phases will occur because of capabilities that hydro-organic mobile phases lack. This article is an overview (for comprehensive reviews see [5, 6]) that emphasizes comparisons between the two phase categories, and areas of interest to bioanalysts.

MICELLIZATION AND SOLUTE RETENTION

The context of Fig. 1 is that above a certain critical concentration (CMC), which is characteristic for each surfactant and set of experimental conditions, surfactant molecules self-aggregate into roughly spherical structures termed micelles. These aggregates are thermodynamically quite stable with the hydrocarbon chains oriented towards the interior and the hydrophilic heads forming the surface of the structure. The number of molecules making up a micelle ('aggregation number') is typically in the range 50-100. Their size is generally within the diameter range 3-6 nm, which enables their macroscopic solution properties to approximate those of a truly homogeneous solution. They cannot be filtered by conventional methods, and they do not cause measurable light-scattering error in UV/visible spectroscopy. The driving force for micellization is the hydrophobic effect, and electrostatic repulsion of the head groups is the counterforce limiting the size of the aggregate. While micelles have been studied for many years by biochemists and physical chemists, analytical chemists have only recently become interested in the effects of micelles on solutes. Guidance has been given [7] on the utility of micelle solutions in chemical analysis.

As surfactant is added to an aqueous mobile phase, a dramatic change in retention of most solutes occurs at a surfactant concentration that corresponds to the CMC. This is shown in Fig. 1, where the retentions of benzene and nitrobenzene are plotted as a function of concentration of Brij-35 (polyoxyethylene lauryl ether), a non-ionic surfactant. Two linear components of these plots are seen, one above and one below the CMC. Extrapolation of the two linear segments should show an intersection at approximately the CMC of the particular system under investigation, and for this plot the value is calculated to be 3.4×10^{-4} M. This value in 3:97 (by vol.) propanol/water at 40° tallies well with a literature value of 1×10^{-4} in pure aqueous solution [7]. The linear portion of the plot for concentrations above the CMC is evidence that the system is behaving in a well-defined RP manner and that the micelle concentration controls retention similarly to organic modifiers. This type of behaviour has been found to hold for cationic, anionic and neutral surfactants, and there is excellent agreement with published CMC values [8, 9]. Micellar mobile phases, with their demonstrable eluting power, are now considered in respect of their chromatographic capabilities.

DIRECT PLASMA/SERUM INJECTION

From a bioanalytical viewpoint, perhaps the most useful application of micellar chromatography that has been shown is the ability to inject serum or urine directly onto a RP column with no protein precipitation or pressure build-up problems. Cline Love and coworkers [10-12] have shown this dramatic benefit in therapeutic drug monitoring - which has become increasingly important and relies on chromatographic methods for the separation step requisite for complex biological matrices. Often a pre-chromatographic deproteinization step is necessary to prevent clogging of the column; besides increasing the time of analysis, this may also reduce accuracy and precision, as the analyte may partially co-precipitate with the protein.

Micellar mobile phases overcome these problems. Aggregates of either SDS or Brij-35 solubilize serum proteins so that they elute with the void volume. Furthermore, the surfactant monomers compete with the analyte for protein binding sites and thereby release it for complete quantitation. Fig. 2 shows urine chromatograms, unspiked and spiked with propanolol and quinidine [11]. DeLuccia et al. have further found excellent agreement between micellar chromatography, with direct serum injection, and EMIT quantitation; they also reported that injection of >250 samples resulted in no increase in system pressure and no noticeable clogging of the injection port [10].

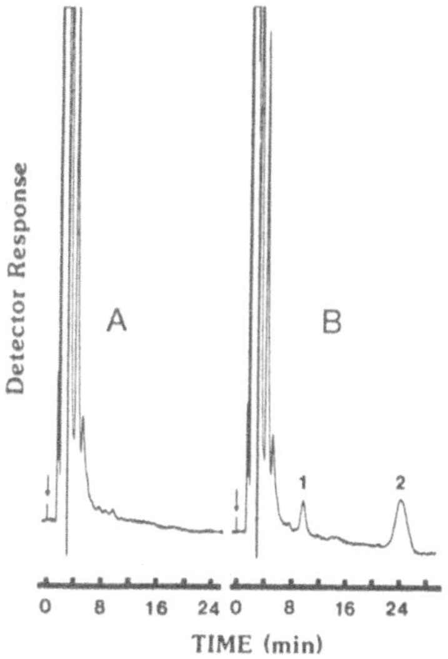

Fig. 2. Chromatograms of (A) urine blank, (B) spiked urine containing 40 ng/ml propanolol (1) and 400 ng/ml quinidine (2). Column: μBondapak C-18; mobile phase 20 mM SDS with 10% (v/v) propanol; flow rate 1.0 ml/min; fluorescence detection (215 nm excitation; emisssion cut-off filter, 300 nm).
From [11], courtesy of the authors and Elsevier.

This technique offers great potential. Direct serum injection has been accomplished previously only by column switching, utilizing short pre-columns for the deproteinization [13, 14], or using specially synthesized stationary phases [15]. [In this book series there are relevant Index entries, under 'HPLC': e.g. pp. 60 & 257 in Vol. 14, p. 311 in Vol. 16.-*Ed.*] Stratton et al. [16] have already shown the usefulness of the micellar technique for the quantitative determination of folylpolyglutamate hydrolase activity in crude tissue extracts.

GRADIENT ELUTION

Gradient elution techniques are the most common solution to the general elution problem in HPLC. Snyder has thoroughly addressed the theory of gradient elution, and has shown the advantages compared with an isocratic separation - faster separations, higher sample capacity, and lower limits of detection [17]. However, especially in repetitive analyses, these techniques have never enjoyed popularity commensurate with the advantages offered. A prime reason is the lengthy column re-equilibration necessary after a hydro-organic gradient separation. RP stationary phases are selectively solvated by the organic component of the mobile phase, to an extent dependent on the mobile phase composition. The solvation structure then changes during a gradient elution program, hence the column must be 're-equilibrated' with the original (weak) mobile phase (maybe needing up to 20 column volumes of the original mobile phase). Consequently, although the gradient may speed the separation, the analysis time, defined as the time between injections, will often be nearly the equivalent of an isocratic separation.

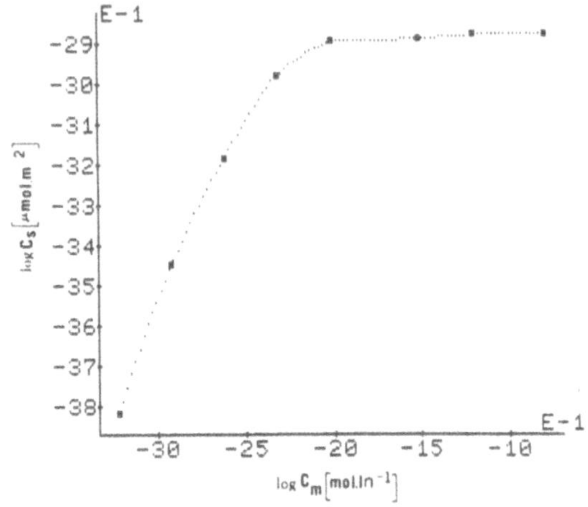

Fig. 3. Log-log representation of adsorption isotherm of SDS on a C-18 column (Altex Ultrasphere) at 30°. Mobile phase: n-propanol/water, 3:97 by vol.
From [19], courtesy of Elsevier.

It is well known from ion-pairing chromatography that surfactants adsorb onto RP stationary phases. Knowledge of the lengthy equilibrations necessary before ion-pairing separations leads to an intuitive belief that gradient elution micellar chromatography would be futile. However, the micelle concentration can be changed without affecting the structure or composition of the stationary phase. Micellar aggregates are in dynamic equilibrium with free surfactant (monomer), and above the CMC there is an approximately constant concentration of free surfactant. As it is the latter that interacts with the stationary phase, this means that after an initial equilibration with any surfactant concentration above the CMC there is no further change in the amount of adsorbed surfactant.

Landy & Dorsey [18] first reported this effect, basing their findings on the precision of retention of an early-eluting solute which was injected after a step gradient from a high to a low surfactant concentration. After a gradient elution separation, failure to re-equilibrate the column leads to irreproducible retention of early-eluting compounds. Dorsey et al. [19] also measured the adsorption isotherm of SDS on a C-18 stationary phase (Fig. 3), and above the CMC found no further modification of the surface, reinforcing the conclusion that no column re-equilibration is necessary after a micellar concentration gradient. Hence RP-HPLC gradient techniques should be applicable to repetitive, routine analyses with dramatic saving of both time and solvent.

Khaledi & Dorsey [20] have also shown that micellar concentration gradients are compatible with amperometric detection, again in contrast to hydro-organic mobile phases. The major limitation of EC detection is its limited applicability to gradient elution techniques. Amperometric EC detectors exhibit both the best and the worst characteristics of solute property and bulk property detectors. While the Faradic current arises only from the solute, the non-Faradic current arises from the bulk mobile phase components. This means that while the detector response is solute-dependent, and therefore selective, the noise (residual current) is controlled by the mobile phase. During gradient elution the mobile-phase composition changes dramatically and with it the residual current. Micellar concentration gradients, however, change the bulk properties of the mobile phase much less than does an organic modifier concentration gradient. The bulk solvent remains constant during a micelle concentration gradient, which makes the control of parameters such as conductance, pH and even mobile phase impurities much easier. Therefore, micellar concentration gradients allow the control of mobile phase conductance and pH, and are closely compatible with amperometric EC detection. With an applied potential of +1.2 V, a gradient from 0.01 to 0.40 M SDS resulted in a baseline shift of only 8 nA. Both solutions were buffered at pH 2.35, and the conductivity of the two solutions was balanced through the addition of 0.226 M $NaClO_4$ to the 0.01 M SDS solution and 0.05 M $NaClO_4$ to the 0.40 M SDS solution. It should be emphasized that this is a 'worst case' experiment, and that smaller gradient ranges and lower operating potentials would result in less baseline shift.

EFFICIENCY

One reason for the popularity of RP chromatography is the plethora of distinct selectivities obtainable by adding a species that will enter into a secondary equilibrium with the analyte in the mobile phase. A problem that must always be addressed when a secondary equilibrium is invoked is the effect of that process on the efficiency of separation. While the secondary equilibrium can favourably affect the *thermodynamics*, and thus the selectivity of separation, it can also have a deleterious effect on the *kinetics* of separation, which controls the peak width.

In any discussion of column efficiencies, it is essential that an accurate measure of plate count be made. This problem is not trivial, as both the commonly used manual measures

$$N = 5.54\,(t_r/w_{0.5})^2 \quad \text{(Eqn. 1; } w_{0.5} = \text{width at half peak height)}$$

and $$N = 16\,(t_r/w_b)^2 \quad \text{(Eqn. 2; } w_b = \text{width at peak base)}$$

can give extremely high positive errors for skewed peaks [21]. Bidlingmeyer & Warren [22] have recently compared many methods for calculating column efficiency and have shown that the most accurate manual method is an equation developed by Foley and Dorsey [23]:

$$N = \frac{41.7\,(t_r/w_{0.1})^2}{B/A + 1.25} \quad \text{(Eqn. 3)}$$ where B/A is the asymmetry ratio measured at 10% peak height.

A major criticism of early work in micellar chromatography was that there was a serious loss of efficiency when compared to traditional hydro-organic mobile phases. If these mobile phases were ever to be useful to the practising chromatographer, the efficiency achieved must at least approach that of conventional HPLC. Taking the initiative, Dorsey et al. [8] applied Knox plots (reduced plate height *vs.* reduced velocity) and showed the dominant problem with hydrophobic stationary phases to be slow mass transfer. Most micellar HPLC studies so far reported had entailed use of totally aqueous mobile phases; under these conditions hydrophobic stationary phases generally exhibit very poor mass transfer characteristics [24]. This is presumably because of poor wetting of the stationary phase, resulting in slow equilibrium across the interface of the two highly dissimilar phases.

The goal for micellar RP-HPLC is, then, to provide wetting with the necessary organic solvent while perturbing the micelle structure as little as possible. Here again, knowledge of RP techniques is helpful. Scott & Simpson [25] have studied modification of C-18 phases by organic modifiers and shown that >90% of the surface is covered with the alcohol at a concentration of 3% (w/v) n-propanol, whereas there is only ~50% coverage with the same concentration of methanol. This modification of the surface should allow fast mass transfer, resulting in improved efficiencies.

The use of 3% n-propanol in the micellar mobile phase and column temperatures of 40° appears to offer a solution to the low efficiency previously reported for micellar mobile phases. These conditions have resulted in reduced plate heights of 3-4 for SDS, cetyltrimethylammonium bromide (CTAB) and Brij-35 [9]. This efficiency optimization strategy appears, then, to be of broad applicability to micellar mobile phases of any surfactant, allowing the surfactant type to be varied so as to improve selectivity with no loss in efficiency.

Foley & May [4] recently confirmed that the significant problem causing low efficiency is poor wetting of the stationary phase. In a study of optimization of pH for separating weak organic acids on hydrophobic stationary phases, they investigated the effects of adding small amounts of methanol, ethanol, n-propanol and acetonitrile as a means of improving efficiency. Much the best effect was with n-propanol: 3-6% (v/v) improved chromatographic efficiencies by factors of 10-15, approaching the efficiencies obtained with traditional hydro-organic mobile phases. The other solvents gave only slight improvements, which is consistent with previous findings that methanol, ethanol and acetonitrile are ineffective at increasing the efficiency of micellar mobile phases [8].

PHYSICOCHEMICAL MEASUREMENTS

Perhaps even more than in other secondary equilibrium-mediated separations, the *value* of the partition coefficients for the partitioning processes in micellar chromatography is of interest. The primary equilibrium is the same as in any RP separation and is represented by the partition coefficient of the solute between the bulk mobile phase (solvent) and the stationary phase. A second equally important solute partitioning process is that between the bulk mobile phase and the micelle. This partitioning coefficient is valuable in understanding retention in micellar chromatography and is also of much interest to micelle chemists. Armstrong & Nome [26] and Arunyanart & Cline Love [27] have derived equations that allow the calculation of solute-micelle partition coefficients and binding constants from chromatographic retention data. These methods offer the usual chromatographic advantages of physicochemical measurements, including the ability to obtain accurate values on small amounts of material, chromatographic separation of any sample impurities, and the capability for simultaneous determination of several solutes. Additionally, Gago et al. [28] have recently shown a correlation of octanol/water partition coefficients with hydrophobicity measurements obtained by micellar chromatography. This method may provide data which could be more biologically relevant than the popular $\log P_{o/w}$ for QSAR analysis.

CONCLUSIONS

A few other practical points need attention in making best use of micellar mobile phases. As with any high ionic strength aqueous mobile phase, precautions need to be taken to preserve the analytical column. Silica has a significant solubility in these mobile phases, and as with aqueous buffer systems a silica saturator column should be placed between the pump and injector. With this precaution there appears to be no difference in column lifetimes between hydro-organic and micellar mobile phases. Secondly, because of the strong adsorption of the surfactants on the stationary phase of C-8 and C-18 columns, a separate column needs to be devoted to each different surfactant. Finally, the strength of common micellar mobile phases is somewhat less than that of traditional organic modifiers. This need not, however, mean longer analysis times. Shorter columns and shorter chain length stationary phases can both be successfully applied to yield acceptable separation times.

Micellar mobile phases are now eight years old, and there is still much left to be learned. Besides the advantage enumerated here, others are sure to follow. Micellar mobile phases will never replace hydro-organic mobile phases, but they do have a role in the laboratory of the practising analyst. At this point in the development of micellar chromatography, it can be stated conclusively that these distinctive mobile phases do offer solutions to problems that cannot be solved by conventional means.

References

1. Armstrong, D.W. & Henry, S.J. (1980) *J. Liq. Chromatog. 3*, 657-662.
2. Hinze, W.L. & Armstrong, D.W., eds. (1987) *Ordered Media in Chemical Separations*, *Am. Chem. Soc. Symp. Ser.*, *342*, 293 pp.
3. Foley, J.P. & May, W.E. (1987) *Anal. Chem. 59*, 102-109.
4. Foley, J.P. & May, W.E. (1987) *Anal. Chem. 59*, 110-115.
5. Dorsey, J.G. (1987) *Chromatog. Mag. 2(4)*, 13-20.
6. Dorsey, J.G. (1987) *Adv. Chromatog. 27*, 167-214.
7. Cline Love, L.J., Habarta, J.G. & Dorsey, J.G. (1984) *Anal. Chem. 56*, 1132A-1148A.
8. Dorsey, J.G., DeEchegaray, M.T. & Landy, J.S. (1983) *Anal. Chem. 55*, 924-928.
9. Landy, J.S. & Dorsey, J.G. (1985) *Anal. Chim. Acta 178*, 179-188.
10. DeLuccia, F.J., Arunyanart, M. & Cline Love, L.J. (1985) *Anal. Chem. 57*, 1564-1568.
11. Arunyanart, M. & Cline Love, L.J. (1985) *J. Chromatog. 342*, 293-301.
12. DeLuccia, F.J., Arunyanart, M., Yarmchuk, P., Weinberger, R. & Cline Love, L.J. (1985) *LC Mag. 3*, 794-800.
13. Roth, W., Beschke, K., Jauch, R., Zinner, A. & Koss, F.W. (1981) *J. Chromatog. 222*, 13-22.
14. van Buuren, C., Lawrence, J.F., Brinkman, U.A.Th., Honigberg, I.L. & Frei, R.W. (1980) *Anal. Chem. 52*, 700-704.
15. Hagestam, I.H. & Pinkerton, T.C. (1985) *Anal. Chem. 57*, 1757-1763.
16. Stratton, L.P., Hynes, J.B., Priest, D.G., Doig, M.T., Barron, D.A. & Asleson, G.L. (1986) *J. Chromatog. 357*, 183-189.
17. Snyder, L.R. (1980) in *High Performance Liquid Chromatography: Advances and Perspectives* (Horvath, C., ed.), V. 1, Acad. Press, N.Y., 207-316.
18. Landy, J.S. & Dorsey, J.G. (1984) *J. Chromatog. Sci. 22*, 68-70.
19. Dorsey, J.G., Khaledi, M.G., Landy, J.S. & Lin, J-L. (1984) *J. Chromatog. 316*, 183-191.
20. Khaledi, M.G. & Dorsey, J.G. (1985) *Anal. Chem. 57*, 2190-2196.
21. Kirkland, J.J., Yau, W.W., Stoklosa, H.J. & Dilks, C.H., Jr. (1977) *J. Chromatog. Sci. 15*, 303-316.
22. Bidlingmeyer, B.A. & Warren, F.V., Jr. (1984) *Anal. Chem. 56*, 1583A-1596A.
23. Foley, J.P. & Dorsey, J.G. (1983) *Anal. Chem. 55*, 730-737.
24. Snyder, L.R. & Kirkland, J.J. (1979) *Introduction to Modern Liquid Chromatography*, 2nd edn., Wiley-Interscience, New York: p. 298.
25. Scott, R.P.W. & Simpson, C.F. (1980) *Faraday Symp. Chem. Soc. 15*, 69-82.
26. Armstrong, D.W. & Nome, F. (1981) *Anal. Chem. 53*, 1662-1666.
27. Arunyanart, M. & Cline Love, L.J. (1984) *Anal. Chem. 56*, 1557-1561.
28. Gago, F., Alvarez-Builla, J., Elguero, J. & Diez-Masa, J.C. (1987) *Anal. Chem. 59*, 921-923.

#E-3

DETERMINATION OF NEFOPAM AND ITS DESMETHYL METABOLITE IN BIOLOGICAL FLUIDS BY PAIRED-ION RP-HPLC

[1]A.J. Woodward, [1]P.A. Lewis, [1]J. Maddock and [2]D. Donnell

[1]Analytical Division, Simbec Research Ltd.
Merthyr Tydfil, Mid Glamorgan CF48 4DR, U.K.
[2]The Medical Department, Riker Laboratories
Loughborough, Leics. LE11 1EP, U.K.

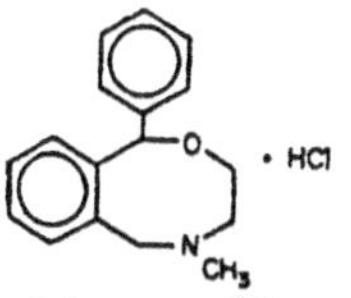

Nefopam hydrochloride (Acupan®), a centrally acting agent used for the treatment of painful conditions, is structurally and pharmacologically unrelated to other analgesics. The assay now described and validated entails solvent extraction steps and final UV measurement after HPLC under conditions that give good peaks even though the analytes are amines. The drug detection limit in plasma is 1 ng/ml.

Most published methods [1-3] for plasma nefopam analysis utilize GC with thermionic detection (N- & P-specific) techniques and are restricted to plasma. Although sufficiently sensitive for pharmacokinetic studies such methodology is unsuitable for the determination of its major metabolite ***N***-desmethylnefopam (NDMN) due to extensive on-column adsorption [4]. The following paired-ion RP-HPLC procedure was developed for the determination of both nefopam and NDMN in human plasma, urine and breast milk and has been used routinely for pharmacokinetic studies.

ASSAY PROCEDURE FOR PLASMA

(1) To 1.0 ml plasma, add internal standard (see below), 0.5 g NaCl and 1.0 ml pH 11 glycine buffer. Extract with 8 ml cyclohexane and centrifuge at 1000 **g** for 10 min.

(2) Transfer cyclohexane layer to a glass tube containing 3.0 ml 0.1 M HCl, back-extract and centrifuge.

(3) Aspirate the cyclohexane layer to waste, add 1.0 ml 0.5 M NaOH and 8 ml cyclohexane, re-extract and centrifuge.

(4) Transfer the cyclohexane layer to a glass vial and evaporate to dryness (room temp., N_2).

(5) Reconstitute in 70 µl acetonitrile/water (50:50 by vol.), and inject 50 µl onto the HPLC column (Dupont Zorbax CN).

(6) Elute with acetonitrile/0.01 M pentane sulphonic acid (PSA; adjusted to pH 3.15), 50:50 by vol., as mobile phase at 1.5 ml/min.

(7) Detect at 215 nm, and quantitate by peak height ratios.

Choice of internal standard (i.s.)

Initially 100 ng ***N***-isopropyldesmethylnefopam (NIPrDMN), an analogue of nefopam, was used as i.s., but due to lack of material 150 ng orphenadrine citrate (OC) was used as an alternative and gave comparable results.

Extraction procedure

Amines tend to adsorb onto glassware. [See Index entries, 'Adsorptive' and 'Vessels', in earlier vols.-*Ed.*] Nefopam and particularly NDMN show this phenomenon especially at low concentrations. It was therefore necessary to silanize all glassware used for extraction.

Quantitative yields of nefopam and a relatively clean extraction were achieved using hexane as extractant. However, in order to get comparable recoveries of NDMN, cyclohexane was substituted for hexane and NaCl was added. This led to the extraction of significant quantities of endogenous compounds which chromatographed both close to the solvent front, interfering with NDMN, and as late-running peaks interfering with subsequent injections.

For therapeutic concentrations of nefopam a back-extraction procedure was incorporated to minimize interferences. This procedure did not significantly affect analyte recoveries. Although nefopam ***N***-oxide, a further nefopam metabolite, has the same retention time as nefopam it did not co-extract and thus did not interfere.

Choice of chromatographic conditions

Compounds containing amino groups are notoriously difficult to chromatograph on conventional RP columns. Peak tailing leads to inadequate resolution with a corresponding loss of sensitivity and reproducibility. An ion-pairing mobile phase modifier, PSA, in conjunction with a nitrile column, as specified above, allowed the rapid separation of the analytes with excellent peak shapes (Fig. 1).

Attempts to introduce a buffer into the mobile phase to control pH impaired chromatographic performance. Retention of the compounds of interest was excessive, with impaired peak shape leading to insensitivity. In the absence of a buffer it was necessary to finely adjust the pH of the PSA to between 3.10 and 3.20 to effect separation of both nefopam and NDMN from co-extracted endogenous and reagent-derived

Fig. 1. Typical chromatograms for normal plasma spiked with [**A**] 100 ng/ml NIPrDMN i.s. (3) and 150 ng/ml orphenadrine (OC) i.s. (4), or [**B**] 50 ng/ml nefopam (1), 100 ng/ml NDMN (2), 100 ng/ml NiPrDMN i.s. (3) and 150 ng/ml orphenadrine i.s. (4).

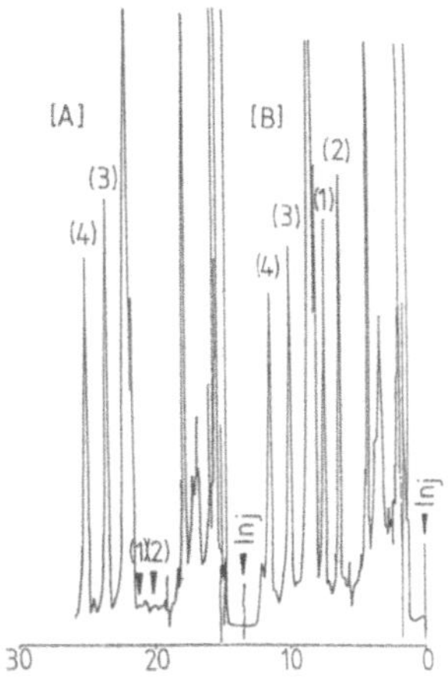

components that could interfere. To achieve optimal sensitivity, a relatively non-specific wavelength (215 nm) was used, with a highly attenuated baseline (0.005 AUFS). Fig. 1 shows typical chromatograms for spiked plasma samples.

Characteristics of plasma assay

With NIPrDMN as i.s., the nefopam/i.s. or NDMN/i.s. peak height ratio was linearly related to concentration over the range studied. For nefopam, 2.5-100 ng/ml, y = 0.0206 x + 0.004, r = 0.9993 (cf. OC i.s., *vs.* NIPrDMN: y = 0.0216 x + 0.009, r = 0.9994). For NDMN, 5.0-200 ng/ml, y = 0.0146 x + 0.015, r = 0.09979 (cf. OC i.s., y = 0.0151 x - 0.009, r = 0.9981). The lower limit for quantitation was 1 ng/ml plasma for nefopam and 2 ng/ml for NDMN. Over the calibration range the mean absolute recovery from plasma was 91% for nefopam and 77% for NDMN. Neither showed any loss during storage of plasma for up to 3 weeks at -20°.

Intra-assay performance was judged by 10 replicate analyses of 3 spiked plasmas containing nefopam and NDMN (ng/ml: 9 & 16; 52 & 95; 93 & 171) besides each i.s. (NIPrDMN and OC). The ranges of the mean recoveries relative to i.s. (as 100%) were 95-106% for nefopam and 99-112% for NDMN *vs.* NIPrDMN (94-105% and 95-107% *vs.* OC). The corresponding mean C.V.'s were 5.1-5.9% and 9.8-12.9% *vs.* NiPrDMN, and 3.3-5.3% and (for NDMN) 10.1-11.8% *vs.* OC. Inter-assay reproducibility was obtained from quality control samples included in subsequent assays, *vs.* OC only: relative recoveries were 95-103% and (NDMN) 98-102%, and mean C.V.'s were 3.6-15.4% and 8.9-18.6%.

Conclusions on plasma assay

The assay for nefopam and NDMN is evidently satisfactory in respects such as sensitivity and precision. It has served for the

Fig. 2. Typical chromatograms for normal urine spiked with **[A]** 750 ng/ml orphenadrine (OC) i.s. (3), or **[B]** 250 ng/ml nefopam (1), 500 ng/ml NDMN (2) and 750 ng/ml orphenadrine (3).

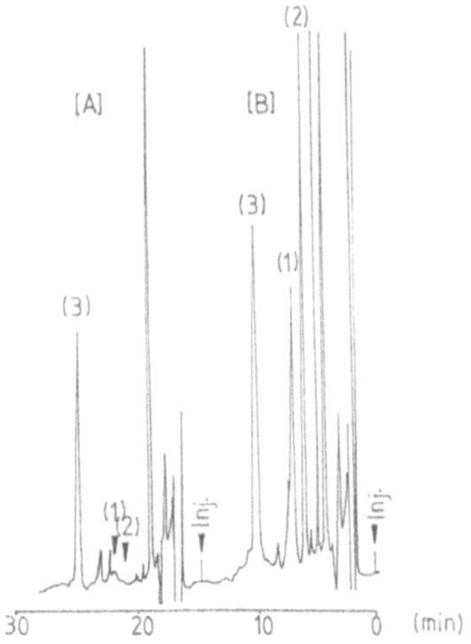

routine analysis of several hundred samples generated by pharmacokinetic studies in human volunteers; both young and elderly subjects have been studied, including urine levels the assay of which is now considered.

URINE ASSAY

Although both nefopam and NDMN are mainly present as conjugates in urine, for the free forms the plasma assay was applicable to urine with minor modifications.- The calibration ranges were adjusted to suit the higher concentrations of both in pharmacokinetic samples; NDMN was adequately recovered without adding NaCl; 750 mg/ml of OC was added as i.s. Fig. 2 shows typical chromatograms.

Characteristics of the urine assay

The assay gave validation data similar to those for the plasma assay, with linear curves over the calibration ranges 0.125-3.0 µg/ml nefopam ($y = 3.66\,x + 0.24$, $r = 0.9966$) and 0.5-12.0 µg/ml NDMN ($y = 2.40\,x + 1.46$, $r = 0.9945$). The mean absolute recoveries over these ranges were 89% and 84% respectively. Both analytes were stable following urine storage at 20° for up to 4 weeks.

Intra-assay precision was judged by replicate analyses of 3 urines spiked with nefopam and NDMN (µg/ml: 0.3 & 1.2; 1.0 & 3.6; 2.7 & 10.0). The ranges of the mean recoveries relative to i.s. were 100-102% and 100-106%, and the mean C.V.'s were 4.7-10.1% and 5.5-9.1% respectively. Inter-assay mean relative recoveries were 89-105% and 100-109%, and C.V.'s were 3.2-10.8% and 2.0-14.4%.

Fig. 3. Typical chromatograms for human milk spiked with **[A]** 50 ng/ml NIPrDMN i.s. (3), or **[B]** 50 ng/ml nefopam (1), 50 ng/ml NDMN (2) and 50 ng/ml NIPrDMN i.s. (3).

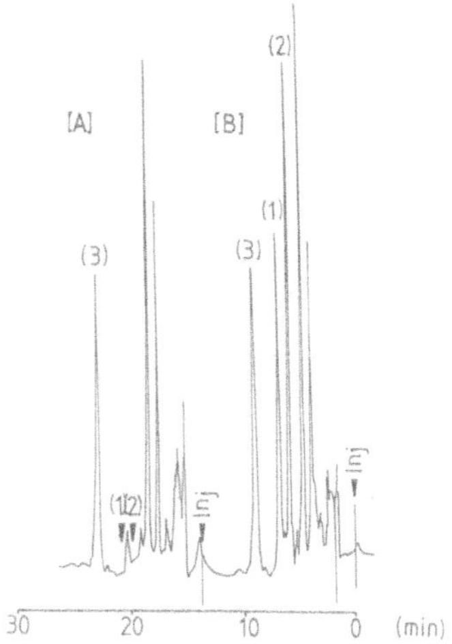

ASSAY IN HUMAN MILK

The plasma assay was applicable with minor changes.- As for urine, NaCl was omitted; 2 ml of sample was taken to allow calibration down to 1.25 ng/ml for both nefopam and NDMN; 50 ng/ml of NIPrDMN was used as i.s. Because insufficient human milk was forthcoming for control purposes, calibration ranges were compared between purchased pasteurized cow's milk and unpasteurized human milk. The former was used to obtain validation data, the following correlations being obtained: nefopam, y = 1.09 x - 0.026, r = 0.99; NDMN, y = 1.04 x - 0.014, r = 0.97.

Fig. 3 shows typical patterns. Results from applying the assay to human milk have been reported [5].

Characteristics of the milk assay

Linear standard curves were obtained over the calibration range 1.25-75 ng/ml for both nefopam (y = 0.0291 x + 0.025, r = 0.9992) and NDMN (y = 0.0323 x + 0.023, r = 0.9977). The mean absolute recoveries over these ranges were 74% and 88% respectively with cow's milk and 75% and 86% with human milk. Both were stable in milk during storage at -20° for up to 3 weeks.

Replicates on 3 samples containing nefopam and NDMN (ng/ml: 5 & 4; 10 & 12; 35 & 35) gave intra-assay relative recoveries, 92-105% and 92-118%, and mean C.V.s, 3.7-12.5%, and 13.2-17.6%. Mean inter-assay relative recoveries were 94-105% and 99-109%, and mean C.V.s were 5.0-10.7% and 3.1-10.2% respectively.

References

1. Ehrsson, H. & Eksborg, S. (1977) *J. Chromatog. 136*, 154-158.
2. Schuppan, D., Hansen, C.S. & Ober, R.E. (1978) *J. Pharm. Sci. 67*, 1720-1723.
3. Chang, S.F., Hansen, C.S., Fox, J.M. & Ober, R.E. (1981) *J. Chromatog. 226*, 79-89.
4. Ebel, S. & Schultz, H. (1978) *Arch. Pharmazie (Weinheim) 311*, 547-552.
5. Lui, D.T.Y., Savage, J.M. & Donnell, D. (1987) *Br. J. Clin. Pharmacol. 23*, 99-101.

#E-4

AN ASSAY FOR IDAVERINE IN PLASMA AND URINE AT PICOGRAM LEVEL

D.J.K. van der Stel, O.A.M. Brockhoff and H. de Bree

Analytical Development Department
Duphar Research Laboratories, P.O. Box 2
1380 AA Weesp, The Netherlands

The method now described was developed to establish the kinetics of idaverine and its metabolite, N-desmethylidaverine, in human plasma in the sub-ng range. The method hinges on a thorough clean-up procedure, which incorporates a solid-liquid extraction of the analytes from plasma, followed by lipid removal by washing with a hydrocarbon solvent. A liquid-liquid extraction is performed to remove acidic plasma constituents. Subsequently NP-HPLC® *is performed to separate the analytes from remaining ballast material, with an i.s. pair which functions also as carriers for the analytes. Finally separation and detection are performed by using automated on-column capillary GC with NPD.*

Fig. 1. Structures of idaverine (n = 3, R = CH_3), its i.s. (n = 2, R = CH_3), its metabolite (n = 3, R = H) and the metabolite i.s. (n = 4, R = CH_3).

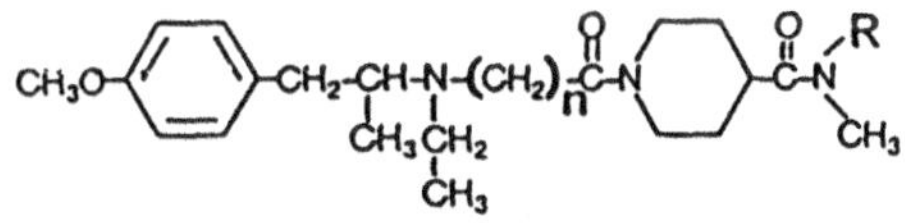

Idaverine (Fig. 1) is a novel antispasmodic compound currently under development at Duphar. To establish the kinetics of idaverine and its *N*-desmethyl metabolite in the animal species used for toxicology, a selective and sensitive method had been developed, However, despite the low determination limit (1 ng/ml plasma), that assay was not sensitive enough to serve for human pharmacokinetic and efficacy studies, since much lower doses are used. Therefore an ultra-sensitive method for the parent drug and metabolite had to be developed.

®*Abbreviations.*- NP[-HPLC], normal- (straight-)phase; NPD, nitrogen-phosphorus detector (N-FID); i.s., internal standard. 'Ether' is diethyl ether.

INSTRUMENTATION AND CONDITIONS

HPLC.- NP-HPLC clean-up of the analytes was performed with a Waters Model 510 Solvent Delivery System equipped with a Kratos Spectroflow 757 UV monitor, using a Zorbax Sil 7 µm column (25 cm × 46 mm i.d.) and, flowing at 2 ml/min, acetonitrile/dimethylformamide /triethylamine (100:5:0.5 by vol.).

GC-NPD (HP 5890A with NPD, and a HP 7672A autosampler and HP 3392 integrator) entailed a modification of the injection port (Fig. 2), to allow automatic on-column injection. The 25 m column was of fused-silica with a chemically bonded phase, Durabond DB1-30W column (film thickness 25 µm; J & W Scientific). From an initial temperature of 60° the column was raised by 60°/min to 100°, then by 40°/min to 250° and 5°/min to 310°.

PRE-GC PROCEDURES

1. To 10 ml plasma add 50 ng of each i.s.
2. Pre-rinse Baker-10 C-18 cartridge: methanol, then water (each 3 × 2.5 ml).
3. Pass sample through cartridge.
4. Wash cartridge with 3 × 6 ml water.
5. Elute with 3 ml methanol containing 2% (w/v) ammonia.
6. Evaporate off the methanol by N_2 stream; add 1 ml 0.01 M HCl.
7. Wash with 9 ml n-pentane.
8. Freeze aqueous layer (dry ice/acetone) and discard pentane layer.
9. Add 25 µl 25% (w/w) ammonia and extract with 9 ml ether.
10. Freeze aqueous layer (dry ice/acetone).
11. Evaporate ether phase with N_2.
12. Dissolve residue in 200 µl HPLC mobile phase (see above).
13. Inject the sample into the NP-HPLC system.
14. Collect the fraction containing the analytes (3.15-6 min).
15. Evaporate off the mobile phase with N_2.
16. Add 2 mg potassium carbonate (granular) and 200 µl acetone.
17. Transfer into mini-vial and evaporate solvent with N_2.
18. Dissolve residue in 15 µl acetone and inject 2 µl on-column.

METHODOLOGICAL RESULTS

Sample clean-up by HPLC.- Fig. 3 shows the chromatograms of a marker run and a typical sample. The whole clean-up operation could be automated employing a Gilson injector/collector system [cf. art. D-5: McDowall]. The collected fraction (5-6 ml) is readily blown to dryness leaving a residue free of GC-interfering substances. The NP-HPLC has excellent stability and hence reliability. In the presence of 50 ng of i.s. even pg-amounts of (radiolabelled) idaverine were completely recovered following this step. Overall,

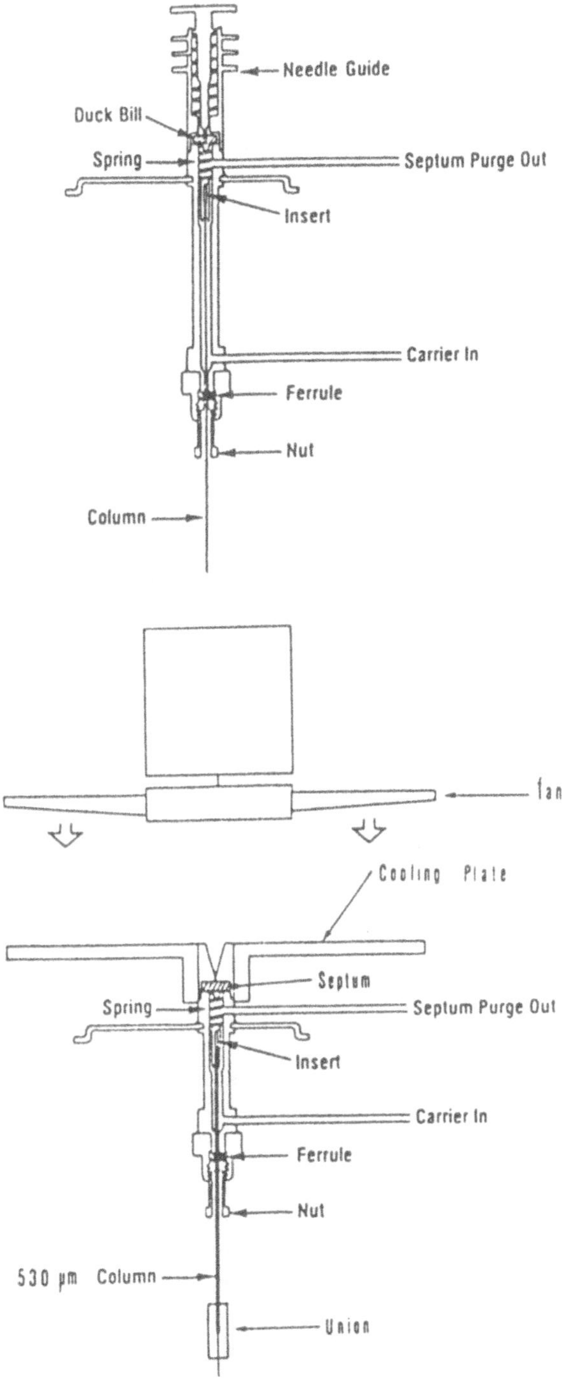

Fig. 2. Original *(upper diagram)* and adapted *(lower diagram)* injection port for the HP 5890 gas chromatograph. The modification furnished a cooled injection port and liner.

Fig. 3. NP-HPLC patterns in the clean-up step, with **a**) a reference mixture, **b**) a test sample. Peak I represents idaverine and its i.s., and peak II represents the *N*-desmethyl metabolite and its i.s *Arrows* indicate the retention window that is collected for subsequent processing. (The idaverine and i.s. levels in the test sample were too low to furnish peaks.)

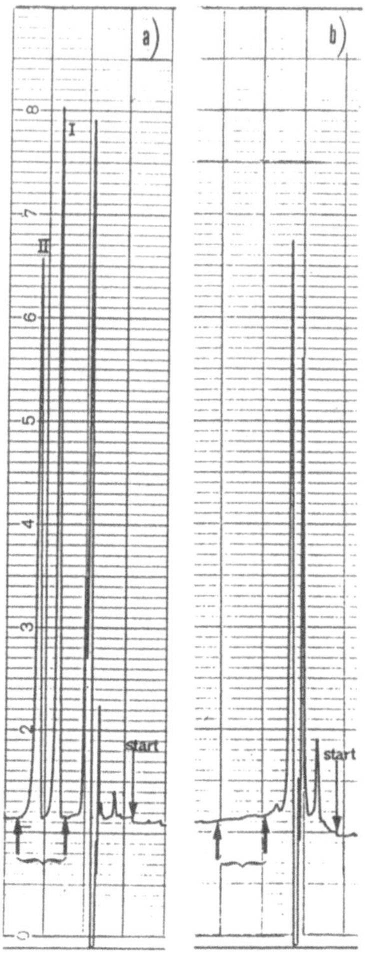

including the initial solid-liquid and liquid-liquid extractions, the absolute recovery was ~60% for both analytes.

GC-NPD.- The bonded-phase fused silica GC technology allows operation at temperatures of up to 350°. At 310° as adopted, these columns were usable at maximum sensitivity for ~4 weeks. Then they start to show tailing peaks and adsorption phenomena, despite the use of a 40-inch pre-column. The chromatograms of a series of additions of both analytes to a blank matrix are shown in Fig. 4. With a 10 ml plasma sample the sensitivity of the method is ~100 pg/ml for both parent drug and metabolite. At a level of 1 ng/ml the reproducibility is 15%.

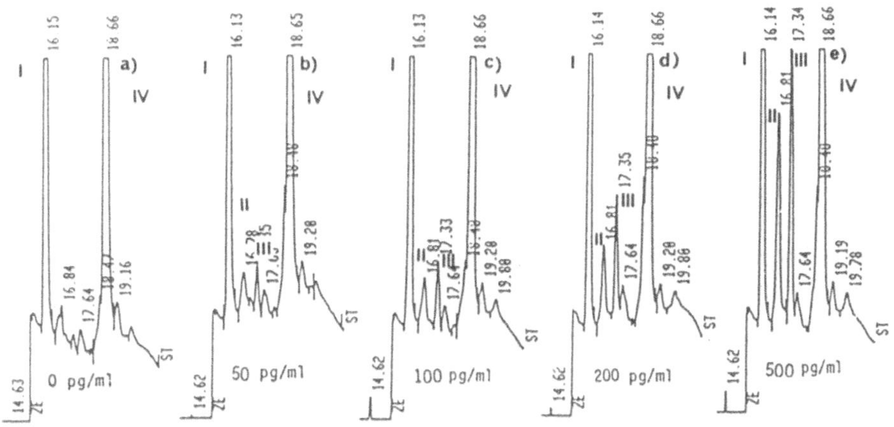

Fig. 4. Capillary-GC patterns for **a**) blank plasma, and **b**)–**e**) blank plasma spiked with each analyte and i.s. with, as marked, increasing amounts of each analyte in the range 50-500 pg/ml. Peak III, idaverine; IV, its i.s.; II, the metabolite; I, its i.s. Each i.s. was in constant amount, 5 ng/ml.

CONCLUSIONS

- The use of an HPLC clean-up step enables a 10 ml plasma sample to be concentrated to a final volume of 15 µl for GC.
- It is possible to inject 10-20% of this concentrate onto the GC column with maintenance of good performance.
- The bonded-phase fused-silica capillaries now available allow GC of relatively polar compounds at temperatures above 300°.
- The determination limits for both parent drug and metabolite are of the order of 100 pg/ml.

#E-5

PITFALLS IN THE ENANTIOSELECTIVE ANALYSIS OF CHIRAL DRUGS

John Caldwell≠ **John F. Darbyshire, Steven M. Winter**† **and** ***Andrew J. Hutt**

Department of Pharmacology and Toxicology, St. Mary's Hospital Medical School London W2 1PG, U.K.

*Department of Pharmacy Brighton Polytechnic Brighton BN2 4GJ, U.K.

There is currently considerable interest in the stereochemical aspects of biochemical pharmacology, largely originating from analytical developments permitting the determination of the enantiomeric composition of chiral drugs in biological media. These new methods allow new types of experiments to be carried out, but there is a need for them to be used critically: a number of pitfalls await those who use them inappropriately. These particularly present themselves in metabolic studies where authentic standards are not available for method validation. Both of the major methods for determining enantiomeric composition can give rise to errors. The formation of diastereoisomers may exhibit stereoselectivity, while the use of chiral columns is fraught with difficulties. In both cases, calibration lines for the two enantiomers are frequently different. Additionally, enantiodifferentiation may occur during sample work-up, as a consequence of achiral processes such as solvent extraction. The importance of proper method validation, with each of a pair of enantiomers, cannot be stressed too highly. If standards of appropriate enantiomeric purity are not available, individual methods cannot be acceptably validated, and data can be accepted only when the same result is obtained with two methods relying on different principles, e.g. a chiral column plus a diastereomeric derivatization, or a chromatographic method plus NMR.⊗

It has been known for more than a century that the animal body is capable of discriminating between pairs of enantiomers since, in bimolecular terms, living matter is a highly chiral environment.

≠to whom any correspondence should be addressed. †now at Dept. of Pharmacology & Toxicology, Coll. of Pharmacy, Univ. of Arizona Health Sciences Center, Tucson, AZ. ⊗*This Forum Abstract has been married up with a provisional text MS., without a proof-correcting stage.- Editor.*

Indeed, Van't Hoff in 1898 quoted this capacity as one of the critical pieces of evidence showing that stereoisomers are different compounds. Early in the history of pharmacology, it was appreciated that pairs of enantiomers often differed substantially in their activities, as first shown with the hyoscine alkaloids by Cushny in 1907 and extended particularly with reference to noradrenaline. The stereoselectivity of drug action provides important evidence for the existence of specific receptors, as was exploited by Beckett & Casy [1, 2] for opiates some 20 years before the discovery of the opiate receptors and their neuropeptide ligands.

Despite this historical background, in the past 30 years there has been a progressive de-emphasis upon stereochemical considerations in biochemical pharmacology and toxicology. Various reasons can be advanced to acocunt for this [3] including the decline in teaching of chemical principles in university biology-based courses and the associated absence of the topic as an index term in virtually every major textbook of pharmacology or toxicology [4]. The explosion in the use of synthetic drugs since the early 1960's has neglected stereochemistry: although some do contain chiral centres, the vast majority are used as the racemic mixtures [see art. #D-1 by I.W. Wainer & co-authors.- *Ed.*]. Last but by no means least, there has been a major deficiency in our ability to determine the enantiomeric composition of chiral drugs in biological media.

The past 5 years have seen the increasingly widespread application of novel analytical methodologies for the separate determination of individual stereoisomers in biological fluids. These studies have shown that the enantiomers can differ widely in biological disposition, and it is now clear that the processes of absorption, distribution, metabolism and excretion can exhibit stereoselectivity. This, together with recent progress, in the fields of stereospecific synthesis and enantiomeric separations, has served to re-emphasize the importance of enantiochemical considerations in drug development and more generally in pharmacology and toxicology.

As with all new advances, the potential contribution which these new analytical methodologies can make will only be realized if their use is informed by a knowledge of their limitations. It is the purpose of this short article to indicate some of the possible problems that may arise, and to highlight the need for extended validation of methods.

Chromatographic methods for determining the enantiomeric composition of chiral molecules depend upon the formation of stable or transient interactions with other chiral centres, and may be considered under three headings: (i) stable diastereoisomeric derivatives formed with one enantiomer of a chiral reagent and separated by TLC, HPLC or GLC; (ii) transient interactions with chiral stationary phases (CSP's) for HPLC and GC, which allow the direct separation of enantiomers; and (iii) transient interactions with chiral agents dissolved

Fig. 1. Calibration curves for the HPLC assay of the enantiomers of 2-phenylpropionic acid.

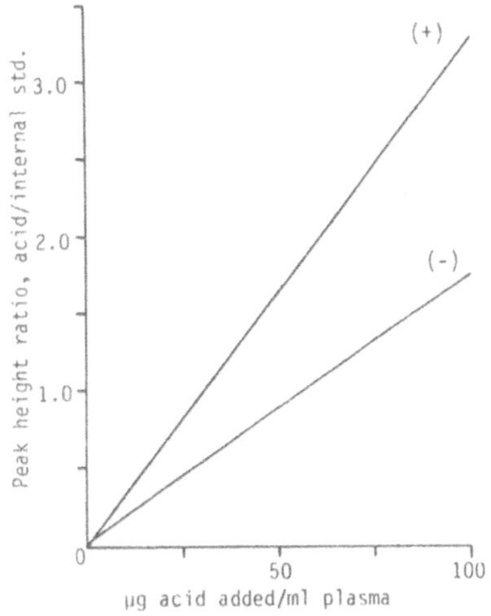

in the mobile phase for HPLC. There exist numerous examples of the applicability of the first two options, but the third is only rarely encountered (see #B-6, pp. 112 & 120, for examples).

It is extremely important that assays making use of these principles are properly validated, and that the inevitable limitations caused by the lack of availability of pure enantiomers as standards are fully appreciated. This is particularly a problem in metabolic studies, where only rarely are the individual enantiomers of metabolites available: indeed, in many cases it is necessary to work without even racemic standards of the metabolites.

Criteria for methods.- Enantioselective analytical methods must be validated according to five separate criteria, as follows.-
(1) Sample work-up: does non-chiral enantiodifferentiation occur?
(2) With chiral derivatization, does 'kinetic resolution' occur?
(3) Assurance of peak identity assigned to the enantiomers is needed.
Also (4) Calibration curves must be set up for each enantiomer (Fig. 1 illustrates a notable difference), and (5) statistical evaluation of the full method for each enantiomer is necessary.

NON-CHIRAL DIFFERENTIATION

Although it is frequently stated that enantiomers can be distinguished only in a chiral environment, exploitation of properties such as lipid solublity, ionization and volatility - a process termed 'non-chiral differentiation' - has been documented recently by Tsai & co-authors [5]. This refers to the progressive enrichment in the more abundant enantiomer of a non-racemic mixture of enantiomers by typical work-up procedures such as solvent extraction, achiral chromatography or chemical reactions, even though these do not involve chiral reagents, columns etc. Since the majority of biological samples for analysis will contain non-racemic mixtures of enantiomers, there must be control of the possible complications introduced by non-chiral enantiodifferentiation during work-up.

CHIRAL DERIVATIZATION

The popular chiral derivatization methods are attractive in producing stable derivatives readily separated by conventional (i.e. inexpensive) HPLC or GC methods, which may be generally applicable to classes of functional group (-OH, $-NH_2$, -COOH) depending upon the derivatizing reagent in question.

Three major problems with these methods must be addressed. (1) It is critical to show that the formation of diastereoisomeric derivatives does not itself exhibit stereoselectivity under the final conditions chosen. The so-called 'kinetic resolution' phenomenon, although fortunately rare, could cause major error. (2) The limits of discrimination between enantiomers in chiral derivatization methods are in part determined by the enantiomeric purity of the derivatizing agent used. It is essential that this be as high as possible and that it be determined as accurately as possible. Even when reagents of acceptable purity are used, these may undergo racemization (at least partial) during the reaction, with consequent misleading results. This has been reported for ketamine [6]. (3) The HPLC separation of diastereoisomers can be highly dependent upon minor changes in mobile phase composition, and even upon the supplier of the solvents used, as we have described elsewhere [7].

Most of these difficulties can be recognized and overcome if optically pure samples of the diastereoisomeric derivatives are available. These greatly facilitate the optimization of reaction conditions etc., as discussed elsewhere (e.g. in #B-6, this vol.).

PEAK IDENTITY

It is obvious that in any analytical method there should be adequate demonstration of peak identity and purity, and this is absolutely critical in the case of chiral assays. Even in the best chromatographic assays, the separation of enantiomers or diastereo-isomers is not notably great: retention differences of 10% are highly acceptable provided that baseline resolution is achieved [8]. Accordingly, structurally related metabolites and/or impurities can readily produce chromatograms resembling those to be expected from pairs of stereoisomers eluting in close proximity from an HPLC or GC column.

We encountered such a problem in the course of studies on the enantioselectivity of aliphatic hydroxylation of 6-n-propylchromone-2-carboxylic acid. In the rabbit, this prochiral model substrate is hydroxylated regiospecifically at the 2'-position of the 6-n-propyl group. The enantiomeric composition of this chiral metabolite, 6-(2'-hydroxypropyl)chromone-2-carboxylic acid, was ascertained by four independent methods (chiral HPLC, HPLC of two diastereoisomeric derivatives, and lanthanide-shift NMR) to be 76% *S* :24% *R* . The isolation procedures used for these studies were laborious, involving first solvent extraction, then methylation of the carboxyl group, a further solvent extraction and two preparative TLC steps. Accordingly, when this project was extended to an examination of the species differences of enantioselectivity of hydroxylation, the work-up procedures were simplified, and the TLC steps were replaced by a single solid-phase extraction.

HPLC chromatograms of diastereoisomeric derivatatives (MTPA esters) of the metabolite isolated from rat urine indicated its enantiomeric composition to be the same as in the rabbit, at least in terms of the UV absorbance of the column eluate. However, we were fortunate to be using ^{14}C-labelled compound, and the radiochromatogram associated with the UV trace gave a very different picture, indicating the enantiomeric composition of the metabolite to be 10% *S*:90%*R*. Subsequent modification of the HPLC mobile phase succeeded in moving an impurity which co-eluted with the *S*-enantiomer in the original system, while when the rat urine samples were prepared for HPLC by the original method the impurity was absent.

Although in this case the problem was relatively easily recognized and overcome, this was true only because of our lengthy experience in working with rabbit urine. It is easy to imagine that gross error can be perpetrated unless peak purity is checked in such situations. This is relatively easy if radiolabelled compounds are used, but is more difficult if cold methods are to be relied upon. Mass spectrometry is obviously able to provide such information, while the laser polarimeter detectors now being introduced for HPLC [see p. 404 - *Ed.*] give the optical purity of peaks.

References

1. Beckett, A.H. & Casy, A.F. (1954) *Nature 173*, 1231-1232.
2. Beckett, A.H. & Casy, A.F. (1955) *J. Pharm. Pharmacol. 6*, 986-999.
3. Smith, R.L. & Caldwell, J. (1988), in course of publication.
4. Ariëns, E.J. (1987) *Med. Res. Rev. 7*, 367-387.
5. Tsai, W.L., Hermann, K., Hug, E., Rhode, B. & Dreiding, A.S. (1985) *Helv. Chim. Acta 68*, 2238-2243.
6. Adams, J.D. jr., Woolf, T.F., Trevor, A.J., Williams, L.R. & Castagnoli, N. jr. (1982) *J. Pharm. Sci. 71*, 658-661.
7. Hutt, A.J., Fournel, S. & Caldwell, J. (1986) *J. Chromatog. 378*, 409-418.
8. Testa, B. (1986) *Xenobiotica 16*, 265-269.

EDITOR'S NOTE.- An intended continuation of the text was not available when the book went to press.

#NC(E)

NOTES and COMMENTS relating to

VARIOUS DRUGS AND SEPARATORY APPROACHES

Comments relating to particular contributions:

#E-1, -2 & -5, and #NC(E)-4: p. 307

#NC(E)-1

A Note on

CHIRAL DIFFERENCES IN THE DISPOSITION OF THE QUATERNARY ANTICHOLINERGIC DRUG OXYPHENONIUM BROMIDE

[1]Karla G. Feitsma, [1]Ben F.H. Drenth, [1]Rokus A. de Zeeuw and [2]Dirk K.F. Meijer

Departments of [1]Analytical Chemistry & Toxicology
and of [2]Pharmacology & Therapeutics
University of Groningen, A. Deusinglaan 2
9713 AW Groningen, The Netherlands

BIOANALYTICAL APPROACH

As for many other anticholinergic drugs, the enantiomers of oxyphenonium bromide exhibit large differences in affinity to muscarinic receptors as well as in therapeutic effect [1]. For detailed study of the fate of the enantiomers in the body an assay is required that determines both enantiomers simultaneously in the biological sample. Although various chiral chromatographic systems have been developed over the years [2], we were unable to separate the enantiomers of oxyphenonium bromide [3]. Recently we succeeded in resolving racemic oxyphenonium bromide on an α_1-acid glycoprotein (EnantioPac®) column; but this system had insufficient efficiency and stability for bioanalytical applications. In order to study differences between oxyphenonium bromide enantiomers in absorption, distribution, metabolism and excretion, another approach was adopted, in which the enantiomers were synthesized separately (Fig. 1).

Fig. 1. Synthesis of the drug from cyclohexylphenylglycollic enantiomers (* = centre of asymmetry; enantiomeric purity determined by differential scanning calorimetry and by chiral HPLC [4]).

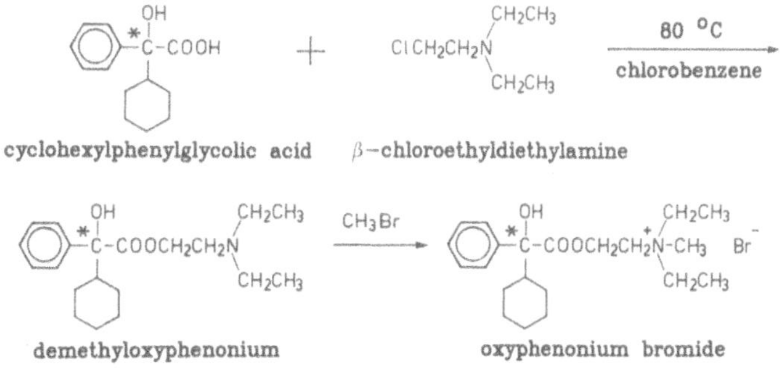

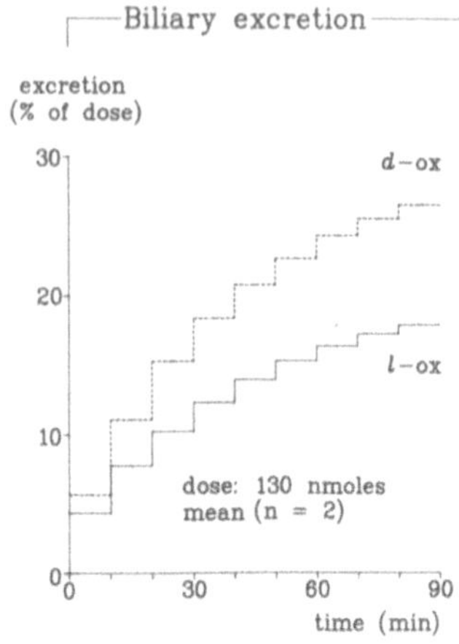

Fig. 2. Cumulative biliary excretion (*in vivo*) after a bolus injection of oxyphenonium enantiomers (two rats, body wt. ~250 g).

For this synthesis, racemic cyclohexylphenylglycollic acid was resolved by crystallization of the diastereomeric salts with (-)- and (+)-ephedrine. The laevo- and dextro-acids were carefully esterified with β-chloroethyldiethylamine and then methylated, resulting in (+)- and (-)-oxyphenonium bromide respectively. The enantiomeric purity of the products was deduced from that of the acid enantiomers, being 99.6% and 99.9% respectively.

Since oxyphenonium is highly potent, the usual dose is relatively low (~25 μg/kg). The enantiomers were therefore radiolabelled for the study of the disposition in the intact rat and the isolated perfused rat liver. This was easily achieved by using $^{14}CH_3I$ in place of methyl bromide for the methylation. These radioactive enantiomers of oxyphenonium were used for the pharmacokinetic experiments.

PHARMACOKINETIC EXPERIMENTS

In the intact rat the bile duct, urinary bladder and carotid artery were cannulated and the lumen of the small intestine was perfused [5]. The radioactivity in the body fluids was measured over 90 min after a bolus injection of (+)- or (-)-oxyphenonium. Recovery was determined from the excretory data and from the radioactivity in some relevant tissues.

For (+)-oxyphenonium the total excretion of radioactivity in bile was considerably higher than for (-)-oxyphenonium (Fig. 2). However, differences in kinetics were partially obscured by the pharmacodynamic effects of the eutomer on renal blood circulation and glomerular filtration, especially after relatively high doses.

In the isolated perfused rat liver [6], in accordance with the *in vivo* results, the biliary output was larger for (+)- than for (-)-oxyphenonium. Hepatic uptake rates were similar, but total biliary

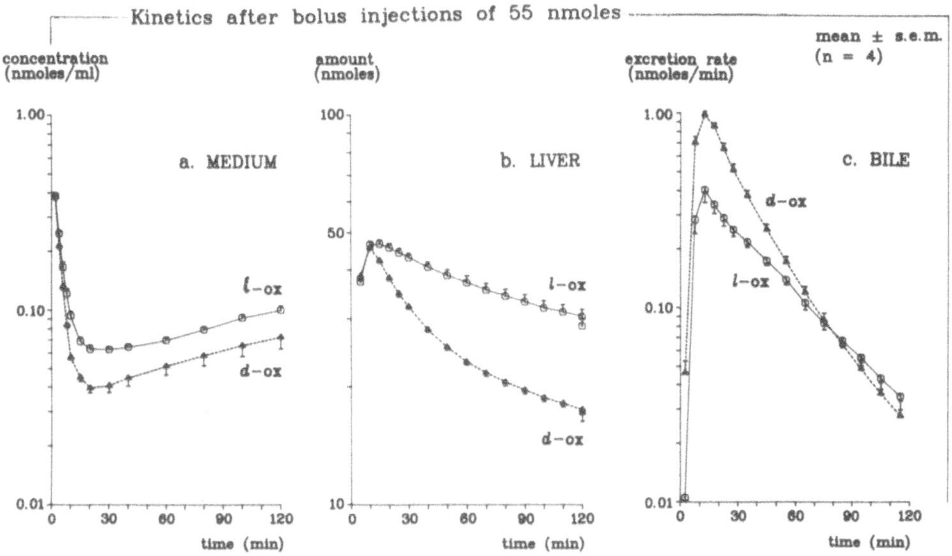

Fig. 3. Kinetics of oxyphenonium enantiomers in the isolated perfused rat liver. Radioactivity measured (Bq) was converted to nmol oxyphenonium disregarding the presence of metabolites. The biliary excretion rate is plotted at each sampling-interval mid-point.

excretion of the laevo-antipode was 55% of the excretion of the dextro-antipode. In line with these data, after 2 h only 30% of the dose of the latter and >50% of the former was still found in the liver (Fig. 3).

The metabolite profile was investigated by TLC using an ion-pair system. In urine only the parent compound was found. In contrast, at least two metabolites were detected in bile for both enantiomers; but unchanged (-)-oxyphenonium persisted longer than the (+)-isomer (Fig. 4).

It is clear that considerable differences in disposition exist between the enantiomers of oxyphenonium, especially in bile. The results indicate a more rapid canalicular transport and/or a more rapid metabolism of (+)-oxyphenonium to cholephilic metabolites.

CONCLUSIONS

For studies of pharmacokinetics of enantiomers, it is preferable to determine them side-by-side in the same sample. Although this was not feasible for oxyphenonium, we were able to gather some information by synthesis from separated enantiomers of a starting compound and individually administering radiolabelled oxyphenonium enantiomers thereby obtained. Their disposition showed differences, especially

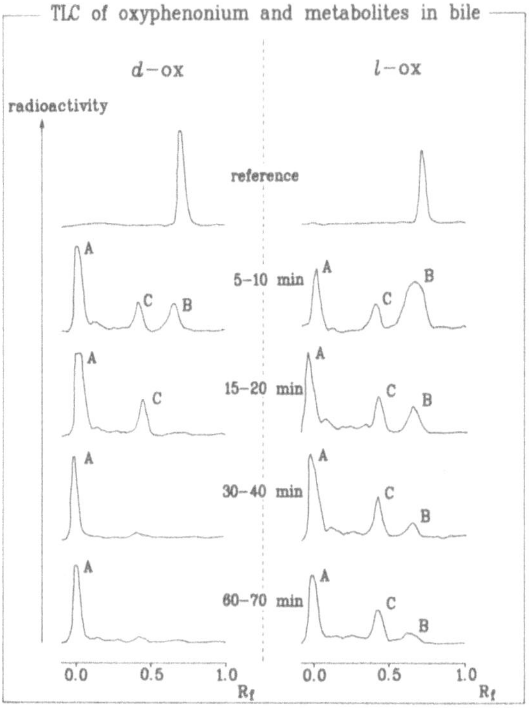

Fig. 4. TLC radioscans of bile samples after introducing ^{14}C-oxyphenonium enantiomers into an isolated rat liver perfusate. Blank bile was spiked with ^{14}C-oxyphenonium. Peak areas are not corrected for differences in scan speed.
TLC system: 20 × 20 cm plates, Silicagel G (Merck), developed with methanol/chloroform (80:20 by vol.) containing 0.2 M sodium perchlorate.

at the hepatic level. It was also evident that the pharmacodynamic characteristics of individual enantiomers may render their pharmacokinetic behaviour unclear.

References

1. Feitsma, K.G. (1987) *Ph.D. Thesis*, Univ. of Groningen, Groningen.
2. Souter, R.W. (1985) *Chromatographic Separation of Stereoisomers*, CRC Press, Boca Raton, FL, 241 pp.
3. Feitsma, K.G., Drenth, B.F.H., Kooi, K., Bosman, J. & de Zeeuw, R.A. (1986) *Method. Surv. Biochem. Anal. 16*, 259-269 [i.e. this series (Reid, E., et al., eds.), Plenum, New York].
4. Feitsma, K.G., Drenth, B.F.H. & de Zeeuw, R.A. (1987) *J. Chromatog. 387*, 447-452.
5. Neef, C., Oosting, R. & Meijer, D.K.F. (1984) *Nauyn-Schmiedeberg's Arch. Pharmacol. 328*, 103-110.
6. Meijer, D.K.F., Keulemans, K. & Mulder, G.J. (1981) *Meths. Enzymol. 77*, 81-94.

#NC(E)-2

A Note on

QUANTITATION OF DANAZOL IN SERUM USING HPLC COLUMN SWITCHING WITH UV DETECTION

H.M. Hill and K. Selinger

Bioanalytical Research and Development
Bio-Research Laboratories
Senneville, Quebec, Canada HQX 3R3

Requirement	A specific HPLC assay sensitive enough to define the pharmacokinetics of danazol (an anterior pituitary depressant) following a 200 mg dose, without interference from other (endogenous) steroids.
End-step	HPLC with column switching and detection at 285 nm.
Sample handling	Danazol is extracted from serum [cleaner for this assay than plasma; cf. #NC(E)-10] using pentane/dichloromethane mixture. The residue from drying down is dissolved in mobile phase and injected onto a 5 µm C-8 column. The fraction containing the internal standard and drug is switched onto a 5 µm C-18 column.
Comments	So as to separate compounds closely related to danazol a mobile phase with a relatively low proportion of component is required. However, this usually results in lengthened chromatography times due to the presence of strongly retained material. The use of column switching provides a solution to this dichotomy. For each new reagent batch it is essential to evaluate the segment time for switching the eluant from the C-8 to the C-18 column. In any system using column switching it is important to use pumps with highly reproducible flow rates and to ensure batch-to-batch reproducibility of mobile phases and columns. The lower limit of quantitation of the assay is 1 ng/ml.

#NC(E)-3

A Note on

SIMULTANEOUS DETERMINATION OF ETRETINATE, ACITRETIN AND 13-*cis*-ACITRETIN IN PLASMA BY GRADIENT-HPLC USING AUTOMATED COLUMN SWITCHING

R. Wyss and F. Bucheli

Pharmaceutical Research, Preclinical Development
F. Hoffmann-La Roche & Co. Ltd.
CH-4002 Basel, Switzerland

Etretinate (Tigason®), an aromatic retinoid ethyl ester, is an oral dermatological drug which is effective in the treatment of psoriasis and other keratinizing disorders [1]. It is hydrolyzed to the corresponding acid *in vivo* [2]. This metabolite, acitretin, (Neotigason®), has a much shorter elimination half-life and shows the same activity in clinical studies; it will succeed etretinate as a drug in the near future. It furnishes 13-*cis*-acitretin by isomerization *in vivo* [2]. Therapeutic drug monitoring is important for this class of compounds because of teratogenicity problems and other side-effects; however, sample handling is difficult in view of the light-sensitivity and the very high protein binding (>99.9%) of the retinoids. Accordingly, an automated method involving gradient elution, column switching and UV detection at 360 nm similar to that described for isotretinoin and its metabolites [3], has been developed for the determination of etretinate, acitretin and 13-*cis*-acitretin.

Plasma proteins are precipitated with ethanol containing two internal standards, an ethyl ester for the quantification of etretinate and 13-*cis*-retinoic acid for the two metabolites. From the supernatant, 0.5 ml is injected onto a pre-column (17 × 4.6 mm) filled with C-18 Corasil (37-53 µm). Polar plasma components are washed out to waste using 1% ammonium acetate and 1% acetic acid/acetonitrile (8:2, by vol.), and the retained components are transferred to the analytical column (125 × 4 mm, protected by a guard column; both filled with 5 µm ODS-silica) in the back-flush mode, and separated by gradient elution (1 ml/min flow-rate) with ammonium acetate/acetonitrile/acetic acid, 300:700:3 for component A and 150:850:1 for B. The separation is achieved by a gradient 100% A to 100% B within 8 min and 100% B for a further 7 min.

The limit of quantification is 2 ng/ml for each analyte. The calibration curve is linear from 2 to 1000 ng/ml at least. The inter-assay precision in the concentration range 20-100 ng/ml is between 0.9 and 4.0% for all compounds. It is for the sake of adequate recovery of etretinate (63%) that ethanol precipitation of protein is performed, rather than inject plasma directly, and the purge mobile phase is not pure water or buffer but contains 20% acetonitrile.

References

1. Bollag, W. & Hanck, A. (1977) *Acta Vitamin. Enzymol. (Milano) 31*, 113-123.
2. Paravicini, U., Stoeckel, K., MacNamara, P.J., Haenni, R. & Busslinger, A. (1981) *Ann.N.Y. Acad. Sci. 359*, 54-67.
3. Wyss, R. & Bucheli, F. (1988) *J. Chromatog. 424*, 303-314.

#NC(E)-4

A Note on

ANALYTICAL CONSEQUENCES OF POTENT DRUGS

Hans de Bree

Analytical Development Department
Duphar Research Laboratories, P.O. Box 2
1380 AA Weesp, The Netherlands

Pharmaceutical companies having their own R & D organization are increasingly confronted with the severe demands of the registration authorities. Besides the proof of safety and efficacy a third criterion is developing: a new drug in a given therapeutic area must be better than existing ones, in respect of therapeutic effect, fewer or no side-effects, or both criteria. This trend has led to the development of a number of rather potent drugs, having low dose regimes. Besides the implications for departments involved in synthesis and pharmacology, there are consequences for other disciplines such as analysis.

This article on bioanalytical strategy is centered on two potential drugs for which an instrumental method had been developed (allowing low ng/ml levels to be determined) during the animal pharmacokinetic studies but for which the appropriate human dose turned out to be 100 times lower than the animal doses, the plasma levels being accordingly lower. Eventually the goal of much greater sensitivity was achieved using GC with selective detection following an obligatory thorough clean-up by HPLC. For both potential drugs, whose structures (both being basic molecules) are on p. 276, the details of the chosen methods are described elsewhere in this book, and are merely outlined here. A simple adaptation of the existing methods to obtain a 10-fold gain in the sensitivity was out of the question; morever, the time factor precluded setting up an ultra-sensitive immunoassay. Therefore it was decided to develop at least two ultra-sensitive instrumental assay approaches for each compound.

Flesinoxan *(formula: p. 276)*

For this antihypertensive compound the earlier assay was described in a previous volume [1]; it was based on RP-HPLC with fluorescence detection using an 'AASP' (cf. [2], and #D-5 in this vol.: McDowall), and with the refinement of a SepPak C-18 pre-enrichment step a sensitivity of ~2 ng/ml was attainable. However, the eventually adopted doses of only a few µg/kg resulted in peak plasma levels of ~1 ng/ml, calling for a determination limit of 100 pg/ml in single-dose human pharmacokinetic studies.

Trial of HPLC with fluorescence detection.- To introduce a strong fluorphore into the molecule, the alcohol moiety was dansylated, or the primary amino group formed by acid hydrolysis of the amide bond was converted to a fluram® derivative. Both fluorescent derivatives improved the sensitivity only marginally, to ~1 ng/ml.

Capillary-GC with selective detection [details: #NC(D)-1].- Since flesinoxan is rather polar to be amenable to GC, we tried derivatization to a more amenable structure: acid hydrolysis of the drug's amide bond resulted in 4-FBz which, after conversion to the PFB-benzoate, was well suited to GC and to ECD®. The initial steps were:
- solvent extraction of drug and isomeric i.s. from plasma or urine;
- NaOH washing to remove metabolites producing 4-FBz on hydrolysis;
- HCl hydrolysis to give FBz's;
- derivatization with PFBB in acetone + triethylamine.

Yields in these steps were >80%; but then began the struggle against the excess of PFBB reagent and co-derivatized endogenous materials. Excess PFBB could be removed elegantly by reaction with ethylamine and acid-extraction of the resulting secondary amine. Removal of the neutral GC-interfering 'rubbish' called for silica-column HPLC clean-up [3], which could be largely automated. The drug and i.s. derivatives co-elute, but the subsequent capillary GC separates the isomers [Fig. 2 in #NC(D)-1]. With the determination limit attained, 100 pg/ml, the assay goal was achieved.

Idaverine *(formula: p. 276)*

Early in the development of this prospective drug for therapy of Irritable Bowel Syndrome, assay of the drug and its ***N***-desmethyl metabolite entailed solid-phase extraction followed by capillary-GC with nitrogen-selective detection. It served for animal pharmacokinetic studies, being sensitive down to ~1 ng/ml. However, with adoption of a dose of ~10 µg/kg in human studies, giving peak plasma levels of a few ng/ml, sensitivity down to 100 pg/ml was needed.

Derivatization and GC-ECD (as for flesinoxan) [details in #E-4].- The initial steps based on the successful assay for flesinoxan comprised amide bond hydrolysis and derivatization of the resulting substituted

---

®*Abbreviations.*- ECD, electron-capture detection; FBz, fluorobenzoic acid; PFB(B), pentafluorobenzyl (bromide); i.s., internal standard; t_r, retention time.

butyric acid with PFBB. Since some metabolites produce the same butyric acid as the drug, they had to be separated initially. This was achieved by HPLC. Acid hydrolysis and PFBB-derivatization could also be accomplished. The overall approach, however, turned out to be non-feasible because of the lack of an i.s. with similar physico-chemical properties and of severe problems in cleaning up the derivatives.

GC-NPD with improved pre-treatment.- Optimization of the existing procedure as an alternative approach offered better prospects, especially since an efficient HPLC isolation step following the initial solid-phase extraction had been developed for the other option. That step separated the drug and metabolites not only from each other but also from the bulk of co-extracted endogenous material that otherwise would have jeopardized GC performance at high sensitivity. The use of a silica column with a rather special mobile phase, acetonitrile/dimethylformamide/triethylamine, yields a very stable system in which even pg amounts can be cleaned up with no evident irreversible adsorptions. Each analyte and its i.s. (a homologue added in ng amount) have the same t_r [Fig. 3 in #E-4]. The analyte-containing HPLC fraction is dried down and dissolved in 15 µl acetone; injection on-column (bonded-phase fused-silica) of 3 µl corresponds to 2 ml plasma, the original plasma sample being 10 ml. For both parent drug and metabolite the determination limit is 100 pg/ml.

CONCLUSIONS

Low plasma levels generally result from the administration of the low doses applicable to novel drugs of high potency. The clinically relevant dose range becomes predictable only after the first human safety and tolerance studies. The instrumental assay developed to assess the animal pharmacokinetics may be too insensitive for assessment of human pharmacokinetics. Important consequences for biomedical assay development are evident from the two cases described here. It appeared that the currently available fluorescence detectors for HPLC are still insufficiently sensitive to quantitate sub-ng levels. On the other hand, the two examples testify to the renaissance of GC with bonded-phase fused-silica technology. That technology permits GC of polar compounds even at temperatures of >300°, allowing sensitive detection by NPD or ECD. For on-column injection of µl amounts an efficient clean-up remained necessary. HPLC clean-up with a silica column is a powerful tool in this respect, especially in the mode of automated injection-collection.

Further development of the procedures that led to the eventually adopted assay for the each of the two drugs took about 6 months. The analysis time in the final method was 30 min longer than for the initial methods, which were for samples with higher drug levels.

References

1. Ceelen, P.R.J., Ruijten, H.M. & de Bree, H. (1986) in *Bioactive Analytes, including CNS Drugs, Peptides and Enantiomers* [Vol. 16, this series] (Reid, E., Scales, B. & Wilson, I.D.), Plenum, N. York, p. 297.
2. Pearce, J.C., Jelly, J.A., Fernandes, K.A., Leavens, W.J. & McDowall, R.D. (1986) *J.Chromatog. 353*, 371-378.
3. de Ridder, J.J., Koppens, P.C.J.M. & van Hal, H.J.M. (1977) *J. Chromatog. 143*, 281-287.

CHEMICAL STRUCTURES OF THE DRUG ANALYTES:

Flesinoxan

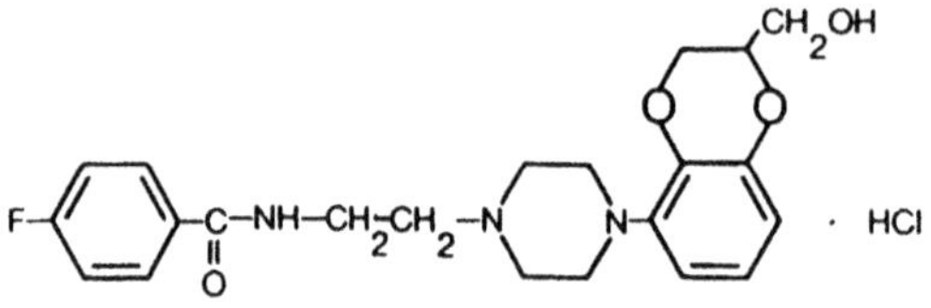

Idaverine

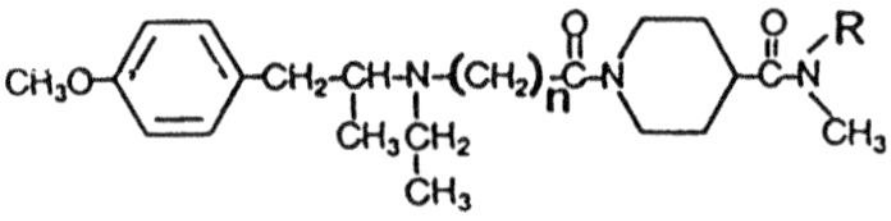

#NC(E)-5

A Note on

OPTIMIZATION OF CAPILLARY GC CONDITIONS FOR THE ANALYSIS OF PSYCHOTROPIC DRUGS AT SUBNANOGRAM CONCENTRATIONS IN PLASMA

H.M. Hill, D. Lessard and L. Letarte

Bioanalytical Research and Development
Bio-Research Laboratories
Senneville, Quebec, Canada HQX 3R3

Following a single therapeutic dose many psychotropic drugs are present in plasma at sub-ng concentrations. As a consequence many of the pharmacokinetic parameters of these drugs have been derived from steady-state data using relatively insensitive analytical methodology. In order to gain pharmacokinetic data from a single dose it is necessary to develop a routine assay with high sensitivity and specificity. This requires optimization of chromatographic and extraction conditions, as now briefly illustrated for two psychotropic agents - thiothixene (TTX), and flurazepam together with its major metabolites desalkylflurazepam (DAF) and hydroxyethylflurazepam (HEF).

The assay for TTX requires quantitation down to 50 pg/ml for cis-TTX and chromatographic separation from trans-TTX, the inactive isomer. DAF, flurazepam and HEF have to be measurable down to 0.5, 0.2 and 1.0 ng/ml respectively, with separation from 5 other metabolites. To achieve these levels of quantitation it was essential to obtain a clean extract. This necessitated the use of a number of back-extraction steps and of clean solvents. For the assay of flurazepam and its metabolites it was necessary to develop extraction procedures which extracted both the relatively non-polar parent compound and its polar metabolite while still ensuring a clean extract.

Both assays entailed use of capillary columns, with NPD for TTX and ECD for flurazepam and its metabolites. For TTX the column type (DB-5), flow rates, sample size, column length and split ratios were optimized to maximize sensitivity. A t_r of 1.56 min for the parent compound and an overall run time of 2 min were achieved. Similar considerations applied to flurazepam assay. The column (DB-210) was chosen to permit separation of polar and non-polar compounds while still providing good peak shape and an overall run time of 8 min.

In attempts to reproduce assays at this level of sensitivity it is essential to pay attention to (and record) every detail of the assay. One aspect rarely evaluated is the interchangeability of equipment for the assay of TTX. On comparing two Hewlett Packard autoinjectors, it was found essential to use Model 7672A; use of the new fast injector HP 7673A produced distorted peak shapes.

#NC(E)-6

A Note on

COMPARISON OF AN ENZYME METHOD WITH A CAPILLARY GC METHOD FOR RED BLOOD CELL SORBITOL ASSAY

R.P.B. Passas and D. Stevenson

Robens Institute, University of Surrey
Guildford, Surrey GU2 5XH, U.K.

The enzyme aldolase reductase is believed to play a crucial role in the long-term complications of diabetes. This enzyme converts glucose to sorbitol and is present in several tissues and also in erythrocytes. At elevated glucose levels (as in diabetes) the pathway is activated and sorbitol accumulates. The determination of erythrocyte sorbitol thus provides a convenient indicator of aldose reductase activity. Several methods are capable of measuring sorbitol including the enzyme sorbitol dehydrogenase (SDH) [1], HPLC [2], both packed- [3] and capillary-column [4] GC; all have been tried in this laboratory.

The HPLC method requires derivatization with *p*-nitrobenzoyl chloride to enhance detection but gives multiple peaks when used with real samples. Packed-column GC of both trimethylsilyl and acetate derivatives did not give adequate resolution of sorbitol, galacitol and mannitol and in any event has little cost benefit in comparison with capillary GC. Both the SDH method and capillary GC of acetate derivatives have been validated and used by us. The methods are now outlined.-

Collect blood, separate and wash erythrocytes; precipitate proteins with perchloric acid, then neutralize supernatant with K_2CO_3 and spin.

SDH method:

- Add glycine buffer (pH 9.4), NAD and SDH.
- Incubate at 25° and read NADH fluorimetrically.

GC method:

- Add internal standard (galacitol), and evaporate down under vacuum.
- Derivatize with acetic anhydride/pyridine, and extract into chloroform after adding water to stop the reaction.
- Evaporate down.
- Re-dissolve residue in butyl acetate for capillary GC (conditions: Fig. 1).

Fig. 1. Specimen chromatogram for sample from a patient. Column: 25 m × 0.3 mm; BP 1, 1 µm. Carrier: He at 20 psi inlet, 1 min splitless; N_2 make-up. 110° (1 min) → 230° (10 min), 40°/ min. Injection at 250°. Detection: FID at 300°.

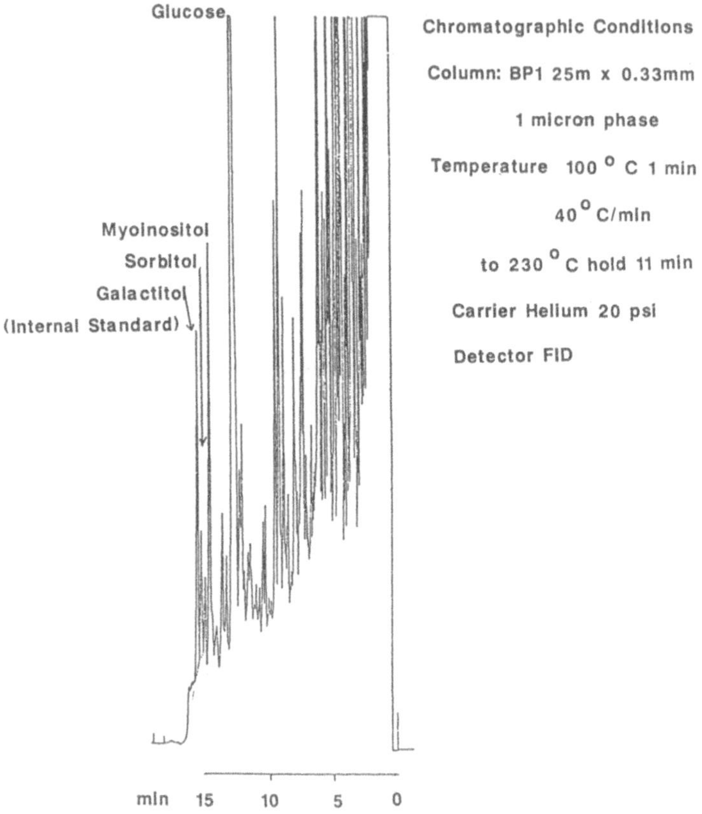

In order to preserve analyte integrity it is important that samples be taken to the perchloric stage within 1 h of blood collection. A typical GC pattern is shown in Fig. 1. The SDH method has the advantages of speed and simplicity but measures only one component, and the enzyme is not highly specific: xylitol, ribitol and xylose all act as substrates. The capillary GC method is more time-consuming but allows measurement of other polyols and carbohydrates.

Fig. 2 shows the results of a comparison of the two methods. The SDH method broadly accords with the more specific GC method whilst, not surprisingly, giving slightly higher values. The correlation between the two methods is 0.87 with a slope of 1.17. When only sorbitol is to be measured, e.g. for screening the efficacy of aldose reductase inhibitors, the SDH method has proven suitable; likewise for routine screening of large numbers of patients. Both methods are sensitive down to ~2 ng/ml.

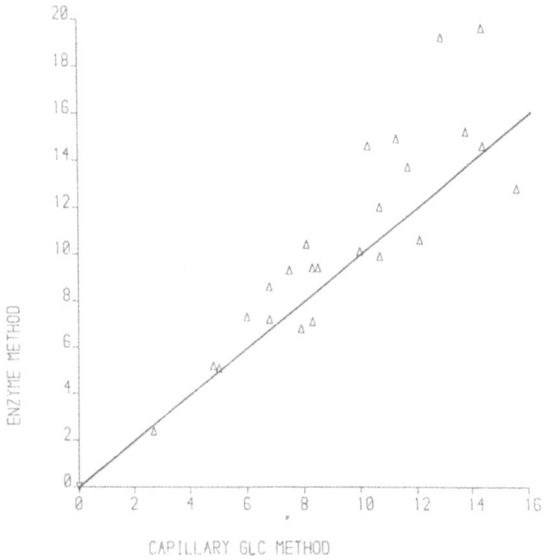

Fig. 2.
Method comparison.
The line through the points has a slope of 1.
The samples were from diabetic patients.

Acknowledgement

The authors gratefully acknowledge the financial support of Pfizer Central Research, Sandwich, Kent, for supporting part of this project.

References

1. Malone, J.I., Knox, G., Benford, S. & Tedesco, T.A. (1980) *Diabetes 29*, 861-864.
2. Nachtmann, F., & Budna, K.W. (1977) *J. Chromatog. 136*, 279-287.
3. Mount, J.N. & Laker, M.F. (1981) *J. Chromatog. 226*, 191-197.
4. Popp-Snijders, C., Lomecky, M.Z. & de Jong, Ad.P. (1983) *Clin. Chim. Acta 132*, 83-89.

#NC(E)-7

A Note on

THE STABILITY OF ANTINEOPLASTIC VINCA ALKALOIDS IN PLASMA AND URINE

D.E.M.M. Vendrig, A. Mekking, J. Teeuwsen and J.J.M. Holthuis

Department of Pharmaceutical Analysis
University of Utrecht, Catharijnesingel 60
3511 GH Utrecht, The Netherlands

Vinblastine and vincristine (Fig. 1) are naturally occurring vinca alkaloids present in the Madagascar Periwinkle plant. Vindesine is a semi-synthetic vinca alkaloid, derived from vinblastine. The vinca alkaloids are active against various types of cancer, e.g. Hodgkin's disease, lymphomas and testicular carcinoma [1].

Until recently, pharmacokinetic studies were mainly carried out using a RIA [2-4]. However, with this technique it is not possible to discriminate between the parent compound and its metabolites. Therefore a sensitive and selective HPLC method was developed, enabling the drugs and a vinblastine metabolite, desacetylvinblastine, to be determined at low ng/ml levels in plasma and urine [5].

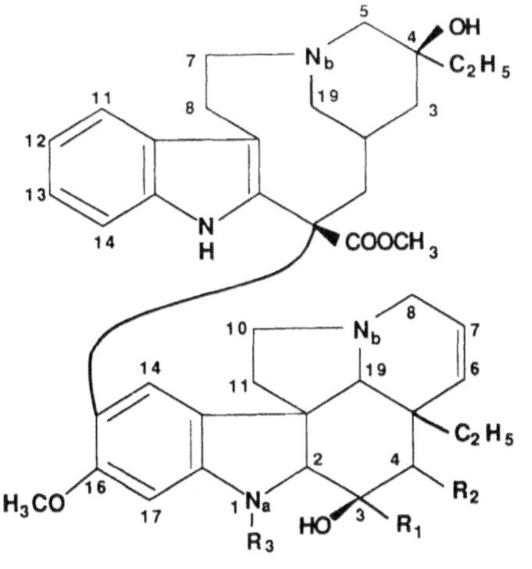

	R_1 (3)	R_2 (4)	R_3
VINCRISTINE	$COOCH_3$	$OCOCH_3$	CHO
VINBLASTINE	$COOCH_3$	$OCOCH_3$	CH_3
VINDESINE	$CONH_2$	OH	CH_3
DESACETYLVINBLASTINE	$COOCH_3$	OH	CH_3

Fig. 1. Chemical structures of vincristine, vinblastine & vindesine.

Because samples are not usually analyzed immediately after collection, the influence of different conditions during periods of storage on the concentration of the 3 compounds was investigated in both plasma and urine. Sometimes samples had to be analyzed more than once on different days; also, frozen samples may inadvertently thaw during transport. Therefore we also studied the influence of repeated freezing and thawing on the concentration of plasma and urine spiked (100 ng/ml) with the different vinca alkaloids.

EXPERIMENTAL

Chromatography (details in [5, 6]).- Briefly, a Hypersil ODS column (5 µm; 100 × 3.0 mm i.d.) was used at ambient temperature, with a methanol/10 mM pH 7.0 phosphate buffer (57.5 : 42.5 by wt.) flowing at 1.0 ml/min. Detection was electrochemical, at +850 mV *vs.* Ag/AgCl (3 M KCl).

Chemicals.- The three alkaloids, as sulphates (Velbe[R], Oncovin[R], Eldisine[R]), were obtained from Eli Lilly Nederlands B.V. Methanol (*pro analyse*, Merck, Darmstadt) and deionized water were distilled and stored in glass. All other chemicals were reagent grade and used as received. Plasma and urine were from healthy volunteers and stored at -20°.

Extraction.- Vincristine was used as internal standard (i.s.) in the determination of vinblastine, desacetylvinblastine and vindesine; for vincristine, vinblastine was the i.s. Before centrifugation (10 min, 3000 **g**) 5-20 µl i.s. in methanol (1-15 ng/µl) were added to 1.0 ml of sample. The alkaloids were extracted using 1 ml BondElut CN columns with a VacElut Processing Station (Analytichem Internatl.). After equilibration with 5 ml methanol and 5 ml 50 mM tetramethylammonium bromide in 10 mM pH 4.5 phosphate buffer, and sample loading, the column was washed with 5 ml methanol/water (20:80 by vol.) followed - ensuring no drying-out - by 2 ml 25 mM pH 7.0 phosphate buffer. Then, after drying the column by air purging, the alkaloids were eluted with 500 or 750 µl methanol, depending on the batch of extraction columns used. The extracts were evaporated to dryness (N_2 stream, at room temp.). After dissolution in 100 µl of the mobile phase and vortexing for 30 sec, 20-30 µl was chromatographed. Figs. 2 (plasma) & 3 (urine) show typical chromatograms. Comparisons of concentration were based on peak-height ratios (*vs.* freshly spiked reference samples).

Stability testing.- Drug-spiked plasma or urine (1 µg/10 ml, in polypropylene tubes) was investigated for analyte stability when kept at 37° (plasma only; n = 3), 4° (refrigerator; n = 3) or -20° (n = 4) or when repeatedly frozen and thawed (n = 4). Aliquots (1 ml) were taken finally or, in the -20° study, were set up at the outset. The aliquots incubated at 37° were put aside at -20° until the assay.

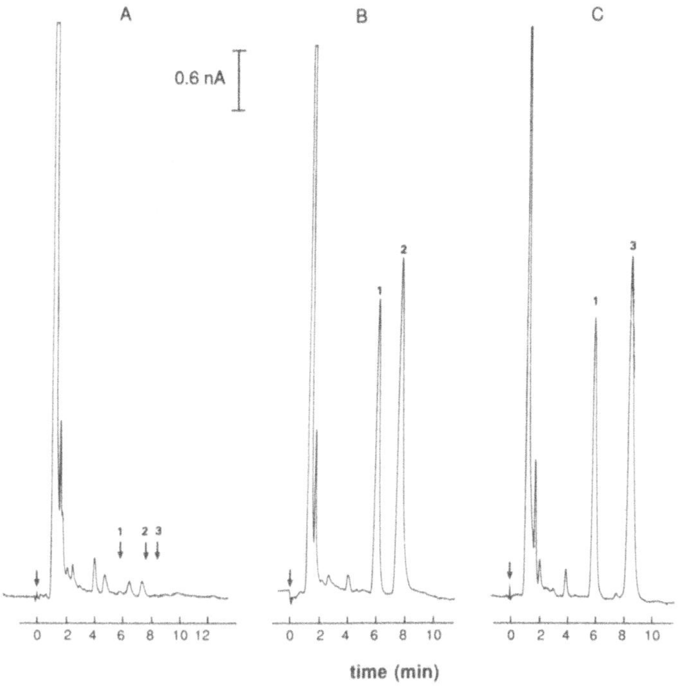

Fig. 2. Chromatograms of extracts of 1.0 ml plasma. **A**, Blank plasma. Plasma spiked to 100 ng/ml: **B**, vindesine; **C**, vinblastine. **1**, vincristine (i.s.); **2**, vindesine; **3**, vinblastine.

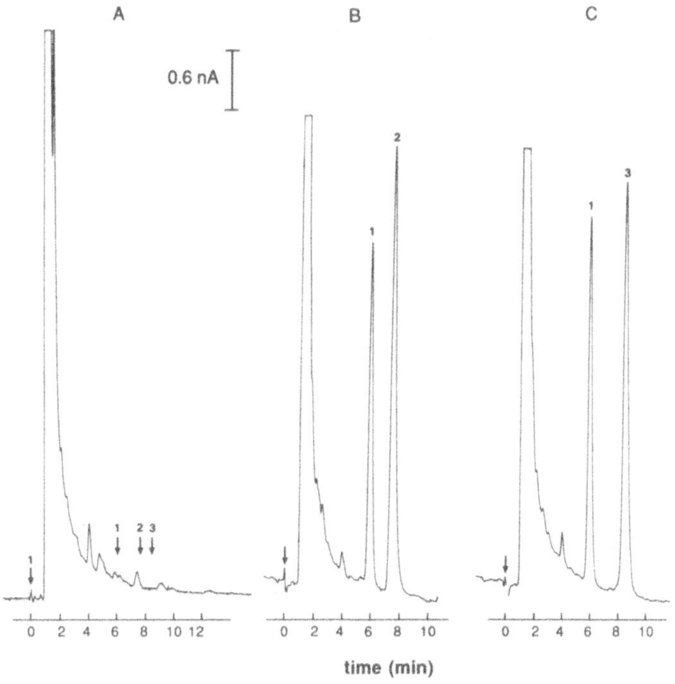

Fig. 3. As for Fig. 2, but urine instead of plasma extracts.

Fig. 4. Recoveries of vinblastine (□), vincristine (■) and vindesine (Δ) in plasma after incubation at 37°.

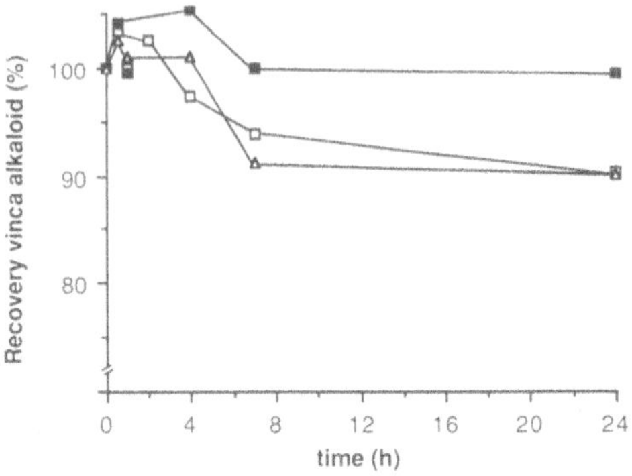

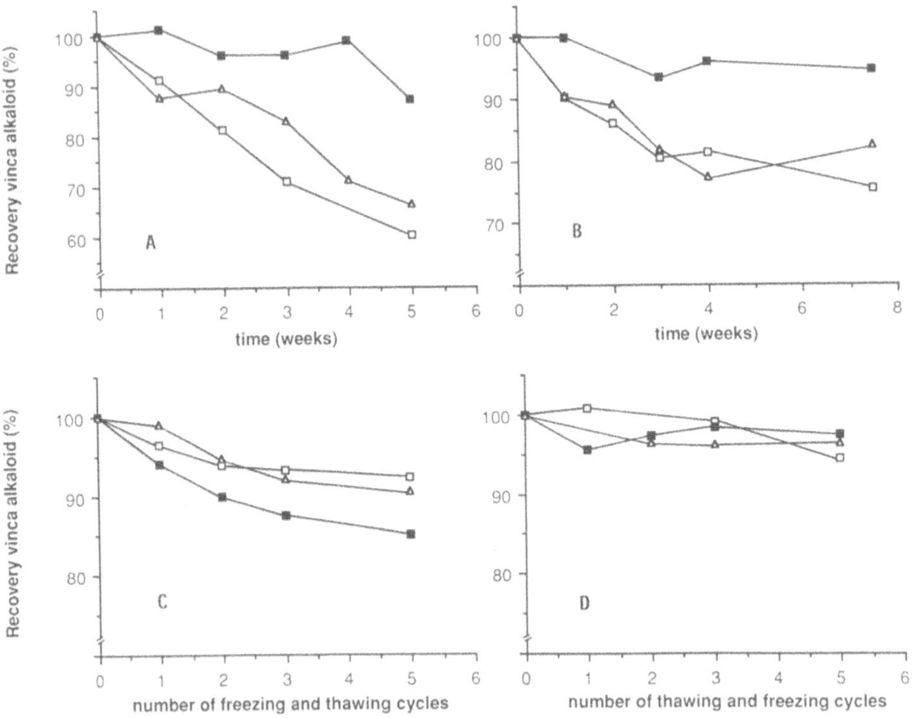

Fig. 5. Recoveries of the analytes (symbols as in Fig. 4) from plasma (**A, C**) and urine (**B, D**) after storage at 4° (**A, B**) or after repeated freezing and thawing of the samples (**C, D**).

RESULTS AND DISCUSSION

Extraction [5].- Briefly, linearity was found with 1-750 ng vinca alkaloid/ml urine and 1-100 ng/ml plasma; recoveries were ~80%. Intra- and inter-day C.V.'s were acceptably low (<5.7%).

Stability in plasma at 37°.- Fig. 4 shows that vinblastine and vindesine decreased to ~90% of the original concentration after 7 h, but vincristine was unchanged during 24 h. Since in aqueous solutions at 37° vinblastine [7] and vindesine degrade more slowly (cf. the 20° half-life, 1050 days, for vinblastine at plasma pH, 7.0 [7]), the concentration decrease could be due to adsorption onto plasma components that are seen to precipitate during the incubation. Since the amount of precipitate depends on the sample, differences between the compounds in the concentration decrease might be thus explained.

Stabilities at 4° and -20°.- In plasma at 4° vinblastine and vindesine but not vincristine fell by 10% within 1 week (Fig. 5A). In urine (pH 6.0) at 4°, vincristine fell by a mere 5% after 4 weeks, whereas vinblastine and vindesine fell by 10% in 1 week (Fig. 5B): evidently urine should be stored frozen. Low extraction recoveries from plasma or urine kept at 4° may reflect evident growth of bacteria or fungi. At -20° for up to 16 weeks all compounds show 100% (±5%) survival.

Thawing and freezing of plasma (5 cycles, over 1 week) led to 92% recovery for vinblastine, 85% for vincristine, 91% for vindesine (Fig. 5C) - probably reflecting adsorption onto precipitated material; with 20 or 500 ng/ml the results were similar. Urine gave higher recoveries - 95%, 98% and 97% (Fig. 5D).

CONCLUSIONS

Following solid-phase extraction of plasma or urine, each alkaloid can be assayed down to 1 ng/ml by HPLC with electrochemical detection. All three alkaloids survive storage of the plasma or urine at -20° for at least 16 weeks. Thawing and freezing (× 5) causes 5-10% loss in plasma but not in urine.

References

1. Creasey, W.A. (1981) in *Cancer and Chemotherapy*, Vol. 3 (Crooke, S.T. & Prestayko, A.W., eds.), Academic Press, N.York, pp. 79-91.
2. Sethi, V.S. & Kimball, J.C. (1981) *Cancer Chemother. Pharmacol.* *6*, 111-115.
3. Van den Berg, H.W., Desai, Z.R., Wilson, R., Kennedy, G., Bridges, J.M. & Shanks, R.G. (1982) *Cancer Chemother. Pharmacol.* *8*, 215-219.
4. Fuks, J.W., Egorin, M.J., Aisner, J., Van Echo, D.A., Ostrow, S., Bachur, N.R. & Wiernik, P.H. (1983) *Cancer Chemother. Pharmacol.* *10*, 104-108.

5. Vendrig, D.E.M.M., Teeuwsen, J. & Holthuis, J.J.M. (1988) *J. Chromatog.* *424*, 83-94.
6. Vendrig, D.E.M.M., Holthuis, J.J.M., Erdélyi-Toth, V. & Hulshoff, A. (1987) *J. Chromatog.* *414*, 91-100.
7. Vendrig, D.E.M.M., Smeets, B.P.G.M., Beijnen, J.H., van der Houwen, O.A.J.G. & Holthuis, J.J.M. (1988) *Int. J. Pharm.* *43*, in press.

#NC(E)-8

A Note on

SOLID-PHASE EXTRACTION: PERSONAL EXPERIENCES

Robin Whelpton and Peter R. Hurst

Department of Pharmacology and Therapeutics
London Hospital Medical College
Whitechapel, London E1 2AD, U.K.

The large number of types of phases and column sizes available for solid-phase extraction (SPE) makes this a very powerful and versatile sample preparation technique. As most of the available phases are based on those used in HPLC, the analyst is likely to be familiar with their behaviour. Some columns have been developed for particular applications, e.g. phenylboronic acid phases for catecholamine separation. We prefer to consider SPE as an important addition to the extraction techniques available, not necessarily as a replacement. Because of their similarities, some of the points discussed here are also applicable to column-switching techniques.

The goals of SPE are to retain 100% of the analyte(s) from the sample and then to elute 100% into the minimum volume of eluent giving maximum recovery and concentration. A high degree of selectivity between the analyte and unwanted components in the sample matrix is required to achieve maximum 'clean-up'. We check retention by passing solutions (1 ml) of analyte down activated/buffered columns and examining the eluent for solute. Those columns that retain the drug are eluted with 1 ml portions of methanol, 0.1 M NH_4OH in methanol, 0.1 M acetic acid in methanol, etc., until elution occurs. In this way, columns and systems worthy of further development are quickly discovered. The conditions for maximum selectivity and sensitivity will require consideration of the finer points as discussed below.

COLUMN SIZE AND SHAPE

Our experiences are confined to BondElut columns (Analytichem International), which are available in different sizes and shapes. The widest choice is the C-18 type which is available in the following sorbent weight/reservoir volume combinations: 100 mg/1 ml; 200 mg/3 ml; 500 mg/2.8 ml; 500 mg/6 ml; and 1000 mg/6 ml. We use the smallest size (100 mg) because of several advantages, e.g. low bulk (easy to handle), small elution volumes, lower cost. A particular advantage is that they can be suspended in centrifuge tubes for sample application and elution. In preliminary investigations, where we want to collect all samples and washes for a balance-sheet, we use centrifugation. (Recommendation: label columns as well as tubes.) In applying the developed method, sample and washes are applied under vacuum and the analyte eluted by centrifugation, suitably with the collection apparatus we described in a previous book (Vol. 16, p. 185). After the last water wash we centrifuge, add 50 µl of elution solvent (say methanol) and centrifuge again. To date no analyte we have tried has eluted in this volume which appears to be primarily water (from the silica pores?). The drug is then eluted and collected in the minimum amount of eluate.

Possible disadvantages of small columns are the small reservoir volume (1 ml) and, because of the small cross-sectional area, a tendency to 'clog'. This area is larger in larger columns, the 6 ml reservoir providing a much greater area and reduced depth of bed than the 2.8 ml column containing the same weight of sorbent. Wide shallow beds may be useful if particulates are a problem, but one should check the effect of faster flow rate on retention.

CHOICE OF BUFFER

For convenience we use salts rather than buffers and generally find 0.1 M H_3PO_4, NaH_2PO_4, Na_2HPO_4 or K_3PO_4 suitable; but problems may arise. For example, to retain the quaternary ammonium compound pyridostigmine on C-18, C-8 or C-2 columns a high pH was required. When the compound was eluted with 0.1 M acetic acid in methanol the eluate contained a crystalline solid. This was overcome by using 1 M NH_4OH rather than 0.1 M K_3PO_4.

The use of alkaline buffers may result in dissolved silica in methanol eluates. This may not be a problem unless the methanol is evaporated to dryness. For this reason we try to minimize the volume so as to avoid the need for further concentration, thereby also obviating other problems that arise in solvent evaporation as documented earlier in this series (especially in Vols. 7, 10 & 12, indexed under 'Evaporation' – *Ed.*).

SAMPLE MATRIX EFFECTS

Pure solutions of the analyte, being directly measurable (UV, HPLC, etc.), are the obvious starting-material in developing an assay; but the biological matrix can have a marked effect on retention. For physostigmine, with nitrile (CN) columns (cf. the earlier work we reported in Vol. 16), plasma samples give the best elution profiles (Fig. 1). Urine, which we had reckoned to be the easiest biological

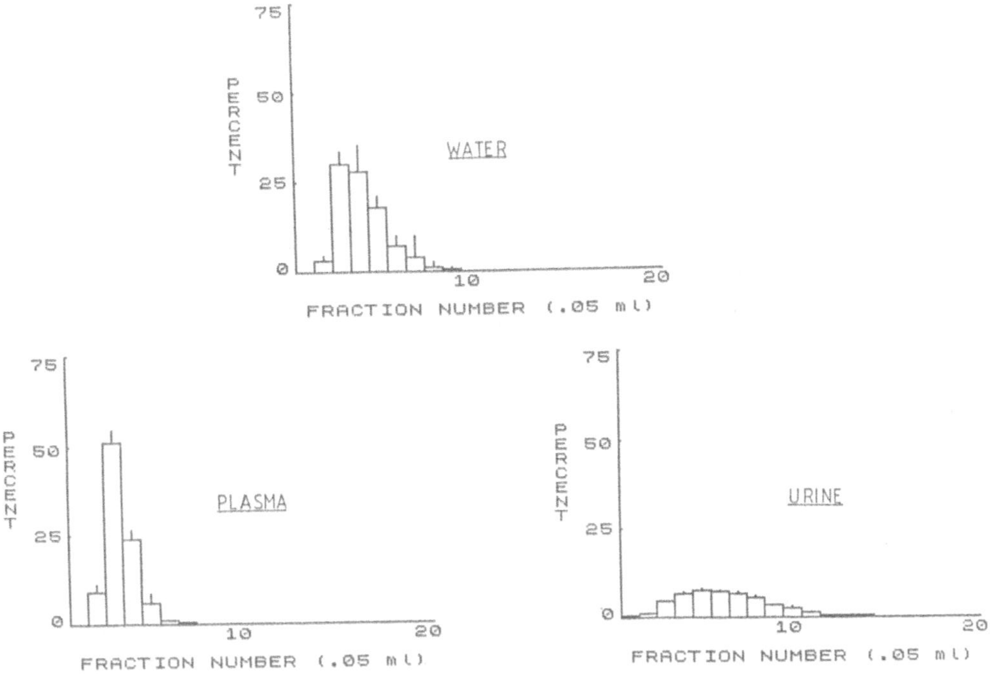

Fig. 1, *above*. Elution of [^{3}H]-physostigmine from 100 mg BondElut CN columns in 50 µl fractions of methanol. (Mean ±S.D., n = 5.)

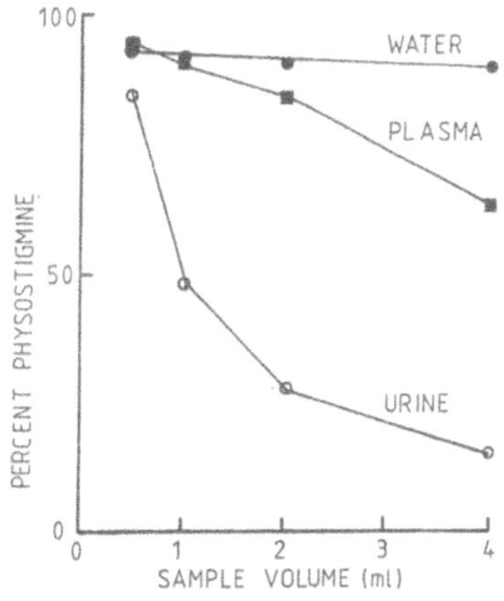

Fig. 2, *right*. Physostigmine recovery as a function of sample type and volume applied to CN columns.

fluid to assay, gave poor profiles and, related to sample size, poor and variable recoveries (Fig. 2). CN columns were chosen as they retained less urinary pigments than C-2, C-8 or C-18. The C-18 columns retained physostigmine and pigments which co-eluted. If these were re-applied in water to a CN column, the drug was retained but not the pigments. The drug-recovery problem was overcome by using only 0.1 ml urine on CN columns. Dilution in water had no effect as the retention on CN was a function of the total *amount* of urine applied, not its concentration.

Table 1. Buprenorphine applied in plasma (or, in parentheses, water) to 100 mg BondElut CN columns: unretained portion *vs.* flow rate.

Flow rate, ml/min	% unretained: C-2	C-8	C-18
10	80.1	55.9	66.5 (3.2)
1.0	62.1 (3.8)	12.5	9.9 (2.5)
0.1	30.5	7.1	5.4

Table 2. Physostigmine/i.s. peak-height ratios determined in 100 µl fractions collected from 5 different BondElut phases.

Fraction, ml	CN	C-18	C-8	C-2	Diol
0 - 0.1	2.3	1.8	1.2	1.3	0.8
0.1-0.2	0.8	0.8	0.8	0.8	1.0
0.2-0.3	1.0	0.3	0.8	0.6	1.5

SPE can be used with whole blood provided that the cells are disrupted first. We pipette blood (1 ml) and water (0.1 ml) into 1.5 ml polypropylene centrifuge tubes and freeze them (e.g. overnight in the freezer). After centrifugation (~15,000 rpm) the supernatant (1 ml) can be applied to prepared columns.

FLOW RATE

In our experience, SPE is tolerant of a wide range of flow rates when analytes are in water. However, problems should be anticipated where drugs are highly protein-bound. The effect of flow rate on the retention of the lipophilic analgesic, buprenorphine, is shown in Table 1. At rates >10 ml/min retention from plasma is poor. On a 100 mg C-18 column only 33% is retained at the fast flow rate, whereas the retention from water is >96% under the same conditions. At 1 ml/min the retention is acceptable on the high-capacity C-18 phase, but even at 0.1 ml/min >30% is lost on C-2 columns. The fact that this is a matrix effect is seen from control experiments using water [(xx) in Table 1]. By careful control of flow rate buprenorphine can be retained on small C-18 columns without protein precipitation.

INTERNAL STANDARDS

Using small eluent volumes to avoid an evaporation step can present problems when using an internal standard (i.s.) as it may be partially resolved from the analyte. The dimethylcarbamate analogue of physostigmine has been shown to be a suitable i.s. when using liquid-liquid extraction [1]. However, it was partially resolved on even the small columns, as evidenced by the peak-height ratios in Table 2 for 3 successive fractions from 5 types of column. On the CN column the i.s. eluted in a narrow band; on C-18, C-8 and

Table 3. Precision (C.V.) of SPE of pyridostigmine without i.s. and with the ethyl or propyl homologue as i.s. Means from n observations.

ng/ml	n	Drug peak ht., mm	C.V.	Ratio, ethyl	C.V.	Ratio, propyl	C.V.
40	6	64.2	7.6%	1.83	7.0%	3.33	5.6%
10	7	16.3	4.6	0.46	4.9	0.86	6.7
5	8	9.3	7.1	0.27	9.5	0.48	7.7

C-2 it eluted after physostigmine and on the diol it eluted first. To obtain a constant and correct ratio, the fraction should be large enough to elute both compounds completely. Obviously a poorly chosen i.s., eluting away from the compounds of interest, will necessitate large elution volumes.

SPE offers high retention with total recoveries >95%. This, coupled with high-precision injection techniques such as loop injection in HPLC, may imply that an i.s. is not required. This is useful when no suitable i.s. is available or there are chromatographic problems: one less compound to resolve and quantify! A comparison of the precision of pyridostigmine assays with and without i.s. illustrates this (Table 3). Ethyl and propyl homologues were each prepared and evaluated as an i.s. but there was little, if any, improvement in the C.V.'s.

TRACE ENRICHMENT*

Concentration of solutes from large volumes of solvent is an obvious advantage of SPE. For example, using 100 mg BondElut C-18 it is usually possible to extract solutes from 10 ml (or more) of urine and elute with no more than 0.5 ml methanol. (WARNING: alkaline urine produces a gelatinous precipitate; as a general rule, samples should be filtered after rather than before pH adjustment.) Thus a 20-fold or greater concentration is achieved in one easy step. As C-18 is not very selective, further purification by SPE or liquid-liquid extraction may be necessary now that the sample has been reduced to a convenient volume. Trace enrichment may be applied to organic solvents, e.g. extracting bases from high-boiling solvents such as toluene onto silica or diol phases. This avoids an evaporation step and the solvent can be recovered for redistillation and re-use if appropriate.

Unfortunately SPE is very good at trace enrichment of impurities. Typically, 6-10 ml of water will be used with each 100 mg reversed-phase column. Impurities with physicochemical properties similar to those of the analyte will be retained and concentrated. This problem, which arose when quantifying sub-ng quantities of physostigmine,

* See this index entry in past vols., notably Vol. 10.- *Ed.*

was solved by passing dist. water through C-18 SepPak cartridges before use. Buffer solutions may need similar treatment.

BATCH VARIATION AND SECONDARY EFFECTS

Considering the differences between C-18 columns from different manufacturers it is not surprising that C-18 cartridges from different sources vary. Just as we might choose Spherisorb ODS-1 to exploit adsorptive effects, one can use secondary interactions of BondElut columns. Indeed, our methods for physostigmine and pyridostigmine require secondary interactions. The pyridostigmine method - retention on alkali-treated columns and elution with acidic eluents - appears to be a form of mild ion-exchange. Retentions of this analyte on C-18, C-8 and C-2 are similar but C-2 columns give the sharpest elution peak and the cleanest extracts from biological fluids.

Intra-manufacturer batch-to-batch variation is another problem as encountered with a batch of CN columns. Physostigmine recovery from water using Batch No. 13119 was <10%, whereas 8 other batches gave >90% total recovery. The 'rogue' batch had a higher organic loading (pers. comm. from R. Calverley, Analytichem Internatl.) which presumably reduced the availability of residual silanols. As manufacturers are now alert to batch-to-batch variation they are improving quality control. The suppliers of BondElut cartridges will endeavour to supply against particular batch numbers.

THROUGHPUT, CONVENIENCE, AND GENERAL ROLE OF SPE

SPE methods can be very much simpler than liquid-liquid techniques; but this is not universally true and choice may depend on the number of samples involved. The Analytichem method for lignocaine in serum is: (1) activate 100 mg C-18 column with 2 × 1 ml each of methanol then water; (2) add buffer with i.s. (0.5 ml), and sample (0.5 ml); (3) wash 3 times with 25% (v/v) methanol; (4) elute with methanol (2 × 0.2 ml); (5) inject aliquot into GC-NPD system. This is a typical simple SPE method and is fine if sample number is low (e.g. 10-20) or if alternative methods are more complicated. With hundreds of samples, batches of 50 can be treated thus: (1) pipette buffer with i.s., serum and toluene (0.5 ml each) into a glass centrifuge tube; (2) stopper, shake mechanically for 10 min, then centrifuge; (3) inject 2 µl organic phase into the GC.

SPE is valuable in sample preparation, either on its own or combined with other methods. It is of major help when extracting 'difficult' compounds: polar, ionized, unstable at pH extremes, etc. SPE developments will help solve such problems [& see Vol. 16 - *Ed.*].

References

1. Whelpton, R. & Moore, T. (1985) *J. Chromatog. 341*, 361-371.
2. Ruane, R.J. & Wilson, I.D. (1987) *J. Pharm. Biomed. Anal.* 5, 723-727.

#NC(E)-9

A Note on

THE USE OF SECONDARY IONIC INTERACTIONS FOR THE SOLID-PHASE EXTRACTION OF SOME 'β-BLOCKING' DRUGS ON C-18 BONDED SILICA

R.J. Ruane, I.D. Wilson and G.P. Tomkinson

Imperial Chemical Industries plc
Pharmaceuticals Division, Mereside, Alderley Park
Macclesfield, Cheshire SK10 4TG, U.K.

A major advance in the preparation of biological samples for chromatographic analysis was the introduction of SPE* [which featured strongly in Vol. 16, this series - *Ed.*]. This methodology involves adsorption of the analyte onto a solid phase (usually a modified silica gel, e.g. C-18 bonded) contained within a disposable cartridge. The compound of interest is then recovered by elution with a suitable solvent. We previously described some unusual and unexpected results obtained with C-18 bonded silica gel SPE cartridges, from two different manufacturers, used for extracting a number of 'β-blocking' drugs from plasma [1]. Here we describe further investigations on a wider range of commercially available C-18 cartridges.

EXPERIMENTAL

Materials.- The ^{14}C-radiolabelled componds, of >95% radiochemical purity, of the β-blocker type (Fig. 1) were synthesized within this ICI Division. The 1cc Bond-Elut C-18 and silica gel SPE cartridges were manufactured by Analytichem International (Harbor City, CA), and C-18 1cc SPE by J.T. Baker (Phillipsburg, NJ). Sep-Pak C-18 cartridges were from Millipore/Waters (Milford, MA), 3cc C-18 Merck 'Adsorbex' from BDH Chemicals (Poole, U.K.), 1cc C-18 cartridges from Supelco (Bellefonte, PA), and 3cc C-18 SPE-ED cartridges from Laboratory Impex (Teddington, U.K.). All other chemicals were of HPLC grade or equivalent.

Cartridge preparation.- All the C-18 cartridges were prepared for use by washing with methanol (2 × 1 ml), then water (1 ml) and

*SPE, solid-phase extraction; TEA, triethylamine. ^{14}C is implied by '(radio)label'.

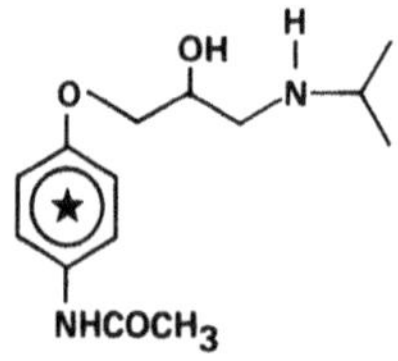

1

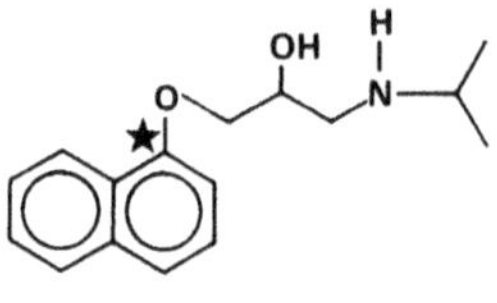

2

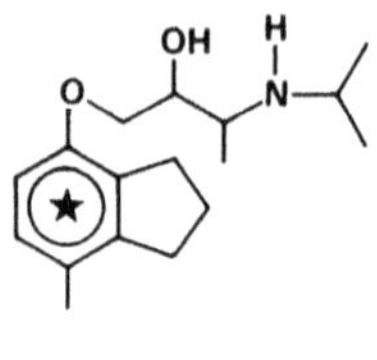

3

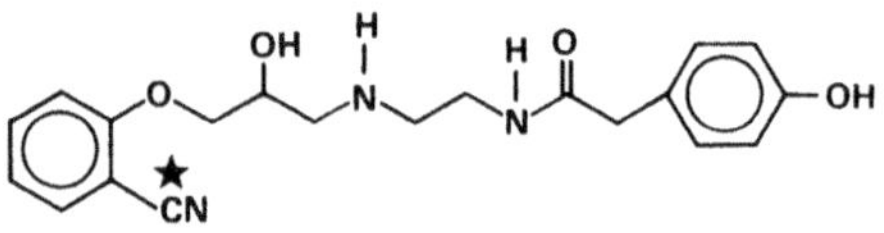

4

Fig. 1. Structures of the test compounds. **1**, ICI 50,172; **2**, ICI 45,520; **3**, ICI 118,551; **4**, ICI 141,292.

finally 0.2 M Na acetate (1 ml) adjusted to pH 5.0 with glacial acetic acid. Silica gel cartridges were treated in the same manner.

Sample preparation.- Samples for extraction were prepared by mixing control dog plasma (0.5 ml) and pH 5.0 Na acetate (0.5 ml) and then spiking with the appropriate radiolabelled β-blocker (10 µg in methanol). The plasma samples were applied to the activated cartridges which were then washed with water (1 ml) followed by acetonitrile (0.5 ml). Analytes were finally recovered from the cartridges with methanol/0.1 M TEA pH 7.0 (80:20 by vol.). Radioactivity present in the various eluates was determined by liquid scintillation counting.

RESULTS AND DISCUSSION

All four test compounds were efficiently extracted from the spiked plasma using the C-18 1cc Bond-Elut cartridges. With the J.T. Baker and Waters C-18 cartridges extraction was less efficient and more variable. However, recovery of radiolabel into acetonitrile when using all these cartridges was much lower than predicted, and efficient recoveries were achieved only with a methanol/TEA buffer mixture. Table 1 summarizes these results. They may be indicative of an ionic interaction between the basic (positively charged) drug molecule and the residual silanol groups (remaining on the silica-gel surface after silylation; negatively charged) rather than the C-18 phase. Support for this hypothesis comes from experiments which showed that one of the compounds (**3**, Fig. 1) was adsorbed onto a silica 1cc Bond-Elut cartridge from either aqueous or methanolic solution and then recovered using TEA acetate.

Table 1. Comparison of three C-18 cartridges with 2 eluents in turn: % recovery (mean ±S.D.; n = 3) of radioactivity applied. See Fig. 1 legend for the test compounds.

	1: Acetonitrile			2: Methanol/pH 7.0 TEA (80:20)		
Cpd.	Waters[⊗]	J.T. Baker[⊗]	Bond-Elut	Waters	J.T. Baker	Bond-Elut
1	27.7 ±6.6	36.5 ±1.1	10.3 ±0.7	62.6 ±12.1	18.1 ±0.8	93.7 ±0.5
2	7.1 ±1.4	31.6 ±2.4	17.0 ±0.6	73.3 ±6.9	48.0 ±4.1	77.3 ±0.5
3	1.7 ±0.7	11.5 ±3.9	0.6 ±0.1	93.8 ±2.6	78.3 ±6.5	94.3 ±2.5
4	6/6 ±0.4	39.7 ±2.0	2.8 ±0.4	83.6 ±1.1	40.0 ±1.1	100.2 ±1.7

[⊗] Where overall recoveries in the 2-step elution were low, the cause was non-retention of radiolabel in loading, not irreversible binding.

Table 2. Comparison of C-18 cartridges from 6 manufacturers with 2 eluents in turn: % recovery of compound **3** (mean ± S.D.; n = 3).

Cartridge	Acetonitrile	Methanol/TEA
Bond-Elut	0.6 ±0.1	94.3 ±2.5
J.T. Baker[⊗]	11.5 ±3.9	78.1 ±6.5
Waters Sep-Pak	1.7 ±0.7	93.8 ±2.6
Merck Adsorbex[≠]	0.4 ±0.2	90.8 ±1.5
Supelco	0.8 ±0.2	88.2 ±3.9
SPE-ED[≠]	0.4 ±0.1	94.0 ±1.1

[⊗] As for [⊗], Table 1. [≠]Only 3cc cartridges available at the time.

When it became clear that the three types of C-18 cartridge differed in 'extraction profile' for these β-blocker-type compounds, we briefly examined the properties of a further 3 commercially available products. We investigated the ability of the cartridges to extract and retain radiolabelled compound **3**. Table 2 gives results for all 6 cartridge types. Within their limitations, the results indicate that, for all the phases examined, acetonitrile is a poor eluent and the best recoveries are obtained with methanol/buffer. The C-18 material produced by J.T. Baker is clearly very different from that in the cartridges from the other manufacturers. Differences amongst C-18 bonded materials can probably be explained in terms of the degree of coverage with silanol groups and the extent of end-capping of the residual silanols. Whatever the explanation, this phenomenon can readily be exploited for the selective extraction of this class of drug by selection of the appropriate commercial product.

Some caution is, however, required in assays devised to exploit such secondary interactions when different batches of SPE cartridges are used, because even small changes in the manufacturing process may affect the final degree of silanization. Indeed, we have noted on one occasion that changing from one batch to another resulted in high irreproducibility; this problem was overcome when a third batch of cartridges was obtained [cf. #NC(E)-8: R. Whelpton, this vol.- *Ed.*]. Such experiences suggest that there may be some merit in obtaining sufficient cartridges of one batch to ensure that all the proposed analytical work can be performed with confidence.

CONCLUSION

SPE is progressively replacing liquid-liquid extraction procedures. However, SPE is evidently a more complex system, where several factors may be responsible for the observed results. Thus, whilst in liquid-liquid extraction it usually suffices to balance the lipophilicity of the analyte against the polarity of the extracting solvent, in SPE a mixed mechanism may be operating. The ability to exploit 'secondary' interactions of the type described in these preliminary experiments should allow the use of SPE to perform sample pre-treatment that would not be possible using simple solvent extraction. However, these experiments also illustrate the need to investigate and understand the mechanisms underlying any given extraction procedure in order to obtain reliable and reproducible results.

Reference

1. Ruane, R.J. & Wilson, I.D. (1987) *J. Pharm. Biomed. Anal.* 5, 723-727.

#NC(E)-10

A Note on

APPLICATION OF SOLID-PHASE EXTRACTION TECHNIQUES TO THE ANALYSIS OF BASIC AND ACIDIC DRUGS IN BIOLOGICAL FLUIDS

H.M. Hill, L. Dehelean and B.A. Bailey

Bioanalytical Research and Development
Bio-Research Laboratories
Senneville, Quebec, Canada HQX 3R3

Use of solid phase (SP) extraction techniques as an alternative to the use of organic solvents is now widely accepted, but is not well documented for assays where quantitation down to ng/ml levels is needed. In order to achieve such levels with a relatively non-specific detector such as a UV detector it is important to have a clean sample. To achieve this the use of the appropriate cartridge and the optimization of the wash procedure are essential; thus, variations in drying time between each step can significantly influence the recovery of the drug and/or endogenous plasma components. The situation is further complicated when automated sample processing techniques are used since the number of sample preparation steps may be limited. Such points have become evident to us in setting up an assay for doxepin (1 ng/ml had to be measurable) and its desmethyl metabolite in serum using the Varian 'AASP' with UV detection, and an assay for fenoprofen (0.3 µg/ml had to be measurable) in plasma using the 'PREP' (DuPont) followed by HPLC with UV detection. The operation of these automated systems has been well described [see index entry 'Automated' in Vols. 12, 14 & 16 - *Ed.*]. Serum is preferable to plasma or even plasma centrifuged prior to processing. Fibrin particles block cartridges, altering flow rates and reproducibility.

The AASP, because of a wider variety of SP's, provides sufficient flexibility to produce an excellent clean-up and consequently lower limits of quantitation, while the use of the PREP is more restricted by the range of available cartridges and preparation steps. Hence, using the AASP it is possible to measure in the ng/ml range, whereas the PREP can be used only for assays where measurement of >50 ng/ml suffices.

In terms of materials the cost of a SP extraction is similar to that of a liquid/liquid extraction procedure. However, analysis time is considerably reduced and, especially with long tedious procedures, reproducibility is improved. One major concern in the use of SP extraction is what constitutes a batch. Our usual procedure is to include one standard curve and 6 quality control samples with each batch of 60-100 samples. In the SP procedure using either the PREP or the AASP a batch can be no larger than 20 samples. We therefore incorporate one standard or quality control sample with each 10-20 test samples.

#NC(E)-11

A Note on

THE DETERMINATION OF DRUG PROTEIN BINDING BY HPLC USING A CHEMICALLY BONDED BOVINE-ALBUMIN STATIONARY PHASE

N. Lammers, H. de Bree, C.P. Groen, H.M. Ruijten and B.J. de Jong

Analytical Development Department
Duphar Research Laboratories, P.O. Box 2
1380 AA Weesp, The Netherlands

The binding of drugs to proteins may play an important role in drug bioavailability and in interactions with other drugs bound to proteins. Determination of drug protein binding by equilibrium dialysis [1] is rather time-consuming; size-exclusion chromatography has been described as an alternative [2], but special precautions are required to prevent dissociation of the drug-protein comples [3]. Yoshida et al. [4] used proteins physically bonded to ODS (C-18) columns to separate drugs. Here we describe the use of a commercially available chemically bonded bovine-albumin stationary phase to determine the degree of protein binding of potential drugs. To obtain the required hydrophobic interaction a buffer concentration of ~0.1 M is necessary, and the addition of 1-propanol lowers retention for all solutes [5]. The retention characteristics of a heterogeneous group of 23 compounds and a group of 11 analogous potential drugs has now been studied. We compared the resulting data with equilibrium dialysis results. The amount injected for chromatography had to be small (~0.5 pmol) to prevent overloading of the chromatographic system.

Fig. 1 shows a chromatogram of some sulfa compounds where the separation is based on the degree of protein binding. As a measure of binding of a compound to the stationary phase, the expression $k'/k'+1$ reflects the degree of protein binding shown by equilibrium dialysis. Table 1 shows the influence of the addition of 1-propanol to aqueous 0.1 M phosphate buffer (pH 7.4) on the coefficient of correlation between chromatographic and equilibrium dialysis data [6, 7]. The Table also ahows the S.D. for $k'/k'+1$ values measured in independent runs on two different columns. Evidently the addition of 1-propanol to the buffer has a slight influence on the correlation

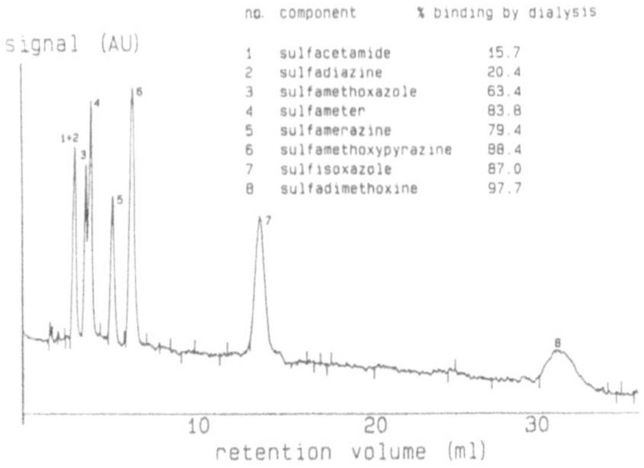

Fig. 1. HPLC of sulfa compounds, on Resolvosil-BSA-7 at 37° (1 µl injected). Mobile phase: 0.1 M pH 7.4 phosphate buffer; 0.5 ml/min. Detection at 254 nm.

Table 1. Influence of 1-propanol addition (% v/v) on the correlation between k'/k' + 1 and protein binding data from equilibrium dialysis and on the spread in chromatographic results, including data for a series of analogue compounds (potential drugs). No. of compounds = n.

Test compounds	Propanol %	Coeff. of correlation	S.D. of k'/k' + 1
Heterogeneous	0	0.84 (n = 19)	0.025
population	1.5	0.88 (n = 22)	0.013
of compounds	3	0.89 (n = 23)	0.009
	4.5	0.88 (n = 23)	0.014
Analogue series	3	0.98 (n = 11)	0.010

between chromatographic and equilibrium dialysis data. However, the mobile phase containing 3% 1-propanol gives a lower spread and is suitable for determining the protein binding of potential drugs in the range 10-99%.

The **advantages** of the method are:
- it is rapid, cheap, and easy to automate;
- the quick answer may aid pharmacological screening of potential drugs;
- it requires only a small amount of technical-grade analyte.
- radioactive labelling is unnecessary;
- in the most relevant range (90-100% binding) retention times are easily assessed, whereas equilibrium dialysis is hampered by poor signals in the non-radioactive assay method or by impurities in the radiolabelled compound.

The **constraints** of the method are:
- as only the binding to albumin is considered, there may be some deviations for drugs which bind strongly to other proteins;
- because the stationary phase can cope only with a low loading a sensitive detector is required.

RECOMMENDATIONS

Studies of the binding of potential drugs to other chemically bonded proteins, e.g. α_1-acid glycoprotein, globulin or melanin, may shed light on phenomena observed in clinical pharmacology. Chemical binding of a standard human plasma on a matrix suitable for HPLC might improve the correlation with equilibrium dialysis results and accordingly with the *in vivo* situation.

References

1. Norris, R.L.G., Ahokas, J.T. & Ravenscroft, P.J. (1982) *J. Pharmacol. Meths.* *7*, 7-14.
2. Sun, S.T. & Wong, F. (1985) *Chromatographia* *20*, 495-499.
3. Sebille, B., Thuaud, N. & Tillement, J.P. (1981) *J. Chromatog.* *204*, 285-291.
4. Yoshida, H., Morita, I., Tomai, G., Masujima, T., Tsuru, T., Takai, N. & Imai, M. (1985) *Chromatographia* *19*, 466-472.
5. Allenmark, S., Bomgren, B. & Born, H. (1984) *J. Chromatog.* *316*, 617-624.
6. Kurz, H. & Friemel, G. (1967) *Naunyn-Schmiedebergs Arch.-Pharmak. -Exp. Path.*, *257*, 35.
7. Lavietes, P.H. (1937) *J. Biol. Chem.* *120*, 265-275.

#NC(E)-12

A Note on

RETENTION MECHANISMS IN REVERSED-PHASE HPLC

John G. Dorsey

Department of Chemistry, University of Florida
Gainesville, FL 32611, U.S.A.

In analytical chemistry HPLC techniques are notably powerful, important and widely used, yet they are still viewed as 'black magic' by many practitioners because of a lack of basic understanding of the chromatographic process. Investigations of retention mechanisms have as their goal a complete and exact understanding of retention, accounting for influences of solute type, mobile-phase composition and stationary-phase structure. This would allow separations to be pre-calculated, largely eliminating trial-and-error development. It would also be feasible to devise a rigorously correct retention index system for HPLC, to obtain exact physicochemical information, and to develop a 'truly expert' system.

It is often assumed that a linear relationship exists between the log of the capacity factor and the % by vol. of organic modifier in the mobile phase. Gradient elution theory as well as many chromatographic optimization schemes and physicochemical measurements are based on this assumed linearity. However, if the relationship is examined over a wide enough range of organic modifier concentrations, a quadratic fit is almost always found. This is particularly true when acetonitrile is used as the modifier instead of methanol. Because the properties of solvent mixtures cannot be predicted accurately by linear addition of the properties of the pure solvents, a number of empirical measures of solvent polarity have been introduced. The most popular single-parameter scale is the $E_T(30)$ scale developed by Dimroth & Reichardt [1]. It is based on the intramolecular charge transfer absorption of a solvent probe:

2,6-diphenyl-4-(2,4,6-triphenyl-*N*-pyridinio)phenolate
- a dye

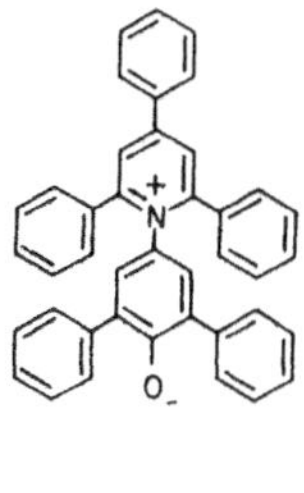

Johnson et al. have reported correlations between log k' and both % organic modifier and $E_T(30)$ polarity for 332 different retention sets [2]; these involved many different solutes on 8 different stationary phases of chain lengths C-2, C-4, C-8 and C-18 with mobile phases of both methanol/water and acetonitrile/water. Plotting log k' *vs.* $E_T(30)$ gave significantly better linearity, with an average r^2 value of 0.9910, compared with 0.9783 when plotted *vs.* % organic modifier. It is interesting, however, to investigate the retention correlations with the two mobile phases separately. For methanol/water (92 data sets) the two log k' plots hardly differed: r^2 = 0.9907 *vs.* $E_T(30)$ and 0.9956 *vs.* % methanol. However, for acetonitrile/water (240 data sets) the former plot (r^2 0.9914) gave significantly better linearity than for log k' *vs.* % acetonitrile (r^2 0.9733).

This linearization will hopefully improve our understanding of retention processes, and aid optimization and physicochemical measurements. The quadratic dependence of plots of log k' *vs.* % acetonitrile has been predicted both by Schoenmakers et al. [3] using the Hildebrand solubility parameter approach, and by Dill [4] using statistical mechanical theory: with binary interaction parameters, χ, he showed that the affinity of a solute for the bonded alkyl chains is determined by the entropy of mixing of the solute, the configurational entropy of the grafted chains, and the contact interactions among solute, solvent and chains:

$$(1/\phi_B)\ln(k'/k'_o) = (\chi_{SB} - \chi_{SA} - \chi_{AB}) + \phi_B(\chi_{AB}) \qquad \text{(Eqn. 1)}$$

where ϕ_B is the vol. fraction of the organic modifier, k'_o is the value of k' when $\phi_B = 0$; S = solvent, A = water, B = organic modifier. Then a plot of $(1/\phi_B)\ln(k'/k'_o)$ *vs.* ϕ_B should be linear for any value of χ_{AB}, provided the regular solution approximation holds. We are now studying this, using the previously reported retention data set [2]. Excellent agreement between linearity of the plots of retention *vs.* $E_T(30)$ and the form derived by Dill suggest that the $E_T(30)$ dye may actually be providing a measure of the binary interaction parameters for the measured solvents.

References

1. Dimroth, K., Reichardt, C., Siepmann, T. & Bohlmann, F. (1963) *Justus Liebigs Ann. Chem. 661*, 1-37.
2. Johnson, B.P., Khaledi, M.G. & Dorsey, J.G. (1986) *Anal. Chem. 58*, 2354-2365.
3. Schoenmakers, P.J., Billiet, H.A.H. & de Galan, L. (1982) *Chromatographia 15*, 205-214.
4. Dill, K.A. (1987) *J. Phys. Chem. 91*, 1980-1988.

Comments on material in #E

Comments on **#E-1**, J.B. Lecaillon - HPLC COLUMN SWITCHING
& **#E-2**, J. G. Dorsey - MICELLAR LC

R. Wyss asked J.B. Lecaillon whether in routine work the stability of samples stored overnight in the autosampler tray causes problems. **Reply.-** Compounds vary in stability, which must be validated and controlled by quality-control samples; a validation (spiked) sample is always added at the end of the series to confirm the accuracy of the preceding results. Pre-adjustment of sample pH may improve stability. **Answer to a further question.-** It was with plasma injection volumes of 400-1000 µl instead of 20-100 µl that column life became short. The use of cartridge switching lengthened the life of the analytical column, as did the heart-cutting technique; but further studies are needed to improve column life where large plasma volumes are needed.

J.G. Dorsey, replying to P.S.B. Minty.- There are very few commercially available phases with high packing density and high selectivity (maybe Vydac offer a polymeric, high-density product). Most manufacturers produce monomeric phases as these are more reproducible. **Brinkman, to Dorsey.-** If indeed one uses C-2 to C-8 rather than C-18 stationary phases with micellar mobile phases, this would not seem to pose any problem, because new stationary phases now coming on the market are quite often of a less apolar (more polar) character. Increasing application of the approach to biological fluids is warranted.

Comments on **#E-5**, J. Caldwell - PITFALLS IN CHIRAL SEPARATIONS
& **#NC**(E)**-4**, H. de Bree - BIOANALYSIS OF POTENT DRUGS

P.S.B. Minty asked J. Caldwell about any examples of acute toxic effects, as in drug overdosage, attributable to a particular stereoisomer. **Replies by Caldwell and Steiner** instanced, respectively, 2 cases of genetic diseases where there is an abnormality in stereospecific metabolism, and the appearance of different isomers in glutethimide overdosage and possibly when neurotoxicity results from giving a cyclophosphamide. **A. Rakhit expressed to H. de Bree** his concern about the large sample volumes used, such that in a pharmacokinetic study it could be difficult to withdraw a sufficient number of samples. **Comment by R. Woestenborghs to de Bree.-** I would have thought the two drugs you studied would have been amenable to RIA, and am surprised that HPLC/GC development took up to 6 months. **Remark by A. Bye** on analysis in general: it could usefully be debated whether development analysts can advantageously be involved in routine analysis, and at what point the developed methods should be transferred.

Citations contributed by Senior Editor

HPLC **column-switching** [cf. arts. **E-1** and **NC(E)-3** & **-4**] has been discussed by Edholm and applied to the assay of urine for **enprofylline** and other xanthines, and of deproteinized plasma for **terbutaline** (a bronchodilator) and **bambuterol** [1,2]. A portion of the eluate from a pre-column was back-flushed into the main column; UV and electrochemical detection were used. For urine from rats given [^{14}C]-**felodipine**, Weidolf [3] used bimodal column switching: NP-HPLC furnished two metabolite groups - carboxylic acid; hydroxylated - each of which, with initial band compression, was separated on a C-8 column using a gradient. Untreated 'blood' (actually plasma) samples could be chromatographed, after pre-column enrichment, by use of a column-switching device ('pre-column venting-plug technique') [4]. Another approach to loading plasma direct is cited below.

Some unconventional approaches warrant mention. To assay plasma for the haemostatic/anti-fibrinolytic drug **tranexamic acid** (down to 10 ng/ml) after deproteinization by perchloric acid, NP-HPLC was followed by on-line derivatization with OPA for fluorescence measurement [5]. Gradient RP-HPLC with UV detection was used for assay of the cardiotonic drug **enoximone** (0.5 μg/ml measurable) and its sulphoxide metabolite (1 μg/ml) by 'automated sequential trace enrichment' with initial automated dialysis [6]. For resolution (not bioanalysis) of 2-arylpropionic acid **enantiomers** (including cicloprofen and ibuprofen) on a preparative scale after conversion to diastereoisomers, centrifugal TLC was, in view of the good resolution, preferable to preparative HPLC unless gram-scale operation was needed [7].

Peptide/protein analytes run on size-exclusion columns (Toya-Soda or DuPont) showed losses onto adsorptive sites, minimized by pre-analytical injection of a basic peptide and especially by use of aqueous sodium azide (0.14 mM) or 10% methanol (never pure methanol) for column storage [8]. A C-18 **packing coated with proteins** (plasma, denatured) allowed use of unprocessed plasma to assay drugs, as shown for doxorubicin and methotrexate [9].

Stereochemistry in drug action is the theme of recently published surveys, wide in coverage but with little attention to analysis [10]. *Reminder:* in Sect. #**D** the first article (Wainer) deals with chiral separations, not entailing pre-generation of diastereoisomers as is generally the case for the entries prefixed **en** in the Table late in **NC(D)** and the similar Table in **NC(A)**. See also #NC(A)-1 and #B-6.

Citations and Forum comments concerning solid-phase extraction

Analyte extraction is considered for leukotrienes in '**NC**' pages that conclude Sect. **A**. For toxicological screening, including forensic investigation of tissue specimens, traditional solvent extraction prior to looking for acidic and basic drugs can be obviated by use, prior to subtle GC, of coupled cartridges (Chem-Elut; Analytichem International), with intervening acidification or alkalinization [11]. Exemplifying applications to endogenous **peptides**, which feature in Vol. 16, C-18 cartridges were used to extract **cholecystokinin** species from biological samples for HPLC followed by RIA [12].

Discussion remarks.- **U.A.Th. Brinkman, to I.D. Wilson** who had spoken about choosing column eluents of appropriate polarity: if 'pre-columns' are used for a number of consecutive runs, and particularly if one is trying elution with aqueous eluents at increasing concentrations, e.g. 20-40-60-80% methanol, one should realize that pre-columns often *do* contain quite a number of plates at the outset, which they tend gradually to lose with each further run. A better approach may be to try a range of pre-column types, from very apolar (e.g. styrene-DVB polymers) on the one hand to metal-loaded phases on the other hand.

*Also (#***NC(E)-7** *is pertinent):* **remarks by R. Whelpton.**- Many people using BondElut-CN are probably using the residual silanols for separation: the combination of cyanopropyls and residual-OH's seems to give the required retention with a good, clean extract. C-2 to C-18 may retain too much background material by partitioning, and silica is likely to be too retentive. We have encountered a batch of BondElut-CN whose 'loading' is erroneously high, and such batches may not work as the Si-O- groups are masked.

References cited

1. Edholm, L-E. (1985) *Current Clin. Pract. Series 19*, 238-44.
2. Edholm, L-E. (1986) *J. Pharm. Biomed. Anal. 4*, 181-189.
3. Weidolf, L. (1985) *J. Chromatog. 343*, 85-97.
4. Daoud, N., Arvidsson, T. & Wahlund, K-G. (1987) *J. Chromatog. 385*, 311-322.
5. Elmworthy, P.M., Tsementzis, S.A., Westmead, D. & Hitchcock, E.R. (1985) *J. Chromatog. 343*, 109-117.
6. Cooper, J.D.H. & Turnell, D.C. (1986) *J. Chromatog. 380*, 109-116.
7. Maitre, J.M., Boss, G., Testa, B. & Hostettmann, K. (1986) *J. Chromatog. 356*, 341-345. [Cf. ref. 7 in #B-6.]
8. Link, G., Keller, P.L., Stout, R.WA & Banes, A.J. (1985) *J. Chromatog. 331*, 253-264.
9. Yoshida, H., Morita, I., Tamai, G., Tsuru, T., Takai, N. & Imai, H. (1984) *Chromatographia 19*, 466-467.
10. Articles (1988) in *Biochem. Pharmacol. 37* [*Part 1*], 1-147.
11. Cordonnier, J., Van den Heede, M. & Heyndrickx, A. (1987) *Int. Analyst 1, May issue*, 28-30.
12. Miller, L.J., Bouska, J.B. & Go, V.L.W. (1986) *J. Chromatog. 378*, 77-84.

also, in this series:

Vol. 16: Reid, E., Scales, B. & Wilson, I.D., eds. (1986) *Bioactive Analytes including CNS Drugs, Peptides, and Enantiomers*, Plenum, New York.

Section #F

DETECTION, IDENTIFICATION AND INSTRUMENTATION

#F-1

ADVANCES IN INSTRUMENTALIZED TLC

I.D. Wilson

Safety of Medicines Department
ICI Pharmaceuticals Division
Mereside, Alderley Park
Macclesfield, Cheshire SK10 4TG, U.K.

Recent advances in instrumentalized TLC are described with emphasis on development techniques, radio-TLC and the spectroscopic characterization of substances in situ *on the TLC plate. Brief descriptions of automated multiple development, overpressure TLC and centrifugal layer chromatography are given. The use of 'linear analyzers' for the detection of radioactivity following 2-D development of TLC plates is discussed, together with the direct identification of TLC spots using mass spectrometry (MS) without having to recover the sample from the plate.*

Despite considerable advances in instrumentation for quantitative TLC over the last decade it is still fashionable to consider the technique to be inferior to HPLC and GC in almost every respect. The popular conception of TLC is that it is a low-resolution chromatographic method, of poor sensitivity, and unsuited to providing quantitative data. Whilst justified in the past this attitude no longer reflects the capabilities of modern instrumentalized HPTLC which can now compete with both HPLC and GC in terms of its performance with regard to sensitivity, specificity, resolution and ability to provide quantitative data.

This transformation of TLC has touched all aspects of the technique, from improvements in the quality and performance of the plates (viz. the introduction of HPTLC plates and bonded phases), to improved autospotters, development techniques and scanning densitometers. Some of these innovations have been covered in previous volumes [bonded phases in #NC(B)-1 by Brinkman, Vol. 14 - *Ed.*]. Here, advances in development techniques, radio-TLC scanners and spot identification by MS will be described.

NEW AND IMPROVED TLC PLATE DEVELOPMENT TECHNIQUES

Traditionally TLC has been performed by merely immersing the lower end of the plate in a suitable solvent and letting capillary action take its course. For particularly difficult applications, multiple or continuous development or perhaps 2-D development would be used. However in 1973 Perry and co-workers introduced 'programmed multiple development' (PMD) [1] whereby the plate was repeatedly and automatically developed in the solvent. At each development the solvent was allowed to migrate slightly further up the plate before drying. This system permitted some impressive separations to be obtained and, by automating the multiple developments, removed most of the tedium associated with performing this type of TLC manually. An instrument based on this approach was, briefly, commercially available; but PMD was clearly ahead of its time and was not widely adopted. Recently, however, Burger has re-introduced PMD in the form of 'automated multiple development' (AMD) [2]. This has also been commercialized and an instrument for AMD is available from Camag. Compared with PMD as described by Perry et al. [1], AMD has certain advantages: firstly, the solvent is removed at the end of each development cycle using vacuum rather than heating (thus removing the risk of degrading thermally labile substances) and, secondly, the design of the AMD instrument allows quite complex gradients to be run.

Common to both PMD and AMD is the fact that with each development the band is re-concentrated. This is because the advancing solvent front comes into contact with the bottom of the band well before it reaches its leading edge. The rear of the band is therefore pushed into the leading edge, providing re-concentration and resulting in very compact bands. Certainly AMD is capable of producing some excellent separations [2]. In our hands AMD provided chromatography superior to that with single and multiple development [3]. We found, however, that with compounds covering a relatively limited polarity range, the ability to run solvent gradients was not particularly useful and multiple isocratic development was best. In its current design the AMD instrument is also limited to 10 × 20 cm silica gel TLC plates, and is unsuitable for reversed-phase (RP) separations on bonded TLC plates.

Forced-flow techniques.- Besides AMD, other new development techniques have been devised. Overpressure-layer TLC (OPTLC or OPLC) is particularly promising: the plate is held, whilst subjected to a hydrostatic pressure, under an impermeable membrane and solvent is forced through the layer using a pump. Indeed, OPLC rather resembles HPLC with flat columns! Because OPTLC is a 'forced-flow' TLC technique some of the limitations of traditional TLC are circumvented. Thus chromatography is no longer limited by capillary action (with development times that increase with the square of the solvent-migration distance). Also, OPTLC begins to acquire some of the benefits of column techniques in that substances can be eluted completely off the plate and detected with a detector of flow-through type. This feature of OPTLC might be particularly useful for the preparative-scale isolation of, e.g., metabolites without the need for laborious (and possibly hazardous to lungs) plate-scraping and subsequent elution.

Besides OPTLC another forced-flow technique, centrifugal TLC (CLC), has become available: the plate is spun at high speed and solvent fed into the centre. The solvent is then forced through the layer by centrifugal force [4]. Impressive separations have been achieved both by OPTLC and by CLC, and it remains to be seen whether the TLC market can support all three rival systems (AMD, OPTLC, CLC) or whether one will achieve a dominant position.

RECENT DEVELOPMENTS IN RADIO-TLC

The introduction of the 'linear analyzers' for detecting radioactivity on TLC plates has revolutionized the use of TLC in drug metabolism studies employing radiotracers. The high sensitivity of these detectors, compared to conventional autoradiography using X-ray film, means that it is possible to obtain results in minutes or hours rather than days or months. This in turn has enabled the development of suitable chromatographic systems for metabolite profiling in a way which was simply not practicable when 2 or 3 days were required in order to see the effect of changing the solvent composition. In addition the data provided by the linear analyzer is often quantitative whereas autoradiography in general supplies only qualitative information.

In the radio-TLC field a major recent innovation has been the development of methods for detecting and quantifying the radioactivity on TLC plates following 2-D chromatography. As alluded to earlier, 2-D TLC can be used to provide greater resolution than is possible using a single development, especially when solvent systems exploiting different chromatographic mechanisms are employed. The major problem associated with 2-D techniques has been that, if obtaining quantitative data from a 1-D plate was difficult, this was virtually impossible following 2-D development. However, with the ready availability of powerful desk-top personal computers (PC's) the acquisition and storage of the large amounts of data required for 2-D-radio-TLC has become feasible [5].

The results illustrated in Figs. 1-3 were obtained using a Berthold model LB2842A linear analyzer fitted with an IBM PC AT. To obtain them the detector acquired a series of 1-D measurements of 1 × 200 mm lanes. By slowly traversing the plate a large number of very narrow 1-D chromatograms were obtained which could be built up into a 2-D display. Thus, over a period of several hours the whole surface of the plate was covered and the radioactivity present detected. The main disadvantage of using a conventional linear analyzer for

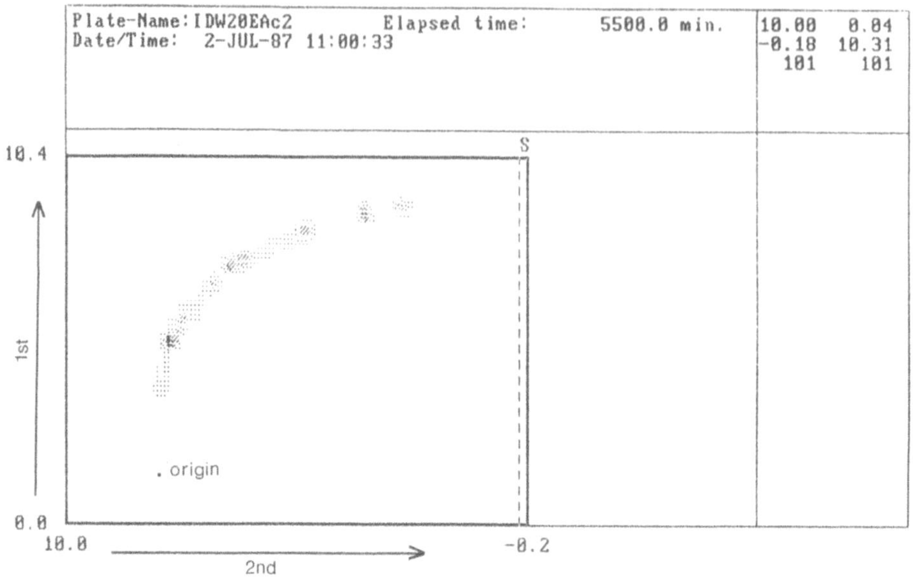

Fig. 1. A 'digital autoradiogram' obtained for acetate derivatives of a ^{3}H polyhydroxylated steroid on a 10 × 10 cm silica gel HPTLC plate using a Berthold LB 2842A linear analyzer.

2-D work is that, because of the design of the detector, the resolution achievable 'across' the plate is not as good as that obtained 'up' the plate. Also, the process is rather time-consuming.

Alternative solutions to this problem have been devised by other manufacturers (as exemplified by the AMBIS 2D TLC scanner [6] and by the TRACY and BERTA instruments produced by Raytest [7]). The output shown in Fig. 1 is analogous to a traditional autoradiogram. However, by utilizing the various subroutines provided by the software it can be examined as a 3-D image from various angles, as illustrated in Fig. 2, and even converted into a 'standard' linear chromatogram as shown in Fig. 3. Similar routines could no doubt be readily devised for UV-Vis scanning densitometers with similar benefits.

TLC-MS WITHOUT PRIOR SUBSTANCE ELUTION

The major practical difficulty in linking most chromatographic separations to an MS instrument is devising an interface compatible with the chromatographic systems needed for a mobile phase and the spectrometer's requirement for high vacuum (see other arts. in this volume). In the case of TLC the solvent is no longer present when MS comes to be performed and the problem does not arise. Instead

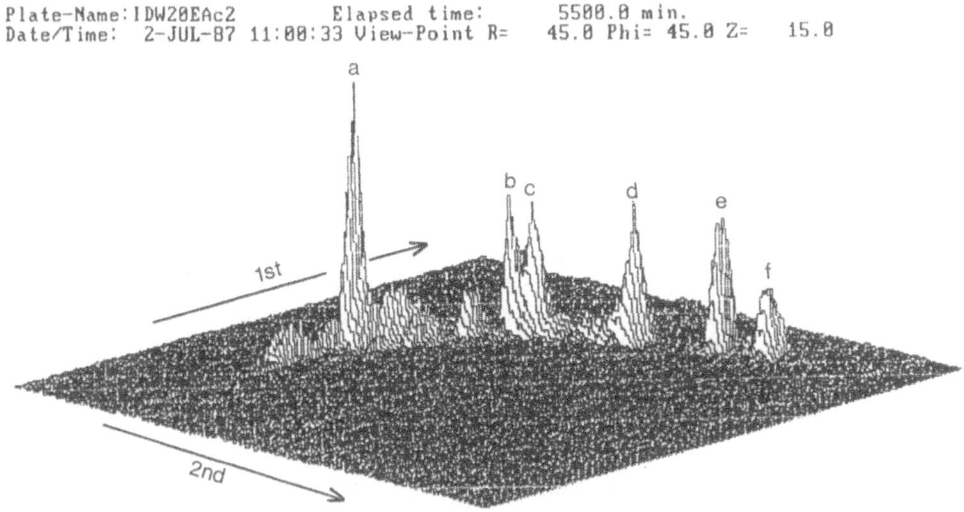

Fig. 2. A 3-D representation of the ^{3}H distribution shown in Fig. 1.

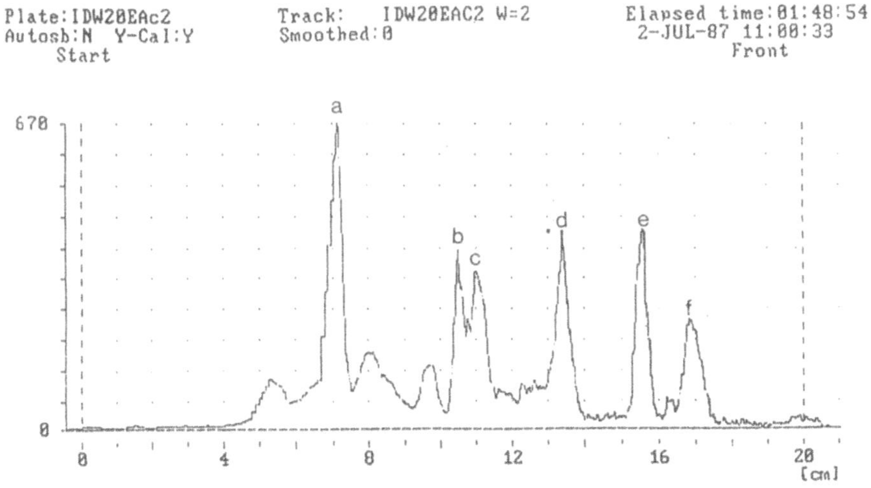

Fig. 3. A linear representation of the data shown in Figs. 1 and 2.

the difficulty is in persuading involatile molecules to volatilize from the surface of a strongly attractive stationary phase such as silica gel. Prior to the introduction of fast atom bombardment MS (FAB-MS) few examples of successful TLC-MS had been reported, and most of these were from TLC phases such as polyamide rather than silica gel [e.g. 8]. With the advent of FAB, however, the difficulties of TLC-MS have been substantially reduced.

The first successful report involved attaching double-sided adhesive tape to the FAB probe which was then pressed down firmly on the area of interest on the TLC plate, thus transferring silica from the plate to the probe [9]. Mass spectra were then obtained in the usual way (following the coating of the silica with a suitable FAB matrix such as glycerol). Since then quite a number of successful trials of TLC-FAB-MS have appeared in the literature [e.g. 10, 11], together with other approaches such as TLC combined with secondary-ion MS [12].

Our own variation on the TLC-FAB-MS technique is to remove the relevant area of the TLC plate and mix with glycerol, then transfer the sample to the probe for spectrometry (P.J. Phillips, pers. comm.). In common with others we have achieved considerable success with this relatively simple approach (there can be few techniques where one can claim to have given the mass spectroscopist his samples 'on a plate'!) [cf. #NC(F)-1, this vol.-*Ed*.]. Considerable innovations continue in this area, and one manufacturer (Jeol) provides a device for slowly moving an appropriate track of a TLC plate through the ion source.

Fig. 4 shows an FAB spectrum for a polyhydroxysteroid taken directly from the plate. In our experience, provided that the compound itself will yield an FAB spectrum then a similar spectrum can probably be obtained directly off a TLC plate, although the amount of material required per spot seems to be compound-dependent. The use of tandem MS and TLC (TLC-FAB-MS-MS) should prove an even more powerful technique for identification.

CONCLUSIONS

Debate about which technique is superior, HPLC or TLC, is both sterile and pointless. The two techniques are complementary, each performing particular types of tasks rather well and others poorly. The main advantage of TLC, particularly in metabolic studies with radiolabelled compounds, is that no matter what the 'polarity' of the metabolites they will be located somewhere between the origin and the solvent front and hence detected. There is therefore no need to make prior assumptions in respect of chromatographic properties in TLC. Thus TLC is ideally suited to tasks such as metabolite profiling. The advances described here clearly indicate that TLC

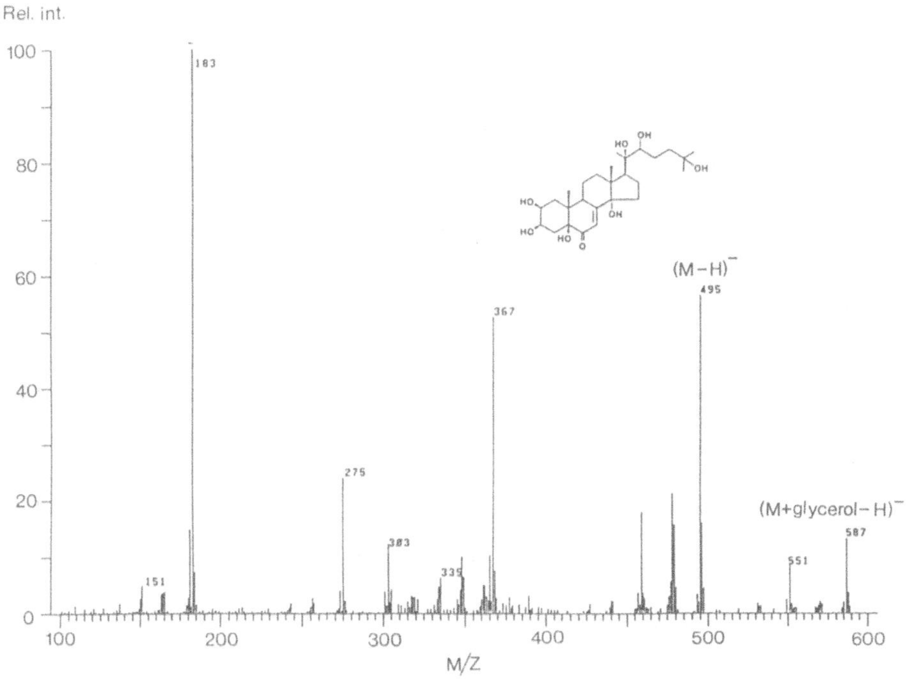

Fig. 4. Negative TLC-FAB-MS of the steroid polypodine B from a silica gel HPTLC plate (to which a plant extract had been applied). The spectrum of the component is identical to that of a standard directly introduced by the probe.

is becoming an increasingly powerful tool for separating, quantifying and identifying substances. Our own experience of using radio-TLC, followed by TLC-MS for metabolic work, suggests that this approach is advantageous: it is very powerful and effective, and technically undemanding besides being relatively cheap, the MS step entailing direct insertion with no need for an interface.

Acknowledgements

Assistance with 2-D radio-TLC from Mr. S. Moss and with the TLC-FAB-MS of polypodine A from Mr. P.J. Phillips is gratefully acknowledged.

References

1. Perry, J.A., Haag, K.W. & Glunz, L.J. (1973) *J. Chromatog. Sci.* *11*, 447-453.
2. Burger, K.D. & Tengler, H. (1986) in *Planar Chromatography*, Vol. 1 (Kaiser, R.E., ed.), Heuthig, Heidelberg, pp. 193-205.
3. Wilson, I.D. & Lewis, S. (1987) *J. Chromatog.* *408*, 445-448.
4. Tyihak, E. & Mincsovics, E. (1988) *J. Planar Chromatog.* *1*, 6-19.
5. Filthuth, H. (1988) in *Recent Advances in Thin Layer Chromatography* (Dallas, F.A.A., Read, H., Ruane, R.J. & Wilson, I.D., eds.), Plenum, New York, in press: pp. 125-130.
6. Smith, I. (1988) *as for* 5., in press: pp. 77-86.
7. Dietzel, G., Kubisiak, H. & Stelzer, H. (1988) *as for* 5., in press: pp. 117-124.
8. Kraft, R., Otto, A., Zopfl, H-J. & Etzold, G. (1987) *Biomed. Environ. Mass Spec.* *14*, 1-4.
9. Chang, T.T., Lay Jnr., J.O. & Francel, R.J. (1984) *Anal. Chem.* *56*, 111-113.
10. Isobe, R., Fujii, I. & Kanematsu, K. (1987) *Trends Anal. Chem.* *6*, 78-81.
11. Hare, K.J. & Read, H. (1987) *Analyst* *112*, 433-439.
12. Iwabuchi, H., Nakagawa, A. & Nakamura, K.I. (1987) *J. Chromatog.* *414*, 139-148.

#F-2

THE USE OF FLUORESCENCE AND CHEMILUMINESCENCE TECHNIQUES FOR SENSITIVE AND SELECTIVE DETECTION IN HPLC

U.A.Th. Brinkman, G.J. de Jong and C. Gooijer

Free University, Department of Analytical Chemistry
De Boelelaan 1083, 1081 HV Amsterdam, The Netherlands

Luminescence techniques, including CL, can be especially useful for HPLC detection when high sensitivity and/or selectivity are required, and they can be nicely adapted to the demands of miniaturized HPLC. In this article four aspects are discussed: (i) on-line post-column reaction detection in HPLC with conventional fluorescence monitoring; (ii) the state-of-the-art of laser-induced fluorescence detection for HPLC; (iii) recent progress in multidimensional fluorescence detection; and (iv) the principles and application of on-line CL (peroxyoxalate system) detection in HPLC.*

The sensitivity and selectivity of detection in HPLC still often do not meet the requirements of modern trace-level determinations in biomedical, pharmaceutical and environmental samples. Among the present-day detection principles, luminescence - notably its best known and most widely applied representative: fluorescence - appears to offer a high potential to achieve (sub)ppb detection limits for analytes in complex matrices. Fluorescence detection has an inherent selectivity which is especially relevant for the analysis of complex samples. Firstly, all compounds that absorb light do not necessarily fluoresce. Secondly, two wavelengths can be selected independently of each other so that, in principle, 3-D chromatograms can be obtained. Both this technique and the on-the-fly determination of fluorescence lifetimes can be utilized for identification purposes. These types of multidimensional detection feature interestingly in recent papers.

Since relatively few compound types display native fluorescence, derivatization plays an important role in HPLC with fluorescence

**Abbreviations (& see later in text).-* CL, chemiluminescence; OPA, *o*-phthalaldehyde; DAS, 9,10-dimethoxyanthracene-2-sulphonate; LIF, laser-induced fluorescence; OTLC, open-tubular HPLC; EEM, emission-excitation matrix; ILDA, intensified linear photodiode array.

detection. Initially, pre-column derivatization (or labelling) was the preferred technique. Today, on-line post-column derivatization is becoming increasingly popular. Relatively fast reactions such as ion-pair formation with fluorigenic counter-ions and true chemical derivatization with, e.g., OPA are especially attractive. An alternative approach is to use a photochemical reactor, wherein UV-vis irradiation converts suitable non- or weakly fluorescing analytes of interest into highly fluorescent products.

Lasers are of growing importance for HPLC detection, viz. RI and thermal-lens besides fluorescence detection [1]. Using laser detection in miniaturized HPLC, laser excitation sources are in fact essential. The beams are highly collimated, so that illuminated volumes of only 10 nl are readily attainable. Detector design is crucial because scattered light and background luminescence limit the sensitivity. Continuous lasers have hitherto predominated. A recent approach to reduce scatter and enhance selectivity entails two-photon excitation, needing pulsed lasers with high peak power.

The limiting factor in the ultimate detectability of fluorescence is the background light, and its noise, that reaches the photomultiplier. In principle, it would therefore be ideal to be able to eliminate the light source from the detection system. One way to achieve this goal is via chemical excitation or CL. Although extensively used in batch analyses, CL as a detection method in HPLC is rather new. Most attention is devoted to the peroxyoxalate system, which has enabled various types of fluorophore and fluorophore-labelled compounds to be determined, besides hydrogen peroxide.

In this article, utilization of the above-mentioned luminescence principles for HPLC detection will be discussed with emphasis on practical aspects and applications.

FLUORESCENCE: POST-COLUMN REACTION DETECTION

On-line post-column reaction detection is a well established technique for improving sensitivity and/or selectivity in HPLC [see Index entry 'HPLC - detection' in previous Vols. - *Ed.*]. Fig. 1 shows a general scheme. Main advantages of such detectors include:

- the analytes of interest are separated in the original form, which often permits adoption of published separation procedures;
- artefact formation plays a minor role in post-column reaction, in contrast with derivatization prior to the separation step;
- the reaction does not need to be complete nor have to be fully defined; the only requirement is reproducibility.

Disadvantages are the need to add reagent solution(s) which generally will require the use of additional pumps, the possibility of band broadening in the reactor which may mar the chromatographic separation, and the presence of excess reagent which may interfere with the (fluorescence) reaction-product signal. Interesting ways to reduce these disadvantages feature in examples that follow.

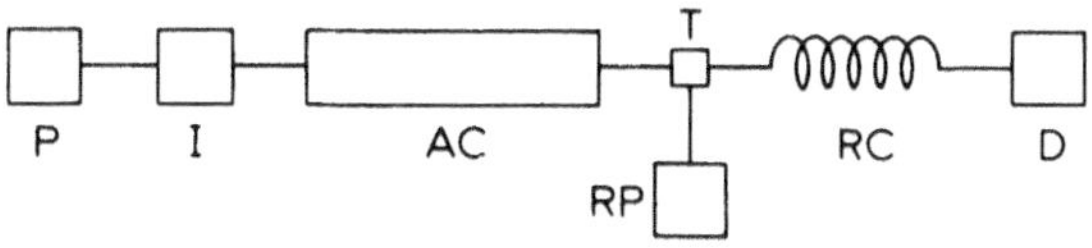

Fig. 1. General set-up of HPLC system with on-line post-column reaction detection unit. P, pump; I, injection valve; AC, analytical column; T, (mixing) T-piece; RP, reagent addition pump; RC, reaction coil (glass, PTFE, stainless steel); D, detector.

Fig. 2. As for Fig. 1 (bonded stationary phase; partly aqueous eluent), exemplifying ion-pair formation, extraction and fluorescence detection. DAS, aq. DAS-Na solution; TCE, tetrachloroethane; O, phase separator; Flu, fluorescence detector.

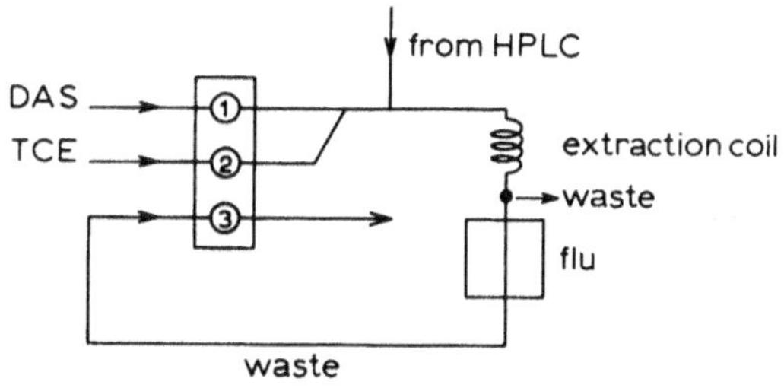

ION-PAIR FORMATION

For basic or acidic analytes an elegant method to minimize post-column broadening is the use of relatively fast ion-pair formation with simultaneous extraction. Under proper pH conditions the analytes are separated and then a reagent solution containing a highly fluorescent counter-ion is added. The fluorescence of the resulting ion-pair is measured. The main problem, of course, is that the excess of counter-ion also fluoresces and therefore has to be separated from the ion-pair to be detected. As Fig. 2 shows diagrammatically, this can be achieved through solvent segmentation, which involves rapid segmentation of the column effluent with plugs of non-miscible solvent, viz. TCE in the example where, after RP-HPLC separation of basic analytes using an aqueous eluent, aqueous DAS (Na salt) is added; the large excess stays in the water phase whilst the non-polar organic ion-pair is extracted into the TCE. After phase separation, the organic solvent stream is monitored continuously in a fluorescence detector while the aqueous flow goes to waste.

Applications include the determination of chlor- and brompheniramine in urine [2], and of drugs such as secoverine (a tertiary amine; [3]), remoxipride (a substituted benzamide; [4]) and a pancuronium-type compound [5] in serum and plasma. Another example involves the detection of linear alkylsulphonates (C-10 to C-18) and related sulphates with acridine-H^+ as counter-ion [6]: separation was achieved by linear gradient elution (acetone-to-water) which did not adversely affect the post-column reaction/extraction procedure.

As an example, the semi-automated procedure developed for the selective dopamine antagonist remoxipride in biological fluids [4] is now outlined. The set-up is essentially as in Fig. 2, plus a disposable pre-column unit to permit the direct injection of diluted plasma (1:1) and urine (1:9). From the separation column, C-18 deactivated with cetrimide, the analyte is eluted with aqueous acetonitrile buffered to pH 3. Aqueous DAS is then used for ion-pairing, and dichloroethane for extraction. A sandwich-type separator (no membrane) of only 30-40 µl internal volume helps minimize band broadening; its upper half is of stainless steel and its lower half of PTFE. Typically, for plasma samples, there is linearity ($r = 0.9999$) over two orders of concentration, and 1 ng/ml is detectable (signal/noise = 3:1); recovery averages 88%, and for 200 ng/ml the within-series and day-to-day C.V.'s are 3.5%.

CHEMICAL DERIVATIZATION

An excess of reagent is readily tolerable when it is only the reaction product, not the reagent itself, that is fluorescent. A good example is OPA, which is non-fluorescent but, at alkaline pH with a strong reducing agent present (e.g. 2-mercaptoethanol), yields highly fluorescent derivatives of primary amino groups. OPA is best known for pg-level assay of amino acids in biological fluids. OPA is also applicable to therapeutic agents such as aminoglycosides (gentamicin, amikacin) and β-lactam antibiotics (penicillin, cephalosporin). In environmental analysis, *N*-methylcarbamates can be quantitated via a dual-reactor approach [7]. They are separated (C-18), catalytically hydrolyzed on-line in a 4-6 cm long stainless steel column packed with anion-exchange resin and heated to ~100° (giving methylamine), and mixed, via a T-piece, with OPA solution for derivatization in the second reactor - a coiled stainless-steel capillary (20-30 sec residence time). The detection limits for 6 carbamates were 0.1-1 ng. With on-line pre-column trace-enrichment, ~20 ppt was detectable in a 20-ml surface water sample spiked with carbaryl [8].

Ligand exchange with its relatively slow kinetics is also effective with a post-column reactor. Trace-level determination of HPLC-separated organosulphur compounds such as penicillamine (e.g. in serum and urine samples) and low-mol. wt. thiols and thioethers has been achieved by an exchange reaction with the non-fluorescent (quenching!) 1:1 Pd(II)-calcein complex: the species detected is the equivalent amount of the highly fluorescent calcein displaced, with no need for an extraction from the water-rich (pH 7) mobile phase since the excess of Pd(II)-calcein does not fluoresce [9]. However, air segmentation is necessary to suppress extra-column band broadening, a large reaction coil being needed because the reaction is rather slow, typically needing 10-15 min in the reactor at 50-60°.

Fig. 3. Clobazam (peak C; R = CH_3) and desmethylclobazam (D; R = H) structures and assay in human serum (0.7 µg/ml each) with a C-18 column and methanol/water (1:1): no UV irradiation (*left*), or 15 sec irradiation (*right*). *Adapted from ref. [10].*

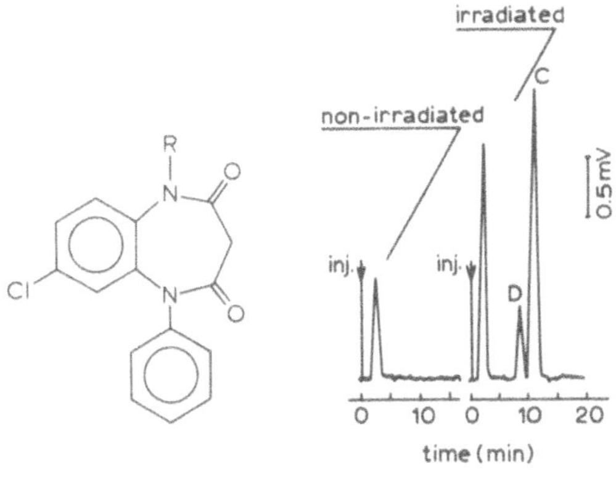

PHOTOCHEMICAL REACTOR

With a photochemical reactor [10] and, in the simplest case, with no reagent added besides photons, non- or weakly fluorescing analytes may be converted by UV irradiation to highly fluorescent reaction products. Photochemical reactors [see arts. in Vol. 14 - *Ed.*] incorporate a lamp, a PTFE reaction coil (200-300 nm transmission comparable to quartz), and a cooled reactor (residence time generally only 10-60 sec). There must be rigid constancy of temperature, because fluctuations will affect flow-rates, reaction kinetics and/or reaction pathways. Typically 50-500 pg is detectable, and selectivity is excellent. This on-line procedure, with the RP-HPLC mode (using aqueous organic eluents) except for clomiphene, has been applied to various analytes in urine, blood or surface water [11, 12](& see Vol. 14): cannabinol, chlorophenols, ciprofloxacin, clobazam (and desmethyl), clomiphene, demoxepam, diethylstilboestrol, LSD, metal ions, methotrexate, phenothiazines, phenylureas, tamoxifen and vitamin K_1.

As an example (Fig. 3) clobazam and its main metabolite, both lacking native fluorescence, fluoresce well if briefly irradiated, with detection limits of 20 and 50 pg/ml serum respectively. Uihlein & Schwab [13] use a very elegant knitted reactor having the UV light source on a sledge adjustable in relation to the reactor in order to vary irradiation time. For methotrexate and metabolites in body fluids [14] a small amount of H_2O_2 is added, and <4 sec suffices for irradiation. Chlorophenols in water are separated as dansyl derivatives and post-column irradiation releases the highly fluorescent dansylsulphonic acid (J.F. Bohle, C. de Ruiter, J. Haslova, J.C. Gluckman, R.W. Frei & present authors, to be published).

DETECTION BY LASER-INDUCED FLUORESCENCE (LIF)

In a 1980 review Yeung & Sepaniak [15] clearly outlined the potential value of LIF detection in HPLC, especially miniaturized. In 1984 Dovichi [16] demonstrated LIF of flowing samples as an approach to single-molecule detection in liquids. Yet, despite the remarkable

Fig. 4. HPLC of a mixture of 15 naphthalene dialdehyde-derivatized amino acids (500 fmol each) using a C-18 column with an aqueous phosphate/ acetonitrile gradient and argon-ion laser detection at 457.9 nm. *From ref. [20], by permission. See note at end of refs. list.*

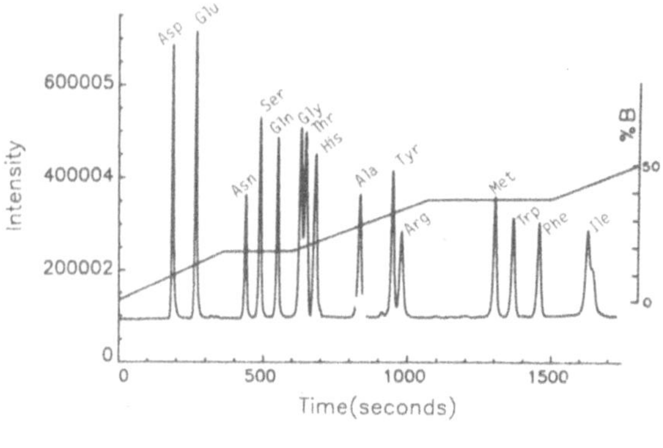

inherent sensitivity of LIF, there are few reports of its use with microbore HPLC or OTLC [17-19]. This is rather surprising, since in HPLC miniaturization is of increasing importance and detection is still a weak point because extremely small detector-cell volumes are required - a problem which can be eased by the intensity and collimation attainable with lasers. Mere replacement of a conventional light source by a laser does not guarantee improved detectability, since background radiation is crucial - especially with very small cell volumes. However, the main disincentive presumably is that lasers are considered too expensive and complicated for routine analytical use.

Besides LIF use with miniaturized HPLC, the excitation source in multichannel detection may advantageously be a nitrogen laser. It has favourable pulse properties and may aid diode array and time-resolution techniques (amplified below). Roach & Harmony [20] developed a pre-column derivatization technique for primary amino acids to make better use of their argon-ion laser; they replaced OPA by naphthalenedialdehyde which, with cyanide present as the nucleophile, reacts with a primary amine to form a 1-cyano-2-alkylbenzisoindole, efficiently excitable by the 457.9 nm argon-ion laser line. They could detect 200-500 amol with a conventional-size HPLC system (see Fig. 4) and 100 amol with a microbore system.

According to Ishibashi and co-workers, lasers are not (yet) practicable for use in commercial HPLC systems because they are bulky, expensive, and relatively difficult to operate [21]. The introduction of semiconductor lasers, well known from videodisc systems, therefore seems propitious. They can be used only for excitation in the near-IR range, but derivatization has been carried out successfully; hence LIF is now possible under conditions where background fluorescence due to the HPLC eluent is negligible. Ishibashi [21] labelled proteins with indocyanide green (ICG), applied 15 mW excitation at 780 nm and detected at ~840 nm. Under these conditions, Raman scattering occurring at 890 nm (small) and 1053 nm (large) can be completely rejected by a fluorescence monochromator, and the noise level depends on the photomultiplier's dark current; cooling will therefore be advantageous. The detection limit for albumin in human serum was 1.3 pmol, which is ~2 orders more sensitive than if using conventional fluorescence detection.

MINIATURIZED HPLC-LIF

Cell design.- With a microbore column, typically 1 mm i.d. with 20-100 µl/min flow-rate, the peak-broadening constraint allows a detector cell of ~0.5 µl. With an open-tubular column (OTLC; only 10-25 µm i.d. and 10-500 nl/min), cell volumes of 0.1-10 nl are required and, for sensitivity (needed in microbore HPLC also), highly sophisticated detection techniques. Compared with conventional columns (3-4.6 mm i.d.), microbore columns show an unfavourable increase (5-fold) in detection limits attainable with UV or fluorescence detectors [19].

Evidently the key property of the laser in miniaturized HPLC is its directionality. Because the angle of radiation is typically only 0.5-1 mradian, the photon flux arriving at the cell is high even though the power of the laser usually is much lower than that of a classical light source. Thus, the frequently used He-Cd laser emits only 1-10 mW at 325 nm, yet it produces a higher intensity in a small detector cell used for OTLC than does a 100 W Hg lamp at 254 nm. The explanation is that the laser can be focused down to 5-50 µm without much loss of photons, whereas the Hg lamp spreads out in all directions, precluding efficient light collection and a spot size <1 mm [9, 17]. Also important in fluorescence detection is reduction of stray light, helped in two respects by laser excitation. Firstly, the Rayleigh and Raman bands are much sharper because of their monochromaticity and, hence, their effect can be more easily minimized. Secondly, a laser-produced scattering pattern is virtually planar, whereas that from an arc lamp is more omnidirectional. Hence background due to laser scatter is easily reduced by careful positioning of the light source, the detector cell and the photomultiplier collecting optics [22, 23].

Notwithstanding the above, detector design is still the main constraint in miniaturized HPLC with LIF. Four interesting principles are now outlined. The **sheath-flow** principle was the basis of a sub-µl flow-through cell [24], the column effluent being surrounded by a second ('sheath') flow having the same composition as the column effluent. If the flows are laminar, no mixing will occur. By adjusting the relative rates of the two streams, the diameter of the sample (inner) stream can be varied and hence different cell volumes can be obtained. The applicability of the cell in conventional and micro HPLC [24] and in OTLC [19] has been demonstrated. Another detection principle depends on **optical fibres**, which are now available down to 10 µm diam. Into the fused-silica column a fibre is inserted, close to the illuminated volume of the detector cell but at a distance where cell-wall interferences such as scattered and reflected light

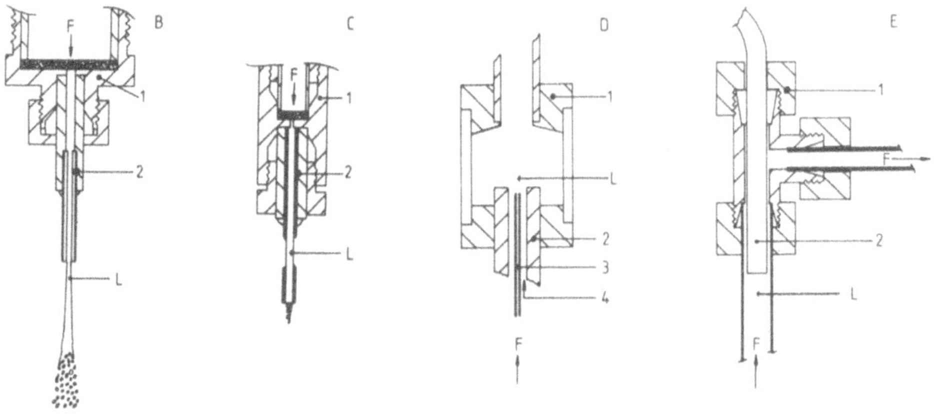

Fig. 5. Cell design for laser fluorescence detection. **B**, Free-falling jet (1, HPLC column; 2, fused-silica capillary glued in stainless-steel tubing). **C**, Fused-silica cell (1, column; 2, fused-silica capillary, with polyimide layer removed over a short distance). **D**, Sheath-flow cell (1, PTFE holder for quartz cuvette; 2, outer tube; 3, terminal part of OTLC column or connection to outlet of packed column; 4, sheath flow). **E**, Optical fibre cell (1, modified T-piece; 2, optical fibre; F, flow direction; L, laser light). *From ref. [19], by permission.*

can be neglected. The luminescence collection efficiency is good and the fibre can be connected to a monochromator without use of additional optics. Gluckman et al. [25] thereby obtained a laser-illuminated volume as small as 3 pl, while the actual cell volume was 90 nl. Unfortunately, even that volume is still too high for use in OTLC.

Thirdly, with **falling jets** the HPLC effluent is led through a fused-silica capillary to increase its linear velocity and obtain a jet stream, i.e. luminescence detection is performed on a liquid eluent cylinder with a smooth and fresh surface which is optically well defined. A zone of intensely reflected light is, however, generated in the laser beam plane, because the liquid cylinder acts as an optical quasi-lens. The collecting optics must therefore be placed outside this area. The possibility of producing free-falling jets with flow-rates compatible with microbore-HPLC columns has been studied by Folestad et al. [23]. With pure methanol at 45 µl/min, a 12 µm i.d. quartz capillary gave a suitable jet. However, so far the jet cell seems to be unsuitable for the extremely low eluent flow-rates (~0.1 µl/min) used in OTLC for which no stable jet can be formed [17, 19]. Fourthly, a cell can be constructed by merely burning away the protective polyimide coating from a **fused silica capillary**, or else the terminal portion of an OTLC column can be

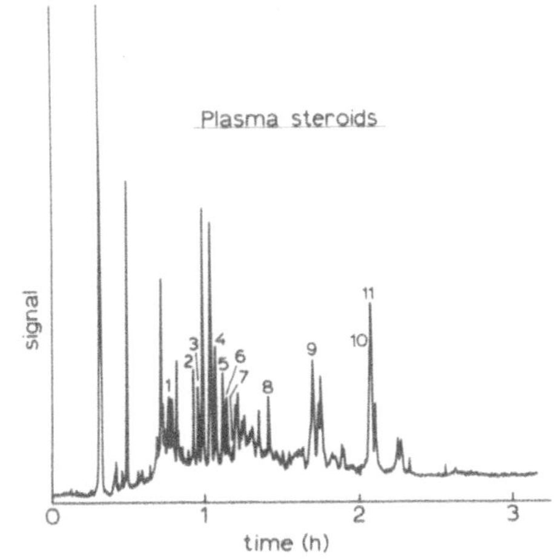

Fig. 6. HPLC-LIF of solvolyzed steroids (~50 pg of each loaded) as found in plasma, viz. (where β**pg** = 5β-pregnane, **as** = -androstan, ^{4}**ol** = 21-tetrol), tentatively:
1, 5α**as**-3α,11β-diol-17-one;
2, 5β**as**-3α-11β-diol-17-one;
3, β**pg**-3α-11β,17α,^{4}**ol**-20-one;
4, β**pg**-3α,17α,20β,^{4}**ol**-11-one;
5, β**pg**-3α,11β,17α,20β,21-pentol;
6, β**pg**-3α,17α,20α,^{4}**ol**-11-one;
7 = 5 but 20α; 8, **as**-3α-ol-17-one;
9, 5-androstene-3^{4}-ol-17-one;
10, β**pg**-3α,20α,21-triol;
11, 5β-**as**-3α,17β-diol.
See text and [25]; by permission.

transformed into the detector cell itself, allowing on-column detection. Unfortunately, capillaries of only 5-10 μm i.d. have relatively thick walls and thus produce a high background signal. The signal-to-noise ratios are comparable with those of the free-falling jet detector [23]. On-column detection in OTLC has been reported [22, 23].

In summary, an ideal cell for miniaturized HPLC does not yet exist, nor can the best type of design be firmly decided. The fused-silica capillary cell, notably simple and practical, warrants preference for OTLC detection because of the absence of any connection between column outlet and detector cell inlet. The main challenge for OTLC is to improve the detection limits which, expressed in units of concentration, are still 2 orders of magnitude higher than in conventional HPLC with 3 μl fluorescence flow cells.

Applications, and laser type.- Amongst the quite few applications (cited in [17, 19]) of LIF in miniaturized HPLC, mostly a low-cost He-Cd laser with its two weak (1-10 mW) lines at 325 and 442 nm was used. Some authors have used a high-power Ar-ion laser, which provides several strong lines in the visible region, e.g. at 458, 476, 488, 496, 514 and 529 nm (1-10 W) and at 334, 351 and 364 nm (only 0.1-1 W). The He-Cd and the Ar-ion laser are both continuous, but the latter can also be operated in a pulsed mode. A tunable dye laser could be effective, but seems not to have been tried in miniaturized HPLC. Generally an alternative approach is preferred, viz. derivatization of the analytes, usually pre-column (but recently post-column [19]), to obtain products with fluorescent characteristics matching the available laser wavelengths. Gluckman et al. [25] showed the usefulness of He-Cd (325 nm) laser fluorescence labelling for biochemically important compounds, e.g. prostaglandins, steroid hormones (Fig. 6) and bile acids. They used reagents containing a coumarin moiety which absorbs maximally at 317 nm. They could detect as little as

8.4 fg of coumarin at a laser excitation power of 6.45 mW, using a 225 cm × 220 μm i.d. packed fused silica HPLC column. Guthrie et al. [22] used OTLC separation columns, 16 or 25 μm i.d., applying their He-Cd laser directly on-column; the collection optics were arranged off the laser-scattering plane. With the 442 nm line they could detect 35 fg of riboflavin.

Until now, pulsed lasers have been used only occasionally for fluorescence detection in miniaturized HPLC. They have the inherent advantage of reducing the background signal by means of time-resolved fluorescence, and some deliver very high peak powers, apt for 2-photon excitation as demonstrated for micro-HPLC by Pfeffer & Yeung [26]. The simultaneous absorption of 2 photons confers extra sensitivity, being determined by selection rules inapplicable to the conventional single-photon absorption process. Besides, long-wavelength excitation suffices, because emission occurs at shorter wavelengths than excitation; hence the background signal is negligible. Compared with single-photon absorption, 2-photon absorption has a much lower probability - which, however, is proportional to the square of the photon density, such that only pulsed lasers can be employed because during the pulses the power can be extremely high. A Cu vapour laser was an appropriate choice [26]: it has excitation lines at 510 and 578 nm, 3 W average power, 30 nsec pulse duration, 5000 Hz pulse repetition frequency, and 2×10^4 W peak power. Effective excitation at 255 or 289 nm permits a choice amongst numerous chromophores. For two phenyloxadiazoles, detection limits of 0.25 and 4 pg were obtained.

MULTI-DIMENSIONAL DETECTION

In HPLC with fluorescence detection, three parameters besides retention time can serve for identification purposes: excitation and emission wavelengths, and fluorescence lifetime. Emission and excitation spectra can be combined to give characteristic matrices (EEM's). Unfortunately, under normal fluid conditions the fluorescence spectra are rather featureless, which limits the selectivity obtainable by means of the EEM's. However, experimental conditions can be created in which the inhomogeneous broadening of the fluorescence transitions is strongly reduced and highly structured spectra are obtained. Solidified samples measured at cryogenic temperatures are required, or else samples introduced into a supersonic jet nozzle, as tried [27] (excitation line-width 0.007 nm) for anthracene and some of its derivatives with, however, rather high detection limits (~40 ng). In our laboratory, micro-HPLC effluents have been immobilized on a slowly moving TLC plate [28, 29], allowing emission and excitation spectra to be obtained using conventional techniques; the TLC plate can also be cooled down to cryogenic temperatures and fluorescence line-narrowing spectra then recorded. Even under rather unfavourable conditions, 30 pg of tetracene could be detected.

Monitoring complete fluorescence spectra during the HPLC run notably helps in identifying peaks, checking their purity and resolving fused peaks. The sensitivity of multi-channel detectors, compared with conventional systems using photomultipliers, is still low, e.g. 1 ng for perylene in a video fluorimetric detector described in 1981 [30]. The situation has been improved by using intensified linear photodiode arrays (ILDA's) and by increasing the number of photomultiplier tubes. With ILDA's, Gluckman et al. [31] fingerprinted polyaromatic hydrocarbons using a xenon light source and a fibre-optic-compatible flow-cell. For such analytes (e.g. in soil extracts) Skoropinski et al. [32] used a nitrogen laser-pumped tunable dye laser with rapid tuning capability so that for each peak excitation could be optimized if a single wavelength (e.g. 365 nm) did not suit. Whereas, with conventional-size columns they could detect 75 pg of benz[a]pyrene by nitrogen-dye laser if with ILDA, 0.75 pg was detectable if with gated photomultiplier - but this was only a 3-fold improvement on the 2.5 pg detectable with a conventional fluorescence spectrometer; however, systems with a single-channel photomultiplier basis rather than ILDA do not provide spectral information. Between-laboratory comparisons cannot readily be made.

As an alternative, Winefordner and co-workers [33] have developed an instrument capable of simultaneously monitoring 4 emission wavelengths, using 4 interference filter photomultipliers, thus increasing both sensitivity and selectivity; as the 4 tubes were driven by a single HV power supply, the system was not very expensive. Fig. 7 shows the system, in which the excitation light from a 300 W xenon arc lamp reaches the flow-cell through optical fibres. Fused HPLC peaks for PAH's in admixture were resolvable (Fig. 8; sub-ng levels). Another approach [34], so selective that 15 PAH's (>0.2ng) could be identified in combustion products, is determination of an analyte's fluorescence lifetime 'on-the-fly' in an HPLC eluent. Excitation (337.1 nm) was by nitrogen laser; pulse repetition frequency was 10 Hz, and pulse width was 1.5 nsec. Simultaneous measurement of only 2 data points on the fluorescence decay curve gave a lifetime estimate.

CHEMILUMINESCENCE (CL) DETEECTION

Peroxyoxalate reaction

The benefit from achieving CL by substituting chemical excitation for photoexcitation was mentioned earlier. The peroxyoxalate reaction is probably the most efficient and versatile CL system currently available. It can be used for exciting and hence determining many classes of fluorophore, and also for detecting small amounts of hydrogen peroxide produced by photochemical or enzymic reactions. The peroxyoxalate reaction, reported in 1967, was first used for HPLC detection in 1980 [35]. Fig. 9 shows the generally accepted reaction mechanism. An energy-rich intermediate, postulated to be 1,2-dioxetanedione,

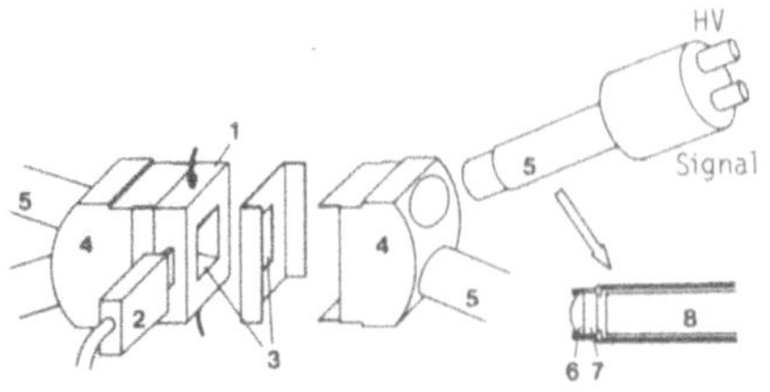

Fig. 7. Detector module used by Winefordner & co-workers. **1**, Black Teflon cell holder; **2**, quartz optical fibre bundle; **3**, side opening for observation of fluorescence signal; **4**, brass detector channel holder; **5**, detector channel; **6**, quartz lens; **7**, interference filter; **8**, photomultiplier tube. *Figs. 7 & 8 from [33], by permission.*

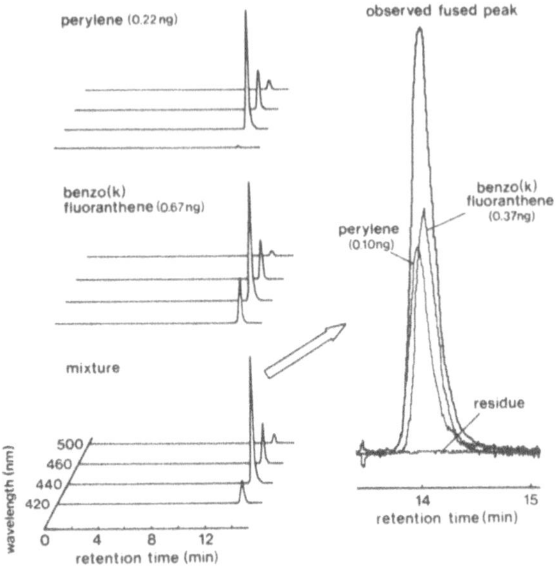

Fig. 8. Resolution of fused HPLC peaks of 2 PAH's (sub-ng amounts); cf. Fig. 7.

Fig. 9. Peroxyoxalate reaction scheme (see text). Possible aryl oxalates: TCPO, bis-(2, 4,6-trichlorophenyl)-oxalate; DNPO, bis-(2,4-dinitrophenyl)-oxalate.

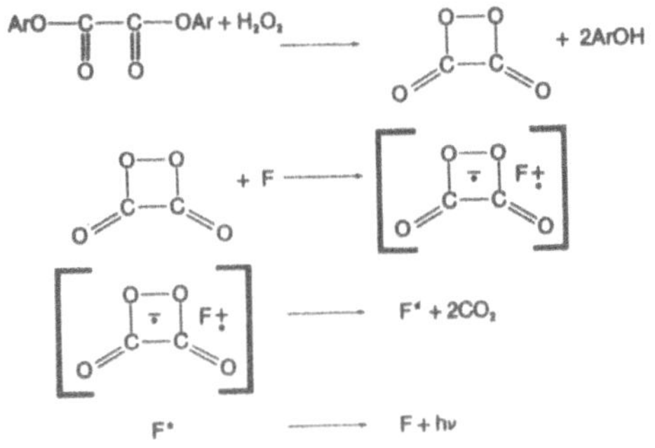

is generated by reaction of hydrogen peroxide and an aryl oxalate such as TCPO or DNPO. The intermediate excites the fluorophore, which returns to the ground state via emission of light. The reported transfer limit, 105 Kcal/mol [36], corresponds to a photoexcitation wavelength of ~270 nm. For high CL intensities the excitation energy should preferably be low [36-38]. A charge-transfer complex is formed between the fluorophore and 1,2-dioxetanedione through electron transfer (Fig. 9), and fluorophores with low oxidation potentials are indeed found to be good CL energy acceptors, as exemplified [38] by results for a series of heterocyclic compounds. The peroxyoxalate reaction is, then, a so-called chemically induced electron-exchange luminescence reaction. The existence of the dioxetanedione intermediate is still debated: e.g. the dependence of the reaction kinetics and CL intensity on the nature of the oxalate aryl groups argues against a common intermediate. Hence Catherall and co-workers [39] have suggested that only one of the two aryls is eliminated from the oxalates, and Alvarez et al. [40] recently concluded that the fluorophore is excited by at least 2 intermediates. A better insight into the reaction mechanism may well improve the analytical potential of this reaction.

The role of various parameters

CL reactions differ significantly from most other detection principles used for HPLC; they display a luminescence growth curve followed by a decay of the signal intensity caused by exhaustion of the light-generating agent(s). In a dynamic system, the half-life of the CL signal is a very important parameter; for given values of the various flow-rates, the dead volume between the mixing-T and the flow-cell and the actual flow-cell volume, the CL half-life determines what % of the emitted light will be measured [41]. The half-life depends strongly on the aryl oxalate structure, and is short with DNPO compared with TCPO [42, 43]. It depends, for a given aryl oxalate, on the (final) solvent composition and invariably decreases - often dramatically - with increase in pH or water content [41, 42, 44]. The CL signal's lifetime is also a function of the concentration of an added catalyst such as imidazole [cf. 45, 46].

Proper solvent selection is also a problem. The aryl oxalate usually chosen is TCPO; esters and ethers are the best solvents for this reagent. However, because ethers react with oxygen to form peroxides, TCPO decomposes rather rapidly if dissolved in an ether. Moreover, most common esters are rather water-immiscible. Hence CL monitoring of partly aqueous solutions - such as typical HPLC eluents - requires the presence of a third solvent to create a homogeneous system. In practice, ethyl acetate is often selected as a solvent for TCPO, and hydrogen peroxide is dissolved in tetrahydrofuran or acetone. Because a pre-mixed TCPO/hydrogen peroxide solution is not very stable, the two reagent solutions are usually mixed on-line for addition to the HPLC effluent.

The maximum allowable concentration is ~0.01 M for TCPO in ethyl acetate, and ~0.02 M for DNPO in acetonitrile. A higher concentration, in principle, means higher detection sensitivity. Imai et al. [47] have found that bis[4-nitro-2-(3,6,9-trioxadecyloxycarbonyl)phenyl] oxalate dissolves very well in, e.g., acetonitrile (1 M). However, such a high concentration also causes a considerable increase of the background signal. In practice, concentrations of 0.001-0.01 M are found to be near-optimal. The hydrogen peroxide level usually is 0.1-1 M, to give high sensitivity and reduce the half-life of the CL signal [35, 46].

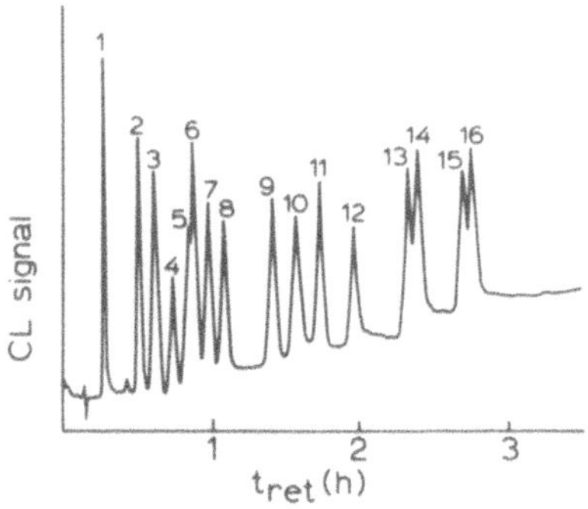

Fig. 10. Gradient-elution HPLC of dansylated amino acids (each 100 fmol) with CL detection. Peaks 1 to 16: Asp, Asn, Gln, Ser, Gly, Thr, Ala, Pro, Lys, Val, Arg, Met, Ile, Leu, Trp, Phe. Conditions: C-18 column; imidazole nitrate-acetonitrile gradient at 0.3 ml/min; CL reagent (TCPO in EtAc/H_2O_2 in acetone) at 1.8 ml/min. *From ref. [49], by permission.*

Applications

The following list illustrates the scope of peroxyoxalate CL for detecting fluorophores and fluorophore-labelled systems and the widespread use of dansylated compounds (derivatizing agent in *italics*).

Dansyl compounds: amino acids, steroids, amines (primary, secondary and tertiary) [35, 41, 43, 46, 48-53].
3-Amino-polynuclear aromatics: carboxylic acids, aldehydes, ketones [38, 54].
OPA, naphthalene dialdehyde: primary amines [52].
Coumarins: carboxylic acids [55].
Fluram/NBD [i.e. 4-chloro-7-nitrobenzo-2,1,3-oxadiazole]: catecholamines, primary amines [35, 56].
[No agent]: polynuclear aromatics (PAH's) [57].

In comparison with the fluorescence detection limit, that for dansylated compounds by CL is often 25- to 100-fold lower: ~200 amol for dansylated amino acids in a microbore HPLC system (gradient-elution run shown in Fig. 10) [50]. In a system developed by Kwakman et al. [53], ion pairs formed between the analytes - protonated tertiary amines - and the CL reagent 5-dimethylaminonaphthalene-1-sulphonate (dansyl-OH) are extracted in an on-line post-column mode into 1,2-dichloroethane containing TCPO. H_2O_2 is added to the organic phase by means of a solid-state perhydrit (H_2O_2 held on a urea support) reactor. Sub-ng amounts of the potential drug secoverine were detectable, the constraint being the background signal due to the co-extraction of the ion-pairing reagent.

CL gives notable sensitivity for amino-substituted PAH's even in the presence of large amounts of other types of polycyclic aromatics [57], as illustrated for a shale-oil extract in Fig. 11 which compares CL with fluorescence detection. Pre- or post-column insertion of a zinc column enables the same system to be used for the sensitive detection of nitro-aromatics. As a further extension, amino-substituted aromatics can be used to convert other analytes into CL-detectable products. Honda et al. [38] were able to detect 100 amol of 3-amino-perylene derivatives of carboxylic acids, and Mann & Grayeski [54] have demonstrated the potential of labelling aldehydes and ketones

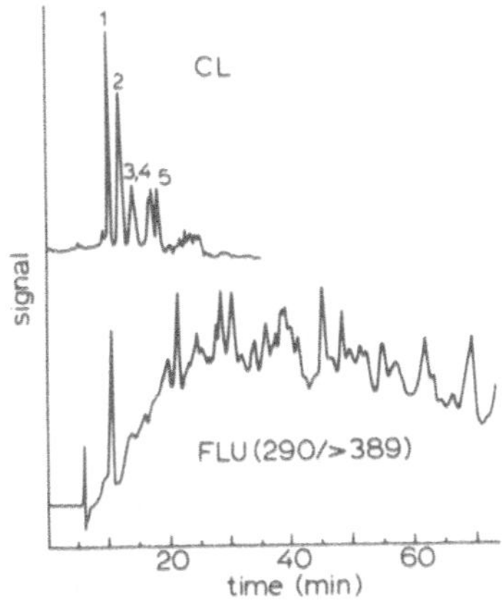

Fig. 11. Comparison of CL and fluorescence detection (HPLC of fractionated shale oil extract). Peaks: 1, aminonaphthalenes; 2, C_1-aminonaphthalenes; 3, aminophenanthrenes; 4, aminoanthracenes; 5, aminopyrenes. Conditions: C-18; acetonitrile/water (3:1) at 0.7 ml/min. *From ref [57], by permission.*

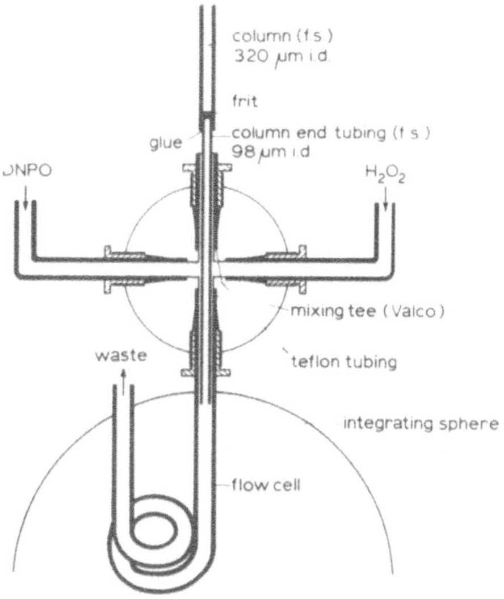

Fig. 12. Mixing of eluent (from a packed capillary column) and DNPO + H_2O_2 (for CL detection; added by syringe pump), in a 70 µl (0.3 mm i.d.) Teflon flow-cell placed in an integrating sphere. *From [46], by permission.*

with 3-aminofluoranthene. Coumarins containing an electron-donating amino functional group serve for determining carboxylic acids [55].

In another approach, quenching of the peroxyoxalate CL signal allowed pmol detection of (e.g.) some anilines and organosulphur compounds and also SO_3^{2-}, NO_2^-, Br^- and I^- [58, 59]. Occasionally compounds which lack native fluorescence are, surprisingly, directly detectable by the peroxyoxalate reaction. Certain compounds enhance the CL background signal from DNPO and H_2O_2, and 2 pmol of ouabain or 20 pmol of urea can be detected [60].

Miniaturization.- The peroxyoxalate system serves well for miniaturized HPLC [41, 47, 50, 51], preferably with DNPO because of its short half-life. However, a satisfactory post-column mixing-cum-detection unit is difficult to construct. The decay of the CL signal is so rapid (often <1 sec half-life) that much emitted light is lost even with a tiny dead time between reagent/effluent mixing and the cell inlet; a remedy is to mix within an integrated sphere which also efficiently collects nearly all the emitted light. Even with a glass coil as large as 450 µl as flow-cell [41], additional band broadening is almost negligible because the CL signal completely decays in a small part of this cell (discussed in [41, 43]). A recent detection system (Fig. 12; see legend) exploits the same principle for HPLC with packed capillary columns (320 µm i.d.) [46], allowing ~400 amol of several dansylated compounds to be detected. With a similar system but with TCPO instead of DNPO, Weber & Grayseki [62] had rather broadened peaks.

Hydrogen peroxide.- Finally, with excess fluorophore the peroxyoxalate reaction serves to determine trace amounts of H_2O_2, such as on-line enzyme reactions may produce [63-66]. Honda et al. [67] found low-pmol amounts of pure choline and acetylcholine to be detectable with a post-column reactor containing immobilized choline oxidase and cholinesterase plus a buffer and a solution containing TCPO and perylene. With the same enzyme system but, more simply, using TCPO and 3-aminofluoranthene chemically bonded to glass beads packed in the flow-cell in front of the photomultiplier [68], choline and acetylcholine in urine and serum have been determined [69], evidencing the high selectivity of the combined use of enzyme reactors and peroxyoxalate CL detection. The same detection system applied to H_2O_2 generated by a post-column L-amino oxidase reactor enabled enantioselective detection of amino acids at pmol levels to be achieved (H. Jansen, U.A.Th. Brinkman & R.W. Frei, to be published). Grayeski et al. [70] have designed a flow-injection system for determining glucose through the detection of H_2O_2 generated by glucose oxidase. A new water-soluble oxalate was employed, with Rhodamine B as fluorophore.

References

1. Yeung, E.S. & Synovec, R.E. (1986) *Anal. Chem. 58*, 1237A-1256A.
2. Frei, R.W., Lawrence, J.F., Brinkman, U.A.Th. & Honigberg, I. (1979) *J. High Res. Chromatog. Chromatog. Comm. 1*, 11-14.
3. Reddinguis, R.J., de Jong, G.J., Brinkman, U.A.Th. & Frei, R.W. (1981) *J. Chromatog. 205*, 77-84.
4. de Ruiter, C., Brinkman, U.A.Th. & Frei, R.W. (1987) *J. Liq. Chromatog. 10*, 1903-1916.
5. Wolf, J.H., de Ruiter, C., Brinkman, U.A.Th., & Frei, R.W. (1986) *J. Pharm. Biomed. Anal. 4*, 523-527.
6. Smedes, F., Kraak, J.C., Werkhoven-Goewie, C.E., Brinkman, U.A.Th. & Frei, R.W. (1982) *J. Chromatog. 247*, 123-132.
7. Nondek, L., Frei, R.W. & Brinkman, U.A.Th. (1983) *J. Chromatog. 282*, 141-150.
8. Low, Kun She, Brinkman, U.A.Th. & Frei, R.W. (1984) *Anal. Lett. 17A*, 915-931.
9. Werkhoven-Goewie, C.E., Niessen, W.M.A., Brinkman, U.A.Th. & Frei, R.W. (1981) *J. Chromatog. 203*, 165-172.
10. Scholten, A.H.M.T., Brinkman, U.A.Th. & Frei, R.W. (1980) *Anal. Chim. Acta 114*, 137-146.
11. Birks, J.W. & Frei, R.W. (1982) *Trends Anal. Chem. 1*, 361-367.
12. Krull, I.S. & LaCourse, W.R. (1986) in *Reaction Detection in Liquid Chromatography* (Krull, I.S., ed.), Dekker, New York, pp.303-352.
13. Uihlein, M. & Schwab, E. (1982) *Chromatographia 15*, 140-146.
14. Salomoun, J. & Frantisek, J. (1986) *J. Chromatog. 378*, 173-181.

15. Yeung, E.S. & Sepaniak, M.J. (1980) *Anal. Chem. 52*, 1465A-1481A.
16. Dovichi, N.J. (1984) *Trends Anal. Chem. 3*, 55-57.
17. Folestad, S., Galle, B. & Josefsson, B. (1985) *J. Chromatog. Sci. 53*, 273-278.
18. Pfeffer, W.D. & Yeung, E.S. (1986) *Anal. Chem. 58*, 2103-2105.
19. van Vliet, H.P.M. (1986) *Laser Induced Fluorescence Detection in Microbore and Open Tubular Liquid Chromatography*, Thesis, U. Amsterdam.
20. Roach, M.C. & Harmony, M.D. (1987) *Anal. Chem. 59*, 411-415.*
21. Sauda, K., Imasaka, T. & Ishibashi, N. (1986) *Anal. Chem. 58*, 2649-2653.
22. Guthrie, E.J., Jorgenson, J.W. & Dluzneski, P.R. (1984) *J. Chromatog. Sci. 22*, 171-176.
23. Folestad, S., Johnson, R., Josefsson, B. & Galle, B. (1982) *Anal. Chem. 54*, 925-929.
24. Hershberger, L.W., Callis, J.B. & Christian, G.D. (1979) *Anal. Chem. 51*, 1444-1446.
25. Gluckman, J.C., Shelly, D.C. & Novotny, M.V. (1984) *J. Chromatog. 317*, 443-453.
26. Pfeffer, W.D. & Yeung, E.S. (1986) *Anal. Chem. 58*, 2103-2105.
27. Imasaka, I., Yamaga, N. & Ishibashi, N. (1987) *Anal. Chem. 59*, 419-422.
28. Hofstraat, J.W., Engelsma, M., van de Nesse, R.J., Gooijer, C., Velthorst, N.H. & Brinkman, U.A.Th. (1986) *Anal. Chim. Acta 186*, 247-259.
29. Hofstraat, J.W., Engelsma, M., van de Nesse, V.R., Brinkman, U.A.Th., Gooijer, C. & Velthorst, N.H. (1987) *Anal. Chim. Acta 193*, 193-
30. Hershberger, L.W., Callis, J.B. & Christian, G.D. (1981) *Anal. Chem. 53*, 971-975.
31. Gluckman, J.C., Shelly, D.C. & Novotny, M.V. (1985) *Anal. Chem. 57*, 1546-1552.
32. Skoropinski, D.B., Callis, J.C., Danielson, J.D.S. & Christian, D. (1986) *Anal. Chem. 58*, 2831-2839.
33. Tanabe, K., Glick, M., Smith, B., Voigtman, E. & Winefordner, J.D. (1987) *Anal. Chem. 59*, 1125-1129.
34. Desilets, D.J., Kissinger, P.T. & Lytle, F.E. (1987) *Anal. Chem. 59*, 1830-1834.
35. Kobayaski, S. & Imai, K. (1980) *Anal. Chem. 52*, 424-427.
36. Lechtken, P. & Turro, N.J. (1974) *Mol. Photochem. 6*, 95-99.
37. Sherman, P.A., Holzbecher, J. & Ryan, D.E. (1978) *Anal. Chim. Acta 97*, 21-27.
38. Honda, K., Miyaguchi, K. & Imai, K. (1985) *Anal. Chim. Acta 177*, 111-120.
39. Catherall, C.L.R., Palmer, T.F. & Cundall, R.B. (1984) *J. Chem. Soc. Faraday Trans. 2*, 823-834 & 837-849.
40. Alvarez, F.J., Porekh, N.J., Matuszewski, B., Givens, R.S., Higuchi, T. & Schowen, R.L. (1986) *J. Am.Chem. Soc. 108*, 6435-6437.
41. de Jong, G.J., Lammers, N., Spruit, F.J., Frei, R.W. & Brinkman, U.A.Th. (1986) *J. Chromatog. 353*, 249-257.
42. Honda, K., Miyaguchi, K. & Imai, K. (1985) *Anal. Chim. Acta 177*, 103-110.

[*Footnote overleaf.

43. de Jong, G.J., Lammers, N., Spruit, F.J., Brinkman, U.A.Th., & Frei, R.W. (1984) *Chromatographia 18*, 129-133.
44. Weinberger, R. (1984) *J. Chromatog. 314*, 155-165.
45. Imai, K., Miyaguchi, K. & Honda, K. (1986) in *Bioluminescence and Chemiluminescence: Instruments and Applications*, Vol. 2 (Van Dijke, K., ed.), CRC Press, Cleveland, OH, pp. 65-76.
46. de Jong, G.J., Lammers, N., Spruit, F.J., Dewaele, C., & Verzele, M. (1987) *Anal. Chem. 59*, 1458-1461.
47. Imai, K., Matsunaga, Y., Tsukamoto, Y. & Nishitani, A. (1987) *J. Chromatog. 400*, 169-176.
48. Melbin, G. (1983) *J. Liq. Chromatog 6*, 1603-1616.
49. Miyaguchi, K., Honda, K. & Imai, K. (1984) *J. Chromatog. 303*, 173-176.
50. Miyaguchi, K., Honda, K. & Imai, K. (1984) *J. Chromatog. 316*, 501-505.
51. Koziol, T., Grayeski, M.L. & Weinberger, R. (1984) *J. Chromatog. 317*, 355-366.
52. Melbin, G. & Smith, B.E.F. (1984) *J. Chromatog. 312*, 203-210.
53. Kwakman, P.J.M., Brinkman, U.A.Th., Frei, R.W., de Jong, G.J., Spruit, F.J., Lammers, N.G.F.M. & van den Berg, J.H.M. (1987) *Chromatographia 24*, 395-399.
54. Mann, B. & Grayeski, M.L. (1987) *J. Chromatog. 386*, 149-158.
55. Grayeski, M.L. & DeVasto, J.K. (1987) *Anal. Chem. 59*, 1203-
56. Kobayaski, S., Sekino, J., Honda, K. & Imai, K. (1981) *Anal. Biochem. 112*, 99-104.
57. Sigvardson, K.W., Kennish, J.M. & Birks, J.W. (1984) *Anal. Chem. 56*, 1096-1102.
58. Van Zoonen, P., Kamminga, D.A., Gooijer, C., Velthorst, N.H. & Frei, R.W. (1986) *Anal. Chem. 58*, 1248-1251.
59. Van Zoonen, P., Bock, H., Gooijer, C., Velthorst, N.H. & Frei, R.W. (198) *Anal. Chim. Acta 200*, 131-141.
60. Capommachia, A.C., Jennings, R.N., Hemingway, S.M., D'Souza, P., Prapaitrakul, W. & Gingle, A. (1987) *Anal. Chim. Acta 196*, 305-310.
61. Grayeski, M.L. & Weber, A.J. (1984) *Anal.. Lett. 17*, 1539-1552.
62. Weber, A.J. & Grayeski, M.L. (1987) *Anal. Chem. 59*, 1452-1457.
63. Riggin, V.I. (1978) *J. Anal. Chem. USSR 33*, 1265-1270.
64. Riggin, V.I. (1979) *J. Anal. Chem. USSR 34*, 619-623.
65. Riggin, V.I. (1981) *J. Anal. Chem. USSR 36*, 1111-1115.
66. Riggin, V.I. (1983) *J. Anal. Chem. USSR 38*, 1328-1330.
67. Honda, K., Miyaguchi, K., Nishino, H., Tanaka, T., Yao, T. & Imai, K. (1986) *Anal. Biochem. 153*, 50-53.
68. Van Zoonen, P., Gooijer, C., Velthorst, N.H., Frei, R.W., Wolf, J.H., Gerrits, J. & Flentge, F. (1987) *J. Pharm. Biomed. Anal. 5*, 485-492.
69. Gübitz, G., Van Zoonen, P., Gooijer, C.. Velthorst, N.H. & Frei, R.W. (1985) *Anal. Chem. 57*, 2071-2074.
70. Grayeski, M.L., Woolf, E.J. & Helly, P.J. (1986) *Anal. Chim. Acta 183*, 207-215.

Concerning ref. 20, from which Fig. 4 is reproduced, Dr. M.D. Harmony as its senior author regrets a publication error by the American Chemical Society: Figs. 1 and 2 were transposed (& **15-40 in text should read **40-15**).*

#F-3

LC-MS AND LC-MS-MS FOR BIOMEDICAL ANALYSES

A.P. Bruins

State University, Department of Pharmacy
A. Deusinglaan 2, 9713 AW Groningen, The Netherlands

The identification of components separated by LC can be done either by preparative chromatography and off-line MS or, with advantages, by on-line LC-MS.*[θ] *The moving belt and DLI are the oldest commercially available interfaces. The belt allows EI, CI or FAB to be chosen for ionization, but samples may suffer from thermal degradation. DLI is limited to CI and suits thermolabile samples, but the 5 μm orifice in the interface is easily blocked. DLI has been replaced by thermospray LC-MS, which has become less elaborate and is now available in nearly all spectrometers on the market.*

The problem of vacuum pumping limitations in LC-MS is circumventable by API, whereby pumping at the ion source is eliminated. Atmospheric pressure operation allows the use of simple, effective pneumatic nebulizers. Analytes of biomedical interest may exist in solution as ions which can be examined by MS if the LC effluent is dispersed into an electrically charged aerosol. The emission of ions from small droplets, induced by an electric field due to the charge on each droplet, creates ions that can be mass-analyzed. In an API source the entire sequence is carried out at room temperature, thus preventing thermal degradation of labile samples. An innovation in vacuum MS is continuous flow FAB, involving bombarding a mixture of the LC effluent and glycerol with fast argon or xenon atoms. The liquid flow through the capillary connected to the source is ~5 μl/min or less. However, at the present time the moving belt and thermospray are the most practical LC-MS interfaces. Where LC-MS information is inadequate because of soft ionization or insufficient LC separation, LC-MS-MS can provide fragmentation reactions for structure elucidation, identification and selective determination.

* *Abbreviations:* LC, liquid chromatography (HPLC); MS, mass spectrometry; API, atmospheric pressure ionization; CI, chemical ionization; DLI, direct liquid introduction; EI, electron impact; FAB, fast atom bombardment; SIM, selected ion monitoring.

[θ]Arts. #NC(C)-4 in V. 14 and #NC(D)-5 in V. 16 (Martin et al.) are pertinent - *Ed.*

The identification of components in a mixture can be done by preparative chromatography and spectroscopic analysis of each component. The on-line combination of chromatography and spectrometry saves time, reduces analyte losses by eliminating sample-handling steps, can deconvolute overlapping chromatographic peaks, and in the case of MS permits quantitative analyses using standards labelled with stable isotopes. On-line GC-MS has been so well perfected that significant improvements seem unlikely. The advent of the flexible fused silica capillary column, which can be slid through a heated transfer line all the way into the ion source, has eliminated the GC-MS interface and its problems of plugging and analyte decomposition. An attached Fourier-transform infra-red detector allows GC-IR and GC-IR-MS as additional tools for the analyst.

Most separations in the biomedical area are performed by HPLC, which obviates thermolability problems and usually needs only simple sample preparation. Although there is an impressive range of detectors, there is no simple universal detector comparable to FID in GC. MS is not simple or truly universal, but a choice of modes for ionization and fragmentation makes MS very versatile. In combining LC with MS, the fundamental problem is the inability of the MS vacuum system to handle the gas load generated by evaporation of the eluate from a 4.6 mm i.d. HPLC column. The system can pump ~20 ml/min of gas (S.T.P.), while 1 ml methanol or water furnishes 550 or 1200 ml of gas (S.T.P.) respectively. Remedies include: (1) use of a moving belt to remove eluent and introduce the sample into the ion source; (2) direct introduction of a small portion of the effluent into the source (DLI); (3) miniaturization of the LC making the eluent flow rate compatible with the pumping speed of the MS vacuum system [1]; (4) connection of an additional vacuum pump to the source together with efficient transfer of a controlled amount of heat to the column eluate (thermospray approach) [2]; (5) no pumping at all at the source: ionization at atmospheric pressure (API) [3].

MOVING BELT

The moving belt and DLI were the earliest commercially available LC-MS interfaces, the belt giving free choice between EI, CI and FAB as ionization methods. Moreover, with CI there is a free choice of reactant ion, which is not the case in DLI, thermospray or API. Thus very reactive quite universal reactant ions such as CH_5^+ and OH^- may be selected [4, 5]. EI operation offers the advantage of reference spectra and library routines that support the identification of general unknowns in environmental and forensic analyses.

Moving-belt operation entails (1) deposition of the eluent, (2) solvent removal in vacuum, and (3) analyte volatilization into the ion source. Originally the eluate was made to flow directly onto the belt, which resulted in noisy total ion current traces.

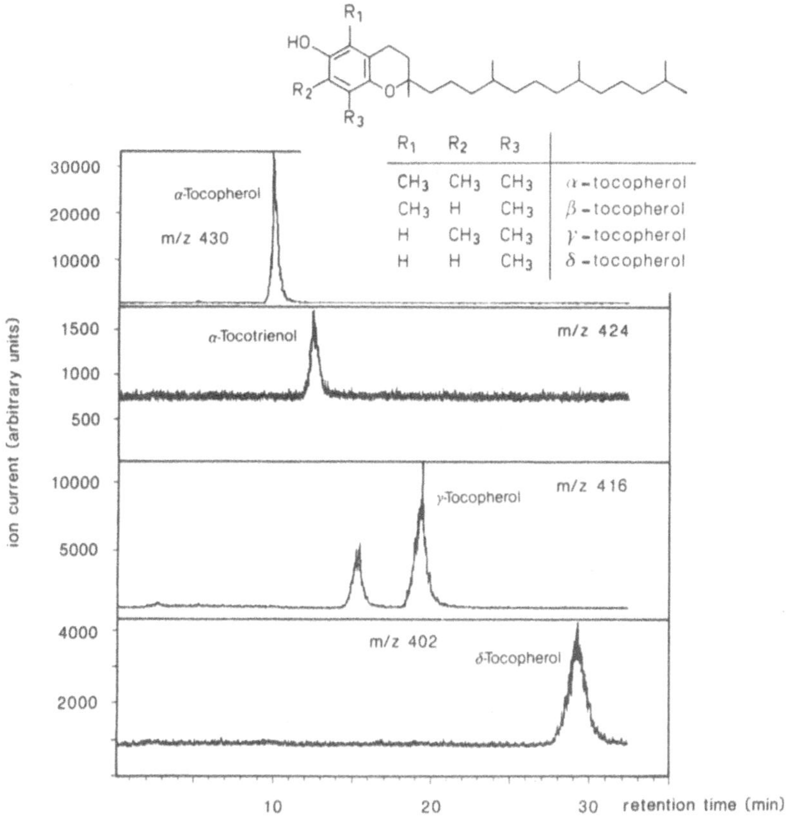

Fig. 1. SIM of the molecular ions of tocopherol standards, with a moving belt and EI. *From [10], courtesy of the authors and Elsevier.*

If the eluent contained a high percentage of water, beads of liquid were formed on the belt; solvent evaporation became difficult and analyte desorption into the ion source was irregular. Spray deposition of the eluate gives a regular layer, resulting in better evaporation of the solvent; hence the irregularities in analyte desorption and the noise level in the LC-MS trace were reduced [6]. Deactivation of the belt with Carbowax facilitates analyte desorption and gives lower detection limits [7]. Thermal degradation of labile analytes is unavoidable but can be reduced if desorption takes place inside the source close to the electron beam [8]. Analyte bombardment on the belt surface inside the source with fast argon or xenon atoms (FAB) generates analyte ions from the condensed phase without recourse to heat and thus avoids thermal degradation [9].

Fig. 1 shows EI results for tocopherol standards [10]. LC-MS with the moving belt is versatile but limited to mass spectrometers manufactured by VG (U.K.) and Finnigan (U.S.A. and W. Germany). It should receive further attention but is at present overshadowed by interest in thermospray LC-MS.

DIRECT LIQUID INTRODUCTION

In the early days of LC-MS the belt and DLI were rival techniques. If a small portion of the LC eluate is forced through a 5 µm diam. pinhole, a jet of liquid is injected into the ion source. The sample is desolvated in the hot source and the solvent vapour serves as the reactant gas for chemical ionization. EI ionization is not possible with commercially available DLI interfaces, and the choice of reactant ions is limited to weak gas-phase acids and bases [4, 5], giving mild ionization conditions. Thermal degradation is strongly reduced in comparison with the moving belt, as demonstrated in a study of ranitidine [11, & #NC(C)-4 in Vol. 14, this series]. DLI is a valuable technique in the hands of a skilled operator although great care has to be taken to avoid the orifice becoming plugged up. DLI was available from Hewlett Packard (U.S.A.) and Nermag (France) but has been virtually superseded by thermospray.

THERMOSPRAY

The evolution from a large scale instrument, with lasers and big pumps designed to accept the full flow from a 4.6 mm i.d. column, into a practical accessory has made thermospray the most popular LC-MS technique available. The heart of the thermospray interface is a heated capillary tube (Fig. 2); the LC eluate is partially evaporated inside the heated section and the remaining liquid is dispersed into a high velocity aerosol beam by the action of the fast-expanding solvent vapour. After ionization the ions of the solvent, additives and the analyte pass through an orifice in a cone and are mass-analyzed by a quadrupole mass filter or a magnetic sector spectrophotometer.

Thermospray is a combination of sample introduction and ionization in one device. Ions can be generated efficiently without the use of electrons emitted from a heated filament or ions produced by an electric discharge as in EI and CI sources; this works well if an electrolyte is added to the eluent. Ammonium acetate is most often used, but other volatile salts, acids or bases may be preferable for certain applications. Desolvation of the ions of the electrolyte (NH_4^+, CH_3COO^-) produces reactant ions for chemical ionization, facilitated if the eluent is rich in water. If the direct formation of reactant ions is not possible, e.g. because of electrolyte lack or a solvent which does not favour the desolvation of ions from solution, an electron emitted by a filament or an electrical discharge is used to support the production of reactant ions.

A clear distinction appears to exist in thermospray LC-MS between analytes that are present as neutrals in solution - and are ionized

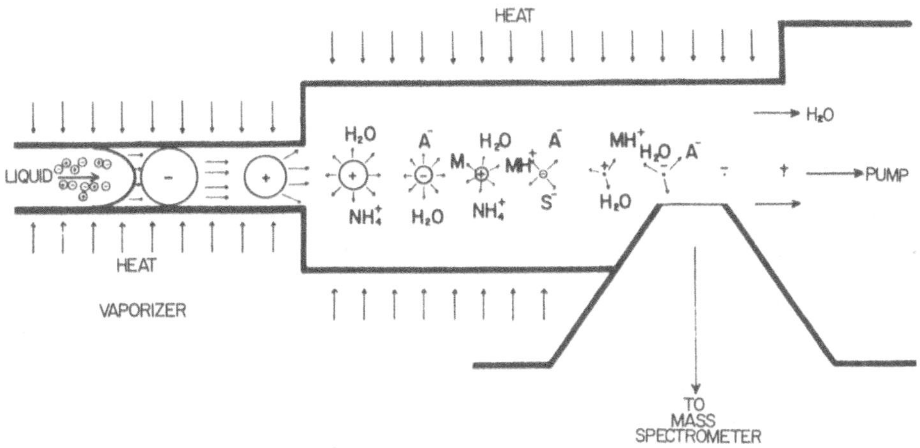

Fig. 2. Schematic diagram of a thermospray LC-MS interface and ion source. *From [12], courtesy of author and publisher.*

by the above-mentioned CI process - and those that are ionized. With the latter the solvation process generates analyte ions in the ion source without the intermediacy of ion molecule reactions [12]. An electrolyte in the eluent is unnecessary and has even been shown to suppress the formation of analyte ions [13].

The reactant ions formed from ammonium acetate are both mild agents [4] that usually yield quasimolecular ions; but some components in a mixture will not become ionized at all. However, since many drugs contain nitrogen atoms, they are strong enough bases in the gas phase to accept a proton from the NH_4^+ ion. Use of ammonium acetate and an electron beam or discharge to facilitate the formation of reactant ions is a good starting point. If ineffective, one must use CI knowledge accumulated since its initiation in 1966, especially the order of acidity and basicity in the gas phase [4, 5]. Triethylamine, a strong base in the gas phase, might be added to the eluent to improve a particular separation. However, it will consume all the positive reactant ions in the ion source so that the sample cannot be ionized and detected.

Despite the heat applied to the thermospray interface, labile molecules can be handled fairly well since the residence time in the heated area is short. Temperature control of the heated capillary is very important: if too low, the generated droplets are too big for the production of reactant and analyte ions. Too high a temperature gives a dry vapour in the ion source without the production of ions from the liquid phase, but ionization can be effected with the electron beam or electrical discharge. Labile analytes decompose under the conditions of a dry vapour. The optimum temperature is limited to a 10° band; since it depends on eluent composition, the interface caters for temperature programming to allow gradient elution.

Schellenberg et al. [14] reported their experience with DLI and thermospray in the analytical laboratory of the Pharma Division of Sandoz. Compared with DLI, thermospray entails no orifice-plugging problem, and is simpler and easier to operate [14] and suitable for automation [15]; but the total ion current traces have greater noise although use of an electron beam or discharge reduces it. Sensitivity fluctuations are not a serious problem in quantitative analyses if standards labelled with stable isotopes are employed. The analyte amount required for full spectra and recognizable peaks in total ion current traces by thermospray is 10 ng in favourable cases but typically 100 ng to 1 μg.

The original thermospray was a simple system for generating an aerosol and ions. The addition of a filament or discharge introduced complexity, and new names added further confusion: plasmaspray (VG Masslab, U.K.) is merely a thermospray with a discharge of construction such that the thermospray vaporizer tip serves as the discharge electrode. The addition of a repeller to the ion source appeared to be useful for generating fragment ions that give insight into the presence of functional groups in the analyte [16, 17].

Thermospray LC-MS of drug metabolites that are ionized in solution, e.g. glucuronic or sulphuric acid conjugates, has been reported [18, 19]. Temperature control is more critical and sensitivity is poorer than for CI of neutral samples. New LC-MS techniques for ionic analytes are continuous-flow FAB [20] and ion spray at atmospheric pressure, as now considered.

ATMOSPHERIC PRESSURE IONIZATION

With no vacuum at all in the ion source the problems of limited vacuum pumping capacity are entirely eliminated. One of the earliest reports of on-line LC-MS [21] described the use of an API source. However, interest in API has been limited because the traditional manufacturers of analytical mass spectrometers do not offer an API source. The original API source was very sensitive but prone to plugging of the ion sampling source (25 μm diam.) and formation of large cluster ions by the association of analyte ions with polar solvent molecules. A modern API source has a wider orifice (100 μm), a gas curtain that keeps dust and polar solvent molecules away from the orifice, extremely fast cryogenic pumping of the mass analyzer and a cluster breaker [22]. Operation at atmospheric pressure allows the use of simple effective pneumatic nebulizers that need several litres/min of gas for the dispersion of the eluate into an aerosol [3].

CI at atmospheric pressure (APCI) can be initiated by an electrical discharge. There is no fundamental difference between APCI and ion molecule reactions in a conventional CI source at 0.3 Torr (40 Pa) or in a thermospray source at ~3 Torr (400 Pa). Again the gas-phase acidity and basicity of all eluent components should be taken into account. LC-MS with APCI is limited to mild reactant ions, as for DLI and thermospray. A simple system is the combination of a pneumatic nebulizer, a heated desolvation region and APCI with a corona discharge [3], in which a dry desolvated analyte is ionized in the gas phase. Since temperature control of the desolvation region is not critical, operation of the heated pneumatic nebulizer is simpler than in thermospray LC-MS; this region need not be temperature-programmed during gradient elution. In the hot region the sample residence time is short; hence there is little thermal degradation of labile analytes. LC-MS with a heated pneumatic nebulizer and APCI is comparable with thermospray but has limitations for compounds that are ionized in solution. [The SCIEX (Canada) installation at Cornell University may be unique in the biomedical area.] The heated nebulizer can accept the full effluent from a 4.6 mm i.d. column. Applications of LC-MS with API are the analysis of tranquillizers and analgesics administered to racehorses and of natural oestrogens in food [3, 23].

Spray techniques in API

As MS is a means of separating and detecting ions, for analytes of biomedical interest that exist as ions in solution the ions present in the LC eluate should be exploitable. This simple approach is indeed possible if the eluate is dispersed into an electrically charged aerosol. The emission of ions from small droplets induced by the electric field generated by the charge on each droplet creates analyte ions that can be mass-analyzed. In API performed at room temperature throughout, thermal degradation of labile analytes is obviated. Three technically different but fundamentally related techniques are available: ion evaporation [24], electrospray [25, 26] and ion spray [27].

In liquid ion evaporation a solution is dispersed through a pneumatic nebulizer into air at atmospheric pressure. Charges are induced on sprayed droplets by means of a small high-voltage electrode placed close to the sprayer. Elevated temperatures are unnecessary and conventional LC flow rates are acceptable, but sufficient sensitivity for trace analyses has yet to be attained.

The production of a fine mist of charged droplets by electrospray was first demonstrated by Zeleny in 1917. LC-MS application, as claimed in patents, has only recently appeared in the literature [25, 26]. A solution of the analyte is injected into dry air or nitrogen at atmospheric pressure through a metal capillary tube that is at a potential of several kV relative to the walls of the ion source. Charge is deposited onto the surface of the emerging liquid, resulting in the production of coulomb repulsion forces sufficient to overcome surface tension so that the liquid is dispersed into a fine spray. The technique works best with flow rates in the range below 5-10 μl/min and hence needs microbore columns; dispersion of

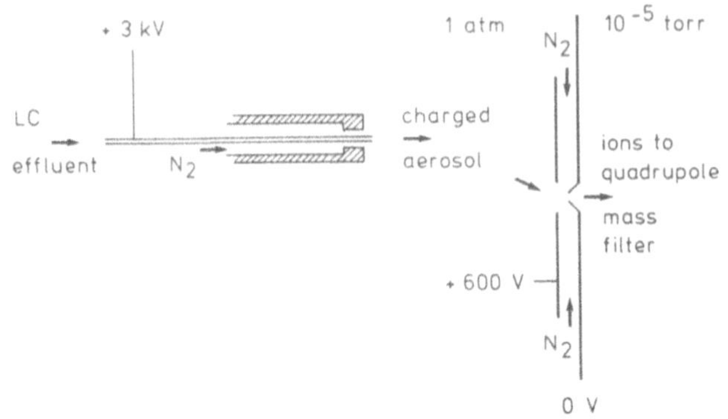

Fig. 3. Schematic diagram of the ion spray LC-MS interface and atmospheric pressure ion source with nitrogen gas curtain. Voltages are given for positive ion formation.

liquids becomes more difficult at flow rates >5-10 µl/min, particularly in the presence of a high percentage of water in the eluate. With its very low flow rates, it is elegantly used in the combination of capillary zone electrophoresis and MS, where the liquid flows at only 1 µl/min [28].

A combination of ion evaporation and electrospray interfaces has been constructed ('ion spray'; Fig. 3): it can be described as electrospray with the assistance of pneumatic nebulization or as ion evaporation with a pneumatic sprayer floating at high voltage. Ion spray works well at a flow rate of 40 µl/min from a 1 mm i.d. microbore column and has been tested at flow rates up to 200 µl/min albeit with some forfeit of sensitivity. It has worked simply and reliably for salts, acids, bases, amino acids, peptides, and sulphuric or glucuronic acid conjugates (Fig. 4). The sample size required for recording of full spectra and recognizable peaks in the total ion current trace is ~10 ng of an ionic analyte. Ion evaporation, electrospray and ion spray cannot be considered universal ionization techniques for many less polar compounds.

CONTINUOUS FLOW FAB

FAB is a simple and effective method for analyzing compounds that are too polar, labile and involatile to be amenable to EI- and CI-MS, but does have shortcomings. First, the significant levels of glycerol or other matrix liquids give rise to a high background level in the spectra. Secondly, conventional FAB using a sample introduction probe is unsuited for the on-line combination with chromatography. To overcome these shortcomings a continuous flow device was reported [29], constructed from a capillary terminated by a mesh frit, or connected to a flat copper surface in the ion source [30].

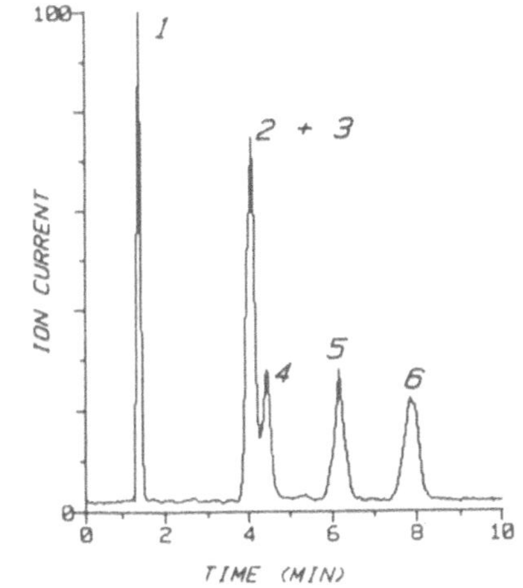

Fig. 4. Ion spray LC-MS of steroidal sulphoconjugates: sum of selected ion current profiles of $(M-H)^-$ ions. Components (1-2 ng of each was taken):
1 17β-Oestradiol disulphate;
2 19-Nortestosterone sulphate;
3 17β-Oestradiol 3-sulphate;
4 Boldenone sulphate;
5 Testosterone sulphate;
6 Dehydroepiandrosterone sulphate.

Courtesy of L.O.G. Weidolf, E.D. Lee and J.D. Henion (unpublished work).

Samples are injected into a solvent mixture containing an involatile component such as glycerol. The liquid surface in the ion source is bombarded with fast argon or xenon atoms; apparently the continuous refreshment of the liquid surface and the minimal need for glycerol compared with normal FAB produce cleaner spectra of sample ions than traditional FAB. The flow rate of 5 µl/min is compatible with packed microbore columns made of 0.32 mm i.d. fused silica tubing. Whilst the approach is attractive for ionizing polar analytes from the liquid phase, it is currently reckoned to be still an art although understanding is growing.

TANDEM MASS SPECTROMETRY, MS-MS

EI mass spectra [31] show abundant fragment ions indispensable for structure elucidation and identification by MS. Soft ionization techniques in LC-MS (DLI, thermospray, APCI, ion spray and continuous flow FAB) typically produce mass spectra that show a peak corresponding with the mol. wt. of the sample, with few or no fragment ions.

Fragmentation can be induced by energetic collisions with neutral gas atoms or molecules such as He, Ar, N_2, air or Xe. To achieve selection of parent ions and mass analysis of fragment ions, two mass spectrometers are connected in series, separated by a collision chamber filled with Ar or another suitable collision gas:

$$\text{ion source} \longrightarrow MS_1 \longrightarrow \text{collision chamber} \longrightarrow MS_2 \rightarrow \text{detector.}$$

A tandem mass spectrometer can be used to record full spectra in three ways:
- selection of a specific ion in MS_1 and a scan of MS_2 gives all daughter ions of a specific ion (daughter ion scan);
- scan of both MS_1 and MS_2 such that MS_2 passes ions with a mass lower

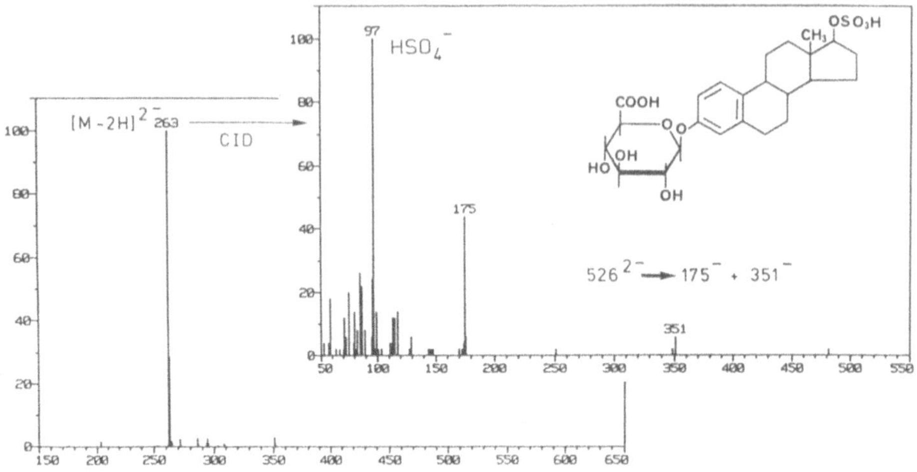

Fig. 5. Ion-spray LC-MS of oestradiol glucuronide sulphate (M_r 528): MS-MS with a collision-induced dissociation of the $(M-2H)^{2-}$ ion and daughter ion spectrum. *From [27], courtesy of authors and publisher.*

than MS_1 at a constant mass difference, gives all ions that lose a common neutral - e.g. CO_2, 44 daltons in the collision chamber (constant neutral loss scan);
- scan of MS_1 and selection of a specific ion in MS_2 gives all precursor ions of a common daughter ion (parent ion scan).

A daughter ion scan is most useful for structure elucidation and confirmation of the presence of certain functional groups in an analyte, as shown in Fig. 5 where the fragment ions at m/z 97 and 175 correspond to the sulphate and glucuronide functions respectively. Constant neutral-loss and parent ion-scans are well suited to the identification of all components in a family of compounds having a common functional group, e.g. drug metabolites, as shown in Fig. 6 where m/z 93 is a common fragment of metabolites of phenylbutazone.

If both MS_1 and MS_2 are set at a fixed but different mass, a selected fragmentation reaction of a specific parent to give a specific daughter ion is observed. This selected reaction monitoring technique is an extension of the well known SIM method in single-stage MS used for sensitive qualitative and quantitative analysis in GC-MS and LC-MS. Selected reaction monitoring is more specific than SIM, shows less background due to ions generated from solvents and contaminants and discriminates more strongly than SIM against components in a mixture eluting together with the compound of interest in a LC separation. The absolute sensitivity of MS-MS is, however, lower than with single-stage MS through the loss of ions by scattering in the collision chamber and through transmission losses in two mass analyzers instead of one.

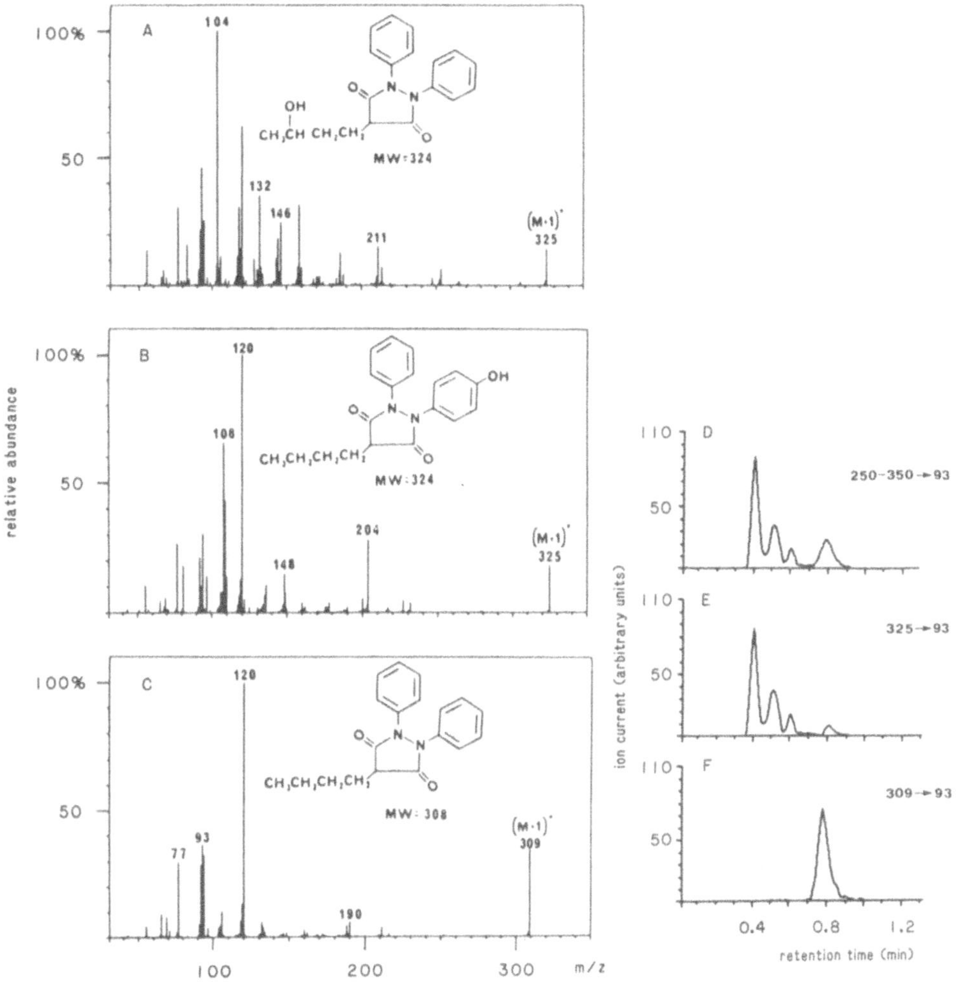

Fig. 6. LC-MS-MS of phenylbutazone and its metabolites. Heated pneumatic nebulizer interface; CI at atmospheric pressure. A-C: daughter ion spectra of MH^+ ions; D: total ion current profile of parent ion scans of m/z 93; E, F: selected reaction monitoring of MH^+ m/z 93 for phenylbutazone (F) and its metabolites (E). *From [23], courtesy of the authors and American Chemical Society.*

To increase speed of analysis, specific detection by MS-MS can be combined with a simplified sample work-up procedure or intentional non-optimization of LC giving poor separation. Saturation effects and reactant ion depletion in the ion source necessitate some chromatographic separation of the analyte from impurities present in swamping amounts [23]; hence MS-MS without chromatography is not a viable alternative to GC-MS, LC-MS or LC-MS-MS.

Fragmentation in the thermospray ion source under the influence of an electrical discharge or a high repeller voltage [16, 17] gives, to some extent, the fragment ions desired for structure elucidation. In the mass region below m/z 100, however, solvent cluster ions and other background ions overwhelm fragment ions of an analyte, and the m/z range between 100 and 150 is not clean enough to identify fragment ions of an analyte with confidence. Thermospray with fragmentation in the ion source is a valuable extension of LC-MS but cannot replace true LC-MS-MS.

CONCLUSIONS

LC-MS with a moving belt interface continues to receive attention because of the ability to generate EI spectra. Thermospray LC-MS is the best choice at present for biomedical research since it can be purchased as an accessory for nearly all makes of mass spectrometer. Continuous flow FAB is still in its infancy as an LC-MS technique but will rapidly find application. API with the ion spray interface is potentially apt for analyzing polar ionogenic analytes without thermal degradation; however, API-MS is available only from SCIEX (Canada) and Hitachi (Japan), and will be as successful as thermospray only when it becomes available from traditional manufacturers of analytical MS-computer combinations. LC-MS-MS provides fragmentation reactions for structure elucidation, identification and selective determination where sufficient LC-MS information is not forthcoming because of soft ionization or insufficient LC separation.

References

1. Bruins, A.P. (1985) *J. Chromatog. 323*, 99-112.
2. Blakeley, C.R. & Vestal, M.L. (1983) *Anal. Chem. 55*, 750-754.
3. Covey, T.R., Lee, E.D., Bruins, A.P. & Henion, J.D. (1986) *Anal. Chem. 58*, 1451A-1461A.
4. Bruins, A.P. (1986) *Adv. Mass Spectrom. 10*, 119-131.
5. Harrison, A.G. (1983) *Chemical Ionization Mass Spectrometry*, CRC Press, Boca Raton, FL, pp. 7-55.
6. Kresbach, G.M., Baker, T.R., Nelson, R.J., Wronka, J., Karger, B.L. & Vouros, P. (1987) *J. Chromatog. 394*, 89-100.
7. Van der Greef, J., Tas, A.C., Rijk, M.A.H., Ten Noever de Brauw, M.C., Höhn, M., Meyerhoff, G. & Rapp, U. (1985) *J. Chromatog. 343*, 397-401.

8. Games, D.E., McDowall, M.A., Levsen, K., Schäfer, K.H., Dobberstein, P. & Gower, J.L. (1984) *Biomed. Mass Spectrom.* *11*, 87-95.
9. Dobberstein, P., Korte, E., Meyerhoff, G. & Pesch, R. (1983) *Int. J. Mass Spectrom. Ion Phys.* *46*, 185-188.
10. Van der Greef, J., Tas, A.C., Ten Noever de Brauw, M.C., Höhn, M., Meyerhoff, G. & Rapp, U. (1985) *J. Chromatog.* *323*, 81-87.
11. Lant, M.S., Martin, L.E. & Oxford, J. (1985) *J. Chromatog.* *323*, 143-152.
12. Vestal, M.L. (1983) *Mass Spectrom. Rev.* *2*, 447-480.
13. Schmelzeisen-Redeker, G., Röllgen, F.W., Wirtz, H. & Vögtle, F. (1985) *Org. Mass Spectrom.* *20*, 752-756.
14. Schellenberg, K.H., Linder, M., Groeppelin, A. & Erni, F. (1987) *J. Chromatog.* *394*, 239-252.
15. Lant, M.S., Oxford, J. & Martin, L.E. (1987) *J. Chromatog.* *394*, 223-230.
16. McFadden, W.H., Garteiz, D.A. & Siegmund, E.G. (1987) *J. Chromatog.* *394*, 101-108.
17. Chapman, J.R. & Pratt, J.A.E. (1987) *J. Chromatog.* *394*, 231-238.
18. Liberato, D.J., Fenselau, C.C., Vestal, M.L. & Yergey, A.L. (1983) *Anal. Chem.* *55*, 1741-1744.
19. Watson, D., Taylor, D.W., & Murray, S. (1985) *Biomed. Mass Spectrom.* *12*, 610-615.
20. Ashcroft, A.E., Chapman, J.R. & Cottrell, J.S. (1987) *J. Chromatog.* *394*, 15-20.
21. Horning, E.C., Carroll, D.I., Dzidic, I., Haegele, K.D., Horning, M.G. & Stillwell, R.N. (1974) *J. Chromatog.* *99*, 13-21.
22. Lane, D.A., Thomson, B.A., Lovett, A.M. & Reid, N.M. (1980) *Adv. Mass Spectrom.* *8*, 1480-1489.
23. Covey, T.R., Lee, E.D. & Henion, J.D. (1986) *Anal. Chem.* *58*, 2453-2460.
24. Iribarne, J.V., Dziedzic, P.J. & Thomson, B.A. (1983) *Int. J. Mass Spectrom. Ion Phys.* *50*, 331-347.
25. Whitehouse, C.M., Dreyer, R.N., Yamashita, M. & Fenn, J.B. (1985) *Anal. Chem.* *57*, 675-679.
26. Aleksandrov, M.L., Gall, L.N., Krasnov, N.V., Nikolaev, V.I. & Shkurov, V.A. (1985) *Zh. Anal. Khim.* *40*, 1570-1580.
27. Bruins, A.P., Covey, T.R. & Henion, J.D. (1987) *Anal. Chem.* *59*, 2642-2646.
28. Olivares, J.A., Nguyen, N.T., Yonker, C.R. & Smith, R.D. (1987) *Anal. Chem.* *59*, 1230-1232.
29. Ito, Y., Takeuchi, T., Ishii, D. & Goto, M. (1985) *J. Chromatog.* *346*, 161-166.
30. Caprioli, R.M., Fan, T. & Cottrell, J.S. (1986) *Anal. Chem.* *58*, 2949-2954.
31. McLafferty, F.W., ed. (1983) *Tandem Mass Spectrometry*, Wiley, New York: e.g. pp. 41-66 (K. Levsen).

#F-4

THE STRUCTURE ELUCIDATION OF DRUG METABOLITES USING THERMOSPRAY LC-MS AND LC-MS-MS

T.J.A. Blake and I.G. Beattie

Department of Drug Metabolism
Smith Kline & French Research Ltd.
The Frythe, Welwyn, Herts. AL6 9AR, U.K.

The isolation of individual metabolites is often laborious and time-consuming, yet in the past was mandatory if structural characterization was required. The advent of directly coupled liquid chromatography [HPLC]-mass spectrometry (LC-MS) has greatly facilitated structure determination of drug metabolites without the need for isolation. This article concerns the thermospray interface, which has proved to be very reliable, and the combined potential of LC-MS and LC-MS-MS as highly specific identification techniques, along with collisionally activated decomposition, as shown for temelastine.

With mobile phases rich in water, as in the RP-HPLC approach favoured for drug metabolites because of their polarity range, severe problems have been encountered in coupling to the MS. A recent remedy, designed specifically to handle aqueous mobile phases at conventional flow rates, is the thermospray interface [1] (& see A.P. Bruins in art. #F-3), which easily accommodates gradient elutions and thus is suited for complex mixtures containing compounds of widely different polarity [2]. It serves as the primary source of ionization: the liquid flow containing a volatile buffer, ammonium acetate, is heated to produce a supersonic vapour jet from which both $(M+H)^+$ and $(M-H)^-$ solute ions are produced. The technique has tended to produce mainly molecular ion data, which has often limited the amount of structural information obtainable. This can be remedied by using LC-MS-MS to select and fragment the thermospray ions. We may consider temelastine:

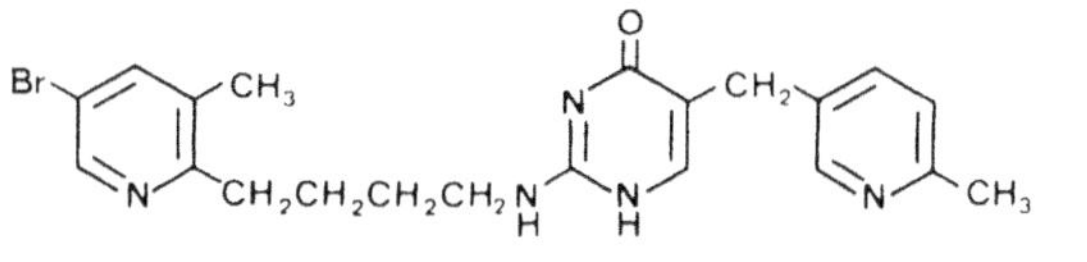

Temelastine (SK&F 93944) is a novel H_1-antagonist, with negligible ability to cross the blood-brain barrier [3]. It undergoes extensive metabolism, notably hydroxylation and glucuronidation: 4 hydroxylated metabolites and one *N*-oxide have been identified to date. Under positive thermospray ionization conditions the latter metabolites generated predominantly protonated molecular ion data [$(M+H)^+$] which made the differentiation of the closely eluting isomeric species difficult. It became necessary to develop an analytical method that could distinguish these metabolites individually. Their HPLC retention times were too similar to allow unambiguous assignments, but by using a combination of conventional LC-MS and LC-MS-MS (daughter ion mode) with collisionally activated decomposition (CAD) these isomeric metabolites could be individually identified.

METHODOLOGY

HPLC.- Runs were performed with a Hewlett-Packard 1090A apparatus fitted with a filter photometric detector, and a Waters C-18 Novapak column (150 × 3.9 mm; 5 µm particles) which was operated at ambient temperature with 1.4 ml/min flow rate. The mobile phases were 0.1 M ammonium acetate (no pH adjustment; filtered though a 0.45 µm filter) and acetonitrile. In gradient runs the first 2 min were with nil acetonitrile, followed by a linear ramp to 40% (v/v) at 20 min and a 2-min hold. UV detection was at 280 nm and was recorded on-line with the MS data.

Mass spectrometry.- Finnigan-MAT equipment was used: a TSQ46 triple quadrupole MS driven by a SuperIncos data system, and a thermospray interface for HPLC coupling. Typically the vaporizer was at 120° (jet at 220°), lowered to ~108° by the end of the gradient run. The instrument was tuned by adding SK&F 93944 to the HPLC buffer and continuously introducing it into the MS via a short loop prior to running the main samples. The MS operating mode was either normal (Q1 scan) or MS-MS (daughter ion scan). In the normal mode, ions were mass-analyzed in quadrupole #1 (Q1) and passed straight through quadrupole #2 (Q2; the collision cell) and quadrupole #3 (Q3; the second mass analyzer) to the detector. In the MS-MS mode selected ions were passed through Q1, collisionally activated by interaction with argon at 1.8 m-torr in Q2 and analyzed for daughter ions in Q3.

ILLUSTRATIVE STRUCTURAL APPLICATIONS

Fig. 1 depicts the UV and thermospray total-ion current (TIC) traces corresponding to a mixture of some synthetic SK&F 93944 metabolites, and shows that 4 eluted with similar retention times, making distinctions difficult. The mixture contained 4 hydroxylated metabolites - SK&F 96294, 95109, 94224 and 94955 - and an *N*-oxide, SK&F 94162 (Figs. 2 & 3 include their structures).

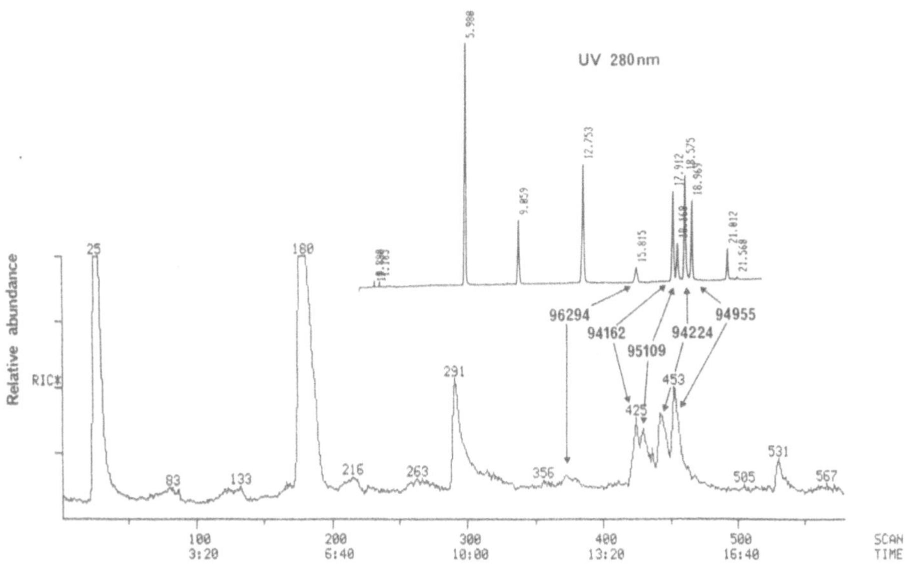

Fig. 1. The TSP(+) TICand UV (280 nm) traces for the SK&F 93944 Standard Metabolite Mix.

Fig. 2 shows the positive ion thermospray [TSP(+)] mass spectra of 4 of the 5 metabolites; that for SK&F 96294 (not shown) was essentially identical with that of SK&F 95109. Clearly the 3 hydroxylated species SK&F 94224, 95109 and 96294 could not be differentiated from their TSP(+) mass spectra alone, each being effectively just the $(M+H)^+$ species; retention times for SK&F 95109 and 94224 were too similar to allow unambiguous identification, although that of SK&F 96294 was different. The *N*-oxide (SK&F 94162) showed extensive loss of oxygen and yielded fragment ions at m/z 242/244 (Fig. 2). SK&F 94955 also appeared to show loss of oxygen and generated fragment ions at m/z 337/339. This enabled SK&F 94162 and 94955 to be distinguished from the other hydroxylated forms using a normal (Q1) mass scan.

Differentiation of the remaining 3 hydroxylated species required LC-MS-MS in the daughter ion mode. The ^{81}Br-containing $(M+H)^+$ ion, at m/z 460, was passed through Q1, collisionally activated in Q2 and analyzed for daughter ions in Q3. Fig. 3 shows the CAD spectra for these 3 metabolites. The molecules fragmented by cleavage between the methylene group and the exocyclic nitrogen atom of the isocytosine ring. The ion current was mostly carried by the Br-containing fragment produced. Thus SK&F 94224 fragmented to give a base peak at m/z 228, while SK&F 95109 generated ions at m/z 244 (16 mass units up) and 226 (additional loss of water). SK&F 96294 and 95109 generated the same ions, under CAD conditions, since their respective sites of hydroxylation were both in the Br-containing half of the molecule.

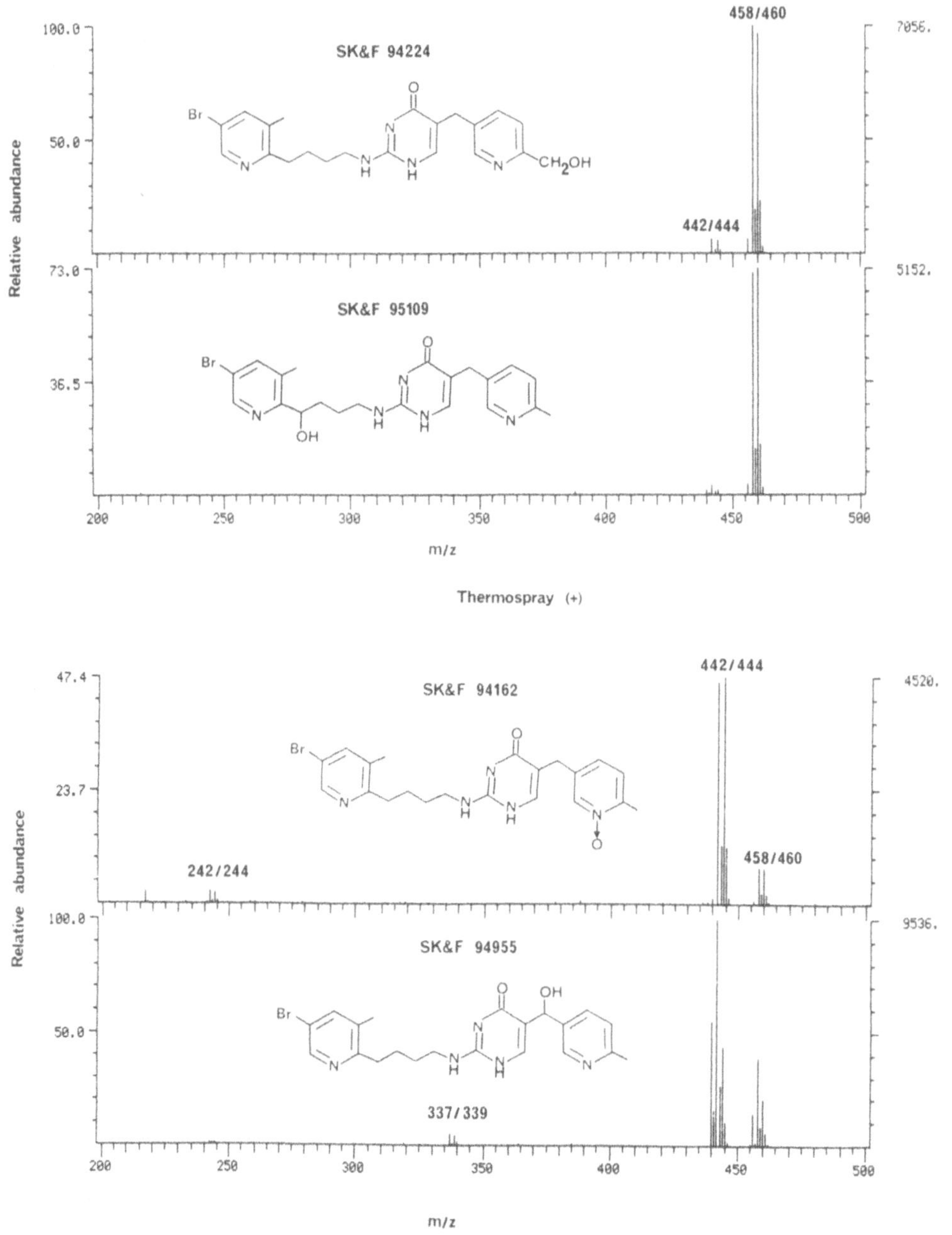

Fig. 2. The TSP(+) mass spectra of SK&F 94224, 95109, 94162 & 94955.

However, the relative abundances of the ions detected at m/z 226 and 244 for SK&F 96294 were reversed with respect to those observed for SK&F 95109. These differences were reproducible and allowed the differentiation of these two hydroxylated forms.

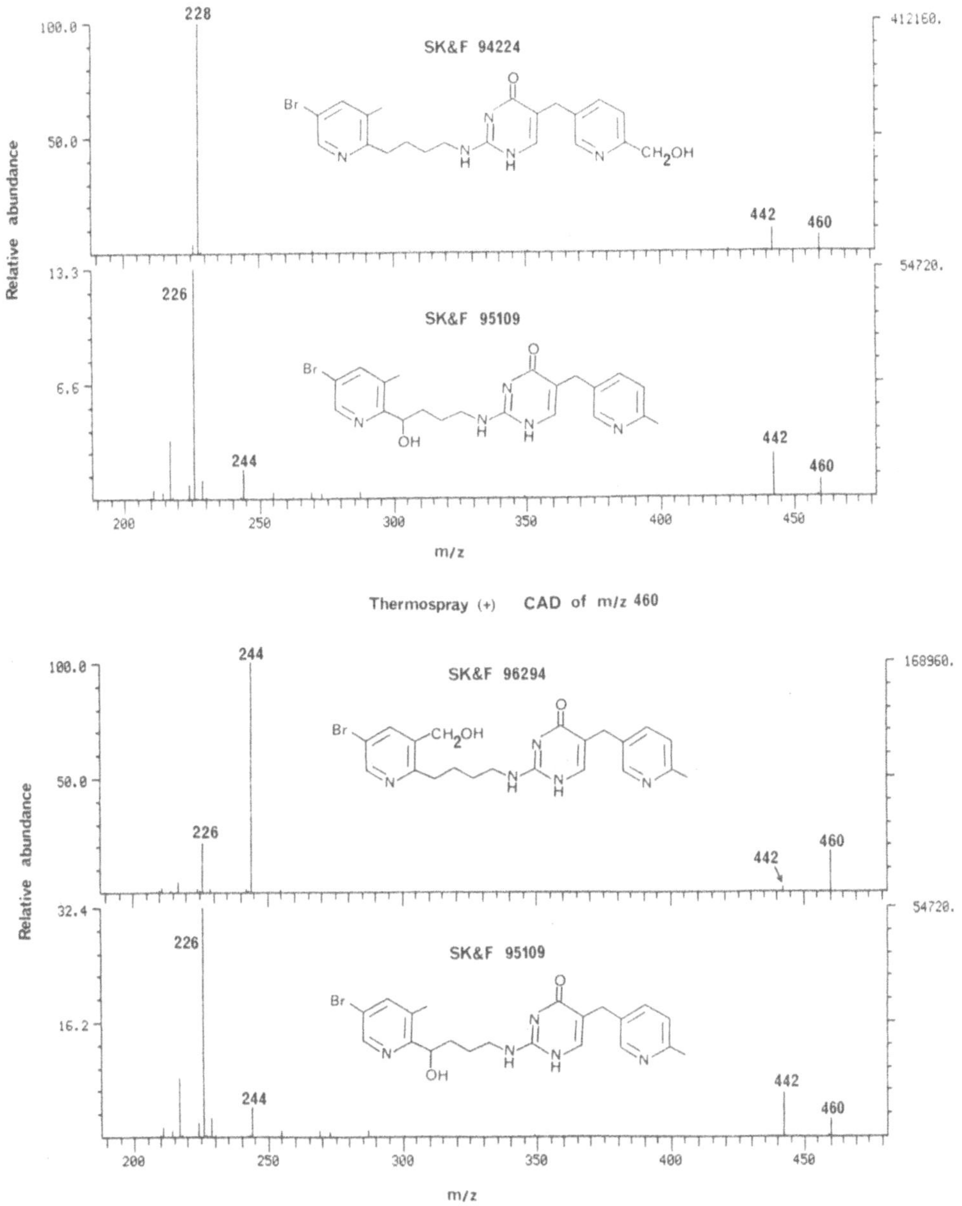

Fig. 3. The TSP(+) CAD mass spectra of m/z 460 from SK&F 94224, 95109 & 96294.

The *N*-oxide (SK&F 94162) and SK&F 94955 (spectra not shown) yielded an ion at m/z 228 similar to SK&F 94224, but were distinguishable in Q1 mode as stated above. Thus by a combination of LC-MS and LC-MS-MS the isomeric metabolites could be differentiated.

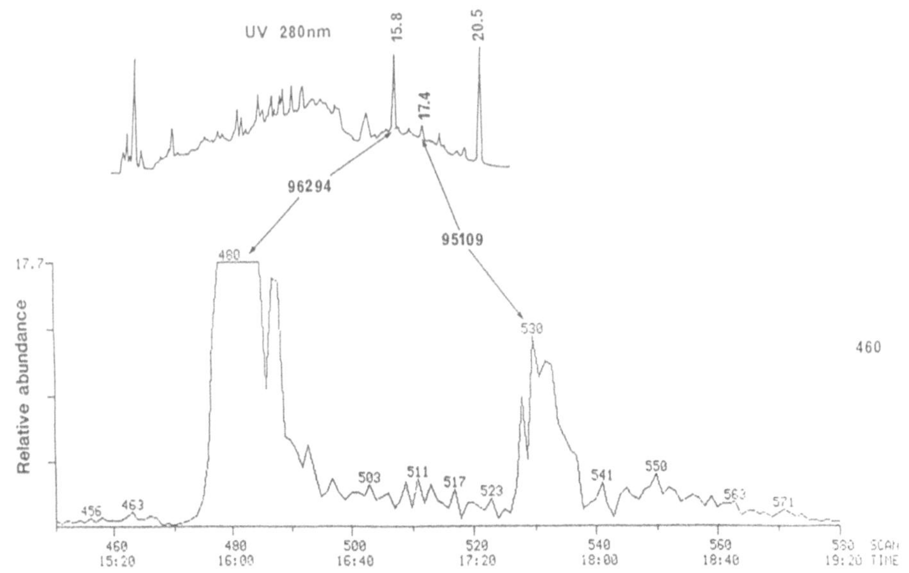

Fig. 4. Ion chromatogram of m/z 460 and UV (280 nm) traces from extracted faeces of a human dosed with SK&F 93944.

An example of use of this methodology is shown in Figs. 4 & 5. The hydroxylated metabolite SK&F 95109 was detected in some human faecal extracts. The problem was to determine whether or not there was any SK&F 94224 eluting under the tail of the SK&F 95109. From the UV trace and the selected ion chromatogram of m/z 460 from the Q1 scan of such an extract (Fig. 4), it was impossible to state with any degree of certainty whether or not any SK&F 94224 was present. Analysis by LC-MS-MS (Fig. 5), however, clearly demonstrated the presence of minor quantities of both SK&F 94224 and 94955 in the sample.

CONCLUSION

The development of the thermospray interface has resulted in a dramatic rise in the use of on-line LC-MS. The technique has allowed valuable structural information to be obtained from complex mixtures without the requirement of individual compound isolation, clearly saving much time and effort. In most cases, however, isolation is still necessary if detailed stereochemical information is required. The added specificity obtainable from the LC-MS-MS approach has been demonstrated and it has shown itself to be a very powerful probe for structural analysis.

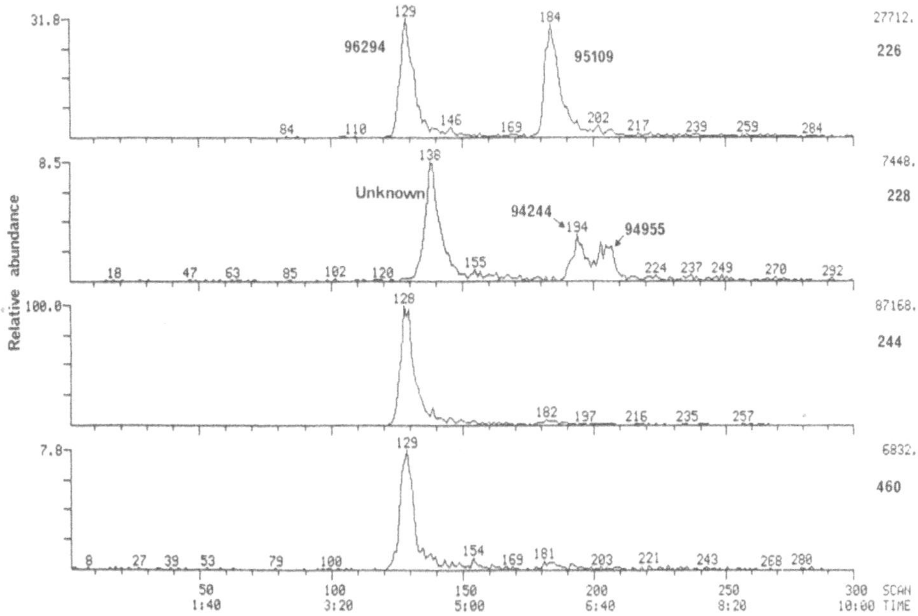

Fig. 5. Ion chromatograms derived from TSP(+) CAD of m/z 460 from a faecal extract as in Fig. 4.

Acknowledgements

The authors are grateful to members of the Investigative Chemistry section of the Department of Synthetic and Isotope Chemistry, SK&F, The Frythe, for synthesis of authentic reference compounds, and to Mr. A.J. Dean for the preparation of the faecal extracts.

References

1. Blakley, C.R. & Vestal, M.L. (1983) *Anal. Chem. 55*, 750-754.
2. Blake, T.J.A. (1987) *J. Chromatog. 394*, 171-181.
3. Durant, G.J., Ganellin, C.R., Griffiths, R., Harvey, C.A., Ife, R.J. & Owen, D.A.A. (1984) *Br. J.Pharmacol. 82*, *Suppl.*, 232P.

#NC(F)

NOTES and COMMENTS relating to

DETECTION, IDENTIFICATION AND INSTRUMENTATION

Comments relating to particular contributions:

#F-1 to -4, and #NC(F)-5, -7 & -8: p. 403.

#NC(F)-1

A Note on

AN APPROACH TO METABOLITE SAMPLE PREPARATION FOR MASS SPECTROMETRY TO AID THE SPECTROSCOPIST

R.D. Brownsill[⊕], A. Gray[⊕] and C.W. Vose[⊗]

G.D. Searle and Co., High Wycombe
Bucks. HP12 4HL, U.K.

Mass spectrometry (MS) is widely applied to the identification of drug metabolites isolated from biological fluids. A wide range of techniques are available, e.g. GC-MS and LC-MS where the MS options include electron-impact (EI), chemical ionization (CI) and fast atom bombardment (FAB). Similarly, many methods are available for metabolite isolation and purification, e.g. liquid-liquid and liquid-solid phase extraction, TLC, HPLC and electrophoresis. Despite this wide range of options, we may frequently experience problems with metabolite identification. These may result in part from certain assumptions:
- metabolite identification is possible on 'anything';
- isolation/purification techniques yield pure samples;
- metabolite identification can be 'fitted in' as an 'also-ran'.

The risk is that the (bio)chemist may isolate samples of the metabolites from a urine extract (Fig. 1) and submit these for MS with no discussions with the spectroscopist and with no agreement on the time frames. When these samples are analyzed the outcome tends to be a mixture of identification of some (generally the most prominent) metabolites with statements regarding the lack of purity or sufficient quantity for most of the metabolites (legend to Fig. 1). The overall result can range from dissatisfaction of both the (bio)chemist and the spectroscopist with the support received from each other, to total failure of metabolite identification and intense mutual recriminations.-

Biochemist: "They've got £250,000 equipment and still can't identify purified metabolites after *x* weeks (months)!"
Mass spectroscopist: "They supply insufficient amounts of impure metabolites contaminated with plasticizers and alkylsilicones and then expect identification in no time at all!!!"

[⊕] now at Servier R & D, Fulmer, Slough, U.K.
[⊗] now at Hoechst UK Ltd., Walton Manor, Milton Keynes, U.K.; the addressee for any correspondence.

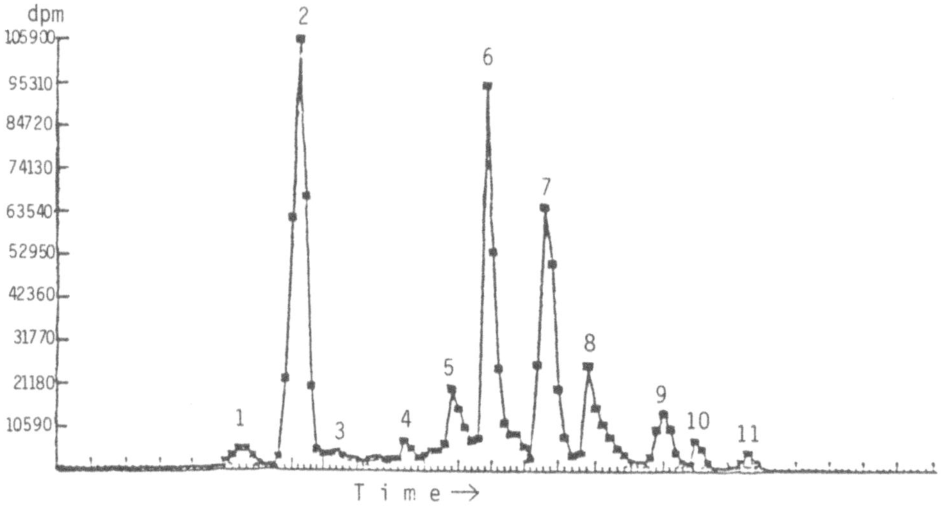

Fig. 1. Objectives and actions in attempting a 'typical' metabolite identification study, for peaks 1-11 in the radiochromatogram shown.
ACTIONS:
- Biochemist isolates samples of each metabolite and submits for MS analysis.
- Mass spectroscopist analyzes samples by MS, GC-MS, LC-MS, etc.

NOTIONAL RESULTS:
- Metabolites 2, 6 and 7 identified.
- Metabolites 1, 3, 4 and 5 impure and insufficient sample.
- Metabolites 9, 10 and 11: insufficient sample to obtain satisfactory spectra.
- Metabolite 8 spectrum 'masked' by interferences from plasticizers and long-chain alkyl material.

To maximize the success of metabolite identification we have found it essential that the (bio)chemist and mass spectroscopist work together as a team from the outset. This requires that they consider the following key items:

- plan the study in detail;
- agree the most appropriate isolation/purification/MS techniques;
- assess the likely sources of contamination;
- agree the amounts of metabolites required;
- agree time schedules for the different activities in the study.

It is not possible to do this from a position of total ignorance and therefore some preliminary information is generally required. We have found the following experiments (pilot study) of help in generating these preliminary data.

- Establish initial isolation/purification procedures using the drug and potential metabolites (if available).

- Obtain mass spectra (EI and CI) of drug and potential metabolites (if available).

- Obtain extracts from control biological samples and fractions in the regions of the chromatograms of these samples where drug and metabolites would be guessed to appear.

- Analyze these blank 'metabolite' extracts by MS with and without the addition of drug and potential metabolites.

These experiments should provide data on which to assess the likely levels of contamination of 'pure' metabolite fractions, decide the amounts of metabolite required and the need for additional procedures, e.g. GC-MS or LC-MS rather than insertion into the ion source directly (not from a column) for the MS investigation. It may be feasible to obtain a preliminary metabolite profile (Fig. 1) and samples of major metabolites in parallel with these experiments. Analysis of these samples can be used to further evaluate the suitability of the established isolation/purification and MS methods.

Based on the results of these experiments, it is possible to agree the details of the metabolite identification procedures and the time-frames. These proposals may appear to increase the time taken to carry our metabolite identification. However, in our experience, they have generally resulted in an improvement in the time taken to identify metabolites and achieved a greater success in terms of the number of metabolites we have been able to identify.

#NC(F)-2

A Note on

APPLICATION OF NMR AND MASS SPECTROMETRY TO THE STUDY OF DRUG HYDROXYLATION

[1]C.M. Walls, [2]A. Gray, [2]R.D. Brownsill and [3]C.W. Vose

G.D. Searle & Co. Ltd., High Wycombe, HP12 4HL, U.K.

Hydroxylation is probably the most common Phase I metabolic pathway for xenobiotics. It is also a pathway readily detectable by MS from the 16 a.m.u. mass shift for the metabolite ions relative to those of the parent compound. However, such information may sometimes be misleading, and MS does not always allow assignment of the site of hydroxylation.

This was evident when two metabolites (M1 and M2) of a phenyl-substituted drug, which had several possible sites for hydroxylation, were isolated by solid-phase extraction (C-18 BondElut) and radio-HPLC with fraction collection, from human urine during a metabolism study. The urine (~10 ml) was applied to an activated C-18 BondElut column, which was washed with phosphate buffer (2 ml; pH 6) and hexane (2 ml); the radioactive material was then eluted with methanol (1 ml). This extract was analyzed by HPLC on a C-18 μBondapak column using a 20% to 100% linear gradient of methanol in water at 4%/min and 1 ml/min flow rate. Scintillation counting of collected fractions allowed the detection of the metabolites, which were isolated and analyzed.

The direct probe electron-impact (EI) mass spectra obtained for both M1 and M2 (50 eV, 220°) showed molecular and fragment ions shifted 16 a.m.u. to higher mass, consistent with the compounds being isomeric monohydroxy metabolites of the drug. However, the relative polarities of M1 and M2 on HPLC were not entirely consistent with the structures indicated by MS. M1 appeared to be much more polar than M2 based on retention times.

From the MS examination there remained 50 and 25 μg of M1 and M2 respectively which were used to obtain proton NMR spectra in hexa-deuterated dimethylsulphoxide (d_6-DMSO), recorded at 250 MHz on a

[1]now at May & Baker Ltd., Medical Dept., Dagenham, Essex; [2]now at Servier R & D, Fulmer, Slough; [3]now at Hoechst UK Ltd., Milton Keynes (the addressee for any correspondence).

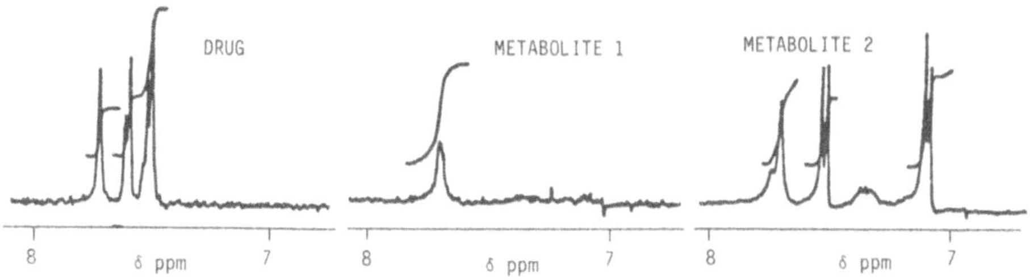

Fig. 1. NMR spectra of the aromatic proton region, pointing to the absence of phenyl ring protons from M1 and to a typical *p*-substituted coupling pattern in M2.

Bruker MH 250 spectrometer. The aromatic proton region of the spectra (6 to 8 δ; Fig. 1) were essentially free of impurity peaks, in contrast to the other regions. The parent drug showed a complex multiplet of a monosubstituted benzene ring at 7.5 to 7.7 δ. In contrast, the spectrum of M1 had no signals in the aromatic proton region. Thus, metabolism had resulted in the loss of aromaticity from the phenyl ring present in the parent drug. In the spectrum of M2, the aromatic proton multiplet was replaced by an A'B' quartet at 6.83 and 7.45 ppm (2H; J = 8.6 Hz) characteristic of a *p*-substituted benzene ring. Metabolites M1 and M2 were therefore both products of metabolism in the drug's phenyl ring, but were *not* isomeric monohydroxy compounds.

The metabolites were recovered from the d_6-DMSO, derivatized with bis-TMStrifluoroacetamide (BSTFA; TMS = trimethylsilyl), and the resulting TMS ethers were analyzed, with on-column injection, by capillary GC-MS with temperature programming (50° to 300° at 30°/min) on a DB-1 fused silica capillary column, to gain further structural information.

The EI mass spectrum showed that the M1-TMS ether had $M^{+\cdot}$ and key fragment ions shifted 178 a.m.u. to higher mass than the parent drug and an $[M-90]^{+\cdot}$ ion characteristic of the loss of $(CH_3)_3SiOH$ — indicating the presence of an aliphatic TMS ether. Thus, the derivative contained two additional oxygen and two additional hydrogen atoms. The M2 spectrum was consistent with that of the TMS ether of a non-aliphatic monohydroxy metabolite, with $M^{+\cdot}$ and key fragment ions shifted 88 a.m.u. to higher mass, and no $[M-90]^{+\cdot}$ ion.

The combined proton NMR and MS data showed that M1 was formed by a metabolic pathway which added two oxygen atoms and two hydrogen atoms to the drug with the loss of the aromaticity of the phenyl group. M1 was therefore the dihydrodiol metabolite (Fig. 2). The equivalent data for M2 showed that it was the phenol formed by *p*-hydroxylation in the phenyl group (Fig. 2). Therefore the combination of NMR and MS allowed the correct assignment of the hydroxylation sites

Fig. 2. Partial structures of drug and metabolites.

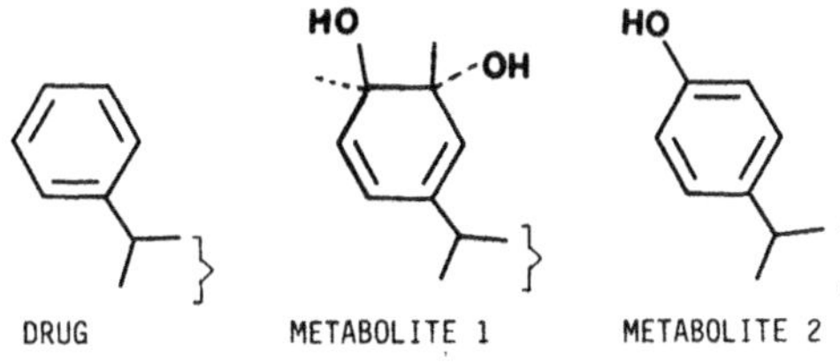

and pathways for the drug. In contrast, the direct insertion mass spectral data alone had given misleading information for the more polar metabolite (M1). This was presumably due to thermal dehydration of the dihydrodiol to yield the spectrum of the resulting phenol.

#NC(F)-3

A Note on

URINANALYSIS BY ^1H NMR: APPLICATION TO CEPHALORIDINE-TREATED RATS

R. Pickford, I.K. Smith and I.D. Wilson

ICI Pharmaceuticals Division
Mereside, Alderley Park
Macclesfield, Cheshire SK10 4TG, U.K.

Although a quite recent development the analysis of biological fluids by high-resolution proton NMR is rapidly becoming an established procedure, especially for urine and plasma (e.g. [1-3], and arts. by J.K. Nicholson & colleagues in this book and in Vols. 16 & 17). A particular advantage is the wealth of additional information furnished, over and above that required for the qualitative or quantitative determination of a single analyte that is manifest in the spectrum from the sample.- Any organic molecule possessing protons, will be detected, provided that the levels are mM or higher. Accordingly, a single assay, on otherwise untreated urine samples, can comprehend drugs, drug metabolites, and endogenous metabolites (e.g. Krebs cycle intermediates, sugars, organic acids, creatinine and various amino acids) that may bear on drug-related toxicicity. No chromatographic method could be as simple and wide-ranging as ^1H NMR. These points are now exemplified by ^1H NMR observations on urine of rats dosed with a nephrotoxic cephalosporin antibiotic, cephaloridine.

Methods.-Male Sprague Dawley rats were given cephaloridine for up to 3 days (0, 1100, 1500 or 1600 mg/kg per day, i.p.); then urine was collected daily and stored frozen. Aliquots (1.6 ml) were taken and mixed with 5 M urea in ^2H_2O (400 µl), and the pH was adjusted to 3.5. ^1H NMR was performed by the water-suppression technique described by Connor et al. [4]. All spectra were recorded at ambient probe temperature (303° ±1°K) on a Bruker spectrometer - AM 400, operating at 400 MHz, or AM 200 at 200 MHz. Spectra were derived from **n** (64-256) free induction decays, which were collected into 32 × 1024 data points. To obtain water suppression the Carr Purcell Mieboom Gill (CPMG) pulse sequence was used:

D_1 = 3 sec T_2 = 0.002 sec n = 64 to 256

On facing page:
Fig. 1. ^1H NMR spectra of control rat urine samples, not the same in **B** as in **A** (the *lower* trace).
A: 400 MHz spectrum following urea addition, pH adjustment to pH ~3.5 and elimination of the water signal as described in the text.
B: 200 MHz spectrum following treatment as for **A** but with the addition of $^2H_6$-DMSO (to 10%, v/v) to provide a field frequency lock for the spectrometer.

Spectra were referenced using, as an internal chemical shift standard (= 0 ppm), trimethylsilylpropionic acid (TSP).

RESULTS AND DISCUSSION

In observing the resonances of analytes present at only mM concentrations in urine, the salient problem is to minimize the interference by the large signal, centred at ~5 ppm, of the water itself. Until recently the only remedies have been either freeze-drying followed by reconstitution in D_2O, or the use of a secondary irradiation at the water resonance frequency in order to saturate the water proton spins and attenuate their signal. Although we have successfully used both approaches, problems can arise. Thus, freeze-drying would cause loss of volatile compounds, whilst irradiation really works only with high-field instruments (i.e. those operating at 400 mHz or above) and also entails loss or distortion of signals for compounds that are close to water in chemical shift.

In more recent methods for eliminating this problem, the T_2 relaxation time of the water signals is increased, resulting in a broadened water peak whereby, combined with the selective attenuation of such broad signals (using spin echo methods), the water signal can be markedly reduced or even completely eliminated. There is the added benefit that signals close to, or even under, that of the water remain unaffected. Fig. 1A shows a typical ^1H NMR (400 MHz) spectrum of control rat urine. Interestingly, the technique also works well at lower field strengths, as manifested by the equivalent spectrum (Fig. 1B) obtained using a medium-field (200 MHz) spectrometer. With the addition of a small amount (~100 µl) of hexadeuterated dimethylsulphoxide ($^2H_6$-DMSO; to provide a field frequency lock for the spectrometer), we have been able to use a 200 MHz instrument fitted with an autosampler for routine analysis of urine.

Fig. 2 shows a spectrum, obtained at 400 MHz, for a urine sample collected over the 24 h following an oral dose of cephaloridine (1500 mg/kg). One difference from the control (Fig. 1A) is the appearance of new resonances due to drug-related material in the aromatic region, whilst those for hippuric acid have been lost. Secondly, large amounts of glucose are apparent, accounting for the bulk of

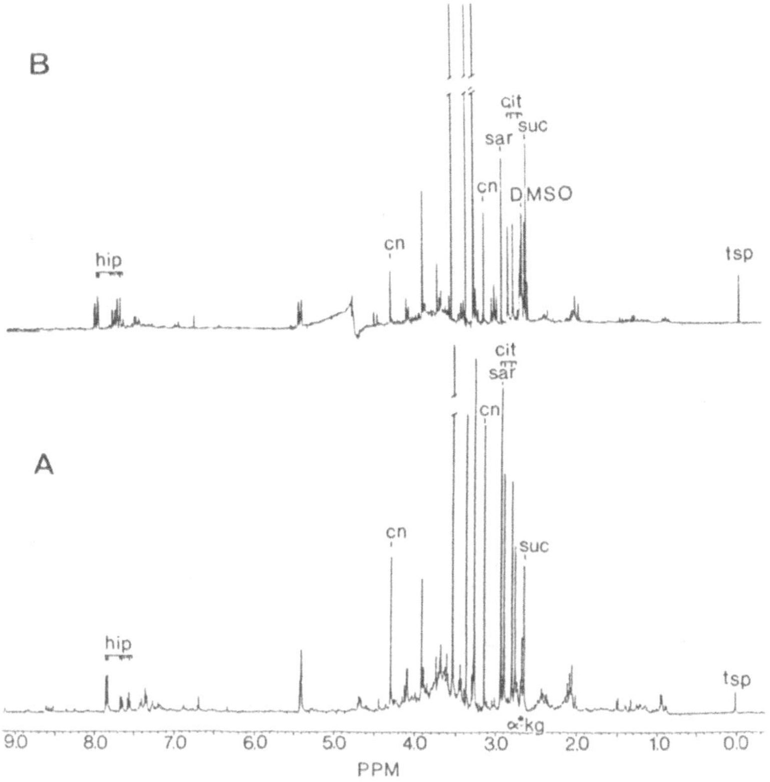

KEY for Fig. 1: cit, citrate; cn, creatinine; hip, hippurate; α-kg, α-ketoglutarate; sar, sarcosine; suc, succinate; tsp, $^2H_6$-TSP (see text, opposite) used as a chemical shift reference.

Fig. 2. ^1H NMR spectrum, 400 MHz, of rat urine obtained, following attenuation of the water signal as described in text, from a 0 to 24 h sample after cephaloridine treatment (see text; 1500 mg/kg).

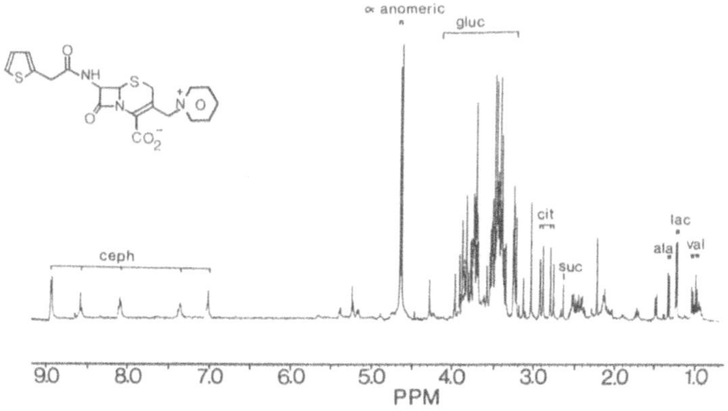

KEY for Fig. 2: ala, alanine; α-anomeric, α-anomeric proton of glucose; ceph, cephaloridine-related material; gluc, glucose; lac, lactate; val, valine. Other abbreviations as for Fig. 1.

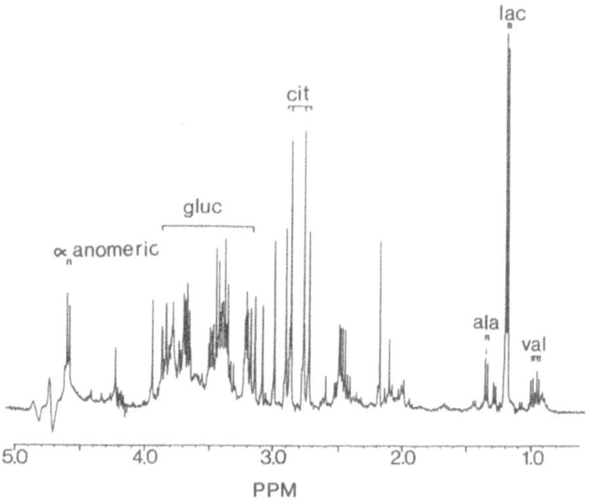

Fig. 3. ¹H NMR part-spectrum (400 MHz) obtained from the urine of a rat dosed with cephaloridine (1600 mg/kg): 24 to 48 h sample. KEY as for Fig. 2.

the material detectable in the spectrum. In addition, lactate, alanine and valine are present. The part-spectrum shown in Fig. 3 is that of the 24 to 48 h urine sample obtained from a rat after 2 daily doses of cephaloridine at 1600 mg/kg. Here the pattern of endogenous metabolites is grossly distorted with very large quantities of lactate present.

CONCLUSIONS

Using ¹H NMR analysis of urine it was possible to detect both the presence of cephaloridine and its nephrotoxic effects as reflected by dramatic changes in concentration for a number of endogenous compounds. In general, ¹H NMR actively complements the established methods for investigating drug excretion and toxicity, and gives much additional useful information not otherwise readily procurable. Indeed, so useful has this methodology proved that it is now routinely used for the qualitative analysis of urine in our laboratories.

References

1. Wilson, I.D., Fromson, J., Ismail, I.M. & Nicholson, J.K. (1987) *J. Pharm. Biomed. Anal. 5*, 157-163.
2. Nicholson, J.K. & Wilson, I.D. (1987) in *Drug Metabolism - from Molecules to Man* (Benford, D.J., Bridges, J.W. & Gibson, G.G., eds.), Taylor & Francis, London, pp. 189-207.
3. Nicholson, J.K. & Wilson, I.D. (1987) *Progr. Drug Res. 31* [Jucker, E., ed.], 427-479.
4. Connor, S., Everett, J. & Nicholson, J.K. (1987) *Mag. Res. Med. 4*, 461-470.

#NC(F)-4

A Note on

PROTON NMR STUDIES ON THE METABOLISM AND BIOCHEMICAL EFFECTS OF HYDRAZINE *IN VIVO*

[1]S.M. Sanins, [2]J.A. Timbrell, [3]C. Elcombe and [1]J.K. Nicholson

[1]Analytical Sciences Group, Deprtment of Chemistry
Birkbeck College (University of London)
Gordon House, 29 Gordon Square, London WC1E 6B7, U.K.

[2]Toxicology Unit, School of Pharmacy (University of London)
29 Brunswick Square, London WC1N 1AX, U.K.

[3]Central Toxicology Laboratories, ICI plc
Alderley Park, Macclesfield, Cheshire SK10 4TG, U.K.

High-resolution ^{1}H NMR spectroscopy is valuable in the bioanalysis of biological fluids (e.g. [1-6], and Vols. 16 & 17 in this series - J.K. Nicholson). Notably, it allows quantitative information on a wide range of endogenous and xenobiotic metabolites to be collected simultaneously. As NMR is non-selective in the detection of metabolites (all compounds present at >0.1 mM levels being detectable), it is particularly appropriate for the investigation of the metabolism of compounds with poorly understood metabolic profiles.

Hydrazine is an important industrial chemical and a metabolite of the drug isoniazid (which is hepatotoxic by an unknown mechanism). However, the metabolic fate and disposition of hydrazine is still poorly understood. Now, using high-resolution ^{1}H NMR methods, we have investigated the urinary excretion and metabolic effects of hydrazine as a model hepatotoxin whose mechanism we aimed to elucidate.

EXPERIMENTAL PROCEDURES

Animal treatment and sample collection.- Male Wistar rats (250 to 350 g; from the ICI laboratory) received hydrazine hydrate, in saline (as given to the controls), as single i.p. doses: 15, 30, 60 or 80 mg/kg. All rats were fasted for the 24-h duration of the experiment but allowed free access to water. Urine was collected over ice during 24 h before dosing and 0-6 and 6-24 h after dosing. Food particles, etc., were removed from urine by centrifugation. It was prepared for ^{1}H NMR analysis merely by adding 100 μl D_2O (as internal field frequency lock), to 400 μl. TSP® was added as a chemical shift reference and concentration standard.

After 24 h the rats were killed and blood was removed from the inferior vena cava into heparinized syringes. Plasma was separated immediately by centrifugation. Liver extracts were made by homogenization in 0.1 M pH 7.4 phosphate buffer (1 ml per 0.25 g). Equal volumes of 6% (w/v) TCA® and tissue homogenates were mixed, kept on ice for 5 min to allow complete protein precipitation, and centrifuged at 3000 rpm for 10 min. The supernatant was freeze-dried and the residue redissolved in 500 μl D_2O (containing TSP) for ^{1}H NMR study.

NMR measurements.- NMR spectra were recorded on a Bruker WH400 spectrophotometer operating at 400.13 MHz ^{1}H frequency at 296°K. For single-pulse experiments, 46-96 transients were collected into 16K computer points using 28° (3 μsec pulses) and an acquisition time of 1.7 sec. A further delay of 3.3 sec was added to ensure that the spectra were fully T_1-relaxed. The total accumulation time for each spectrum was less than 5 min. The intense signal from water was suppressed by applying a secondary irradiation field at the resonance frequency of water. Chemical shifts were referenced to internal TSP (δ = 0 ppm). For 2-D homonuclear COSY NMR spectra (see [3] for the 2-D COSY parameters) as recorded on selected urines and tissue extracts the pulse sequence used was:

$$[90° - t_1 - 90° - \text{collect Fid}(t_2)] - \text{relaxation delay}.$$

RESULTS AND DISCUSSION

NMR urinalysis following administration of a hydrazine

As is evident from Fig. 1 which shows, for the aliphatic regions, the dose-related changes in the ^{1}H NMR metabolic 'profile' of rat urine, there were many well resolved resonances. Each spectrum took <5 min to obtain and represents the sum of 48 transients. Following hydrazine treatment, proton resonances from several hydrazine metabolites were detected. These included acetyl- and diacetylhydrazine (δ = 1.95 and 2.05 ppm respectively), a cyclic hydrazine derivative of 2-oxoglutarate (THOPC; δ = 2.50 and 2.80 ppm) and methylamine (δ = 2.70 ppm).

Hydrazine-induced elevations in endogenous metabolites included [see Fig. 1 legend; cf. #NC(F)-3, I.D. Wilson & co-authors] hypertaurinuria and lactic-aciduria, and represent novel observations

®*Abbreviations*.- COSY, ^{1}H-shift correlated; Fid, free induction decay [cf. GC use of 'FID' - *Ed.*]; HOPC, 1,4,5,6-tetrahydro-6-oxo-3-pyridazine carboxylic acid; TCA, trichloroacetic acid; TSP, sodium d_4-(trimethylsilyl)propionate.

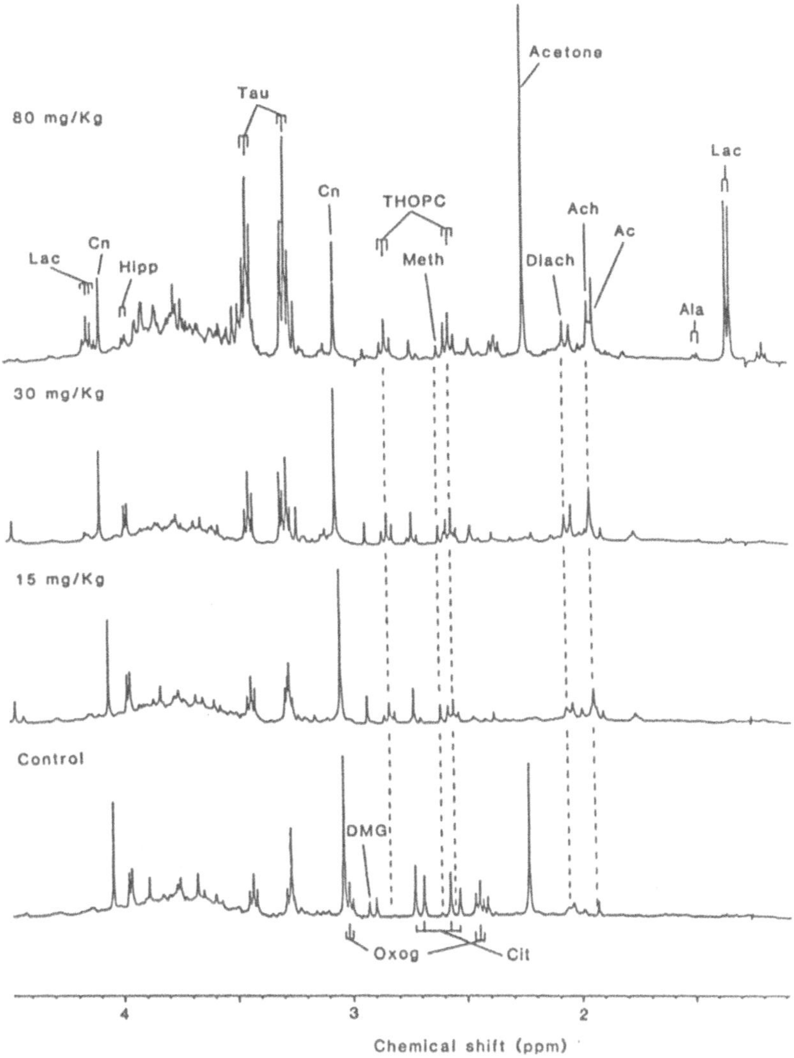

Fig. 1. 400-MHz ^{1}H NMR spectra of the aliphatic region of urine collected 0-6 h after treating rats with hydrazine. Its acetyl (Ach) and diacetyl (Diach) and other metabolites (THOPC; methylamine, Meth) are seen, and endogenous components some showing dose-related rises: Lac, lactate; Ala, alanine; Ac, acetate; DMG, dimethylglycine; Cn, creatinine; Tau, taurine; Hipp, hippurate; Cit, citrate; 2-oxoglutarate, Oxog, was absent).

and display the power of NMR urinalysis in the detection of novel markers for toxicity. The hypertaurinuria may represent one or more of the following hepatotoxin-related events: (i) cholestatic changes, (ii) a perturbation in pyrimidine metabolism producing elevations in serum β-alanine which is reabsorbed preferentially over taurine by the renal transport system for β-amino acids, (iii) increased protein catabolism, or (iv) a direct inhibitory effect on taurine breakdown in the liver.

The hydrazine-induced lactic-aciduria may represent effects on oxidative metabolism. Elevated levels in plasma from hydrazine-treated rats (unpublished observations) suggest a pre-renal effect rather than an effect on renal tubular reabsorption. Related to this may be other changes in metabolites concerned in oxidative cycles, e.g. 2-oxoglutarate. Its significant depletion in urine was initially thought to be due to an effect of hydrazine on its renal uptake. However, as hydrazine is very reactive towards carbonyl groups and as some hydrazine is excreted unchanged in the urine [8], the urinary depletion of 2-oxoglutarate and the production of THOPC may be the result of a direct chemical reaction in the urine.

The simultaneous observation of both drug and endogenous metabolites is a very useful feature of ^{1}H NMR urinalysis. However, problems arise when drug-metabolite signals overlap with those of endogenous compounds, prohibiting assignment of resonances; for such problems 2-D NMR methods are now being applied. COSY spectroscopy was intially used here to simplify the overlapping resonances in the 2.4-3.0 ppm region of the spectrum. For urine from a hydrazine-dosed rat Fig. 2 shows the 2-D COSY contour map and, beneath, the 1-D ^{1}H spectrum corresponding to the 2-D central diagonal contour densities; the symmetrical off-diagonal cross-peaks correlate spin-spin coupled nuclei. In the 2.4-3.0 ppm spectral region there are 3 pairs of cross-peaks correlating to β-alanine, citrate and THOPC. These resonances overlap in the 1-D spectrum and are difficult to distinguish in the absence of 2-D information. Furthermore, 4 pairs of cross-peaks (U_1 to U_4) correlated through spin-spin couplings were observed in the urine from hydrazine-treated rats but were not present in controls. These signals are due to the presence of a hydrazine metabolite not yet identified.

^{1}H NMR studies of protein-free liver extracts

Although there have been a few ^{1}H NMR studies of liver metabolism [1,6,9,10], mostly ^{13}C and ^{31}P NMR have been applied to study metabolism in the isolated perfused liver and in suspensions of hepatocytes. Excellent ^{1}H NMR spectra can be obtained in protein-free TCA extracts of liver especially when metabolites are concentrated by freeze-drying, as illustrated in Fig. 3 for a hydrazine-treated rat. Characteristic resonances can be seen from free amino acids (alanine, glutamine, glutamate, glycine), polyamines (spermine, spermidine), lipotropic compounds (betaine, choline), glucose, lactate, 3-D-hydroxybutyrate and succinate. Observed dose-related changes in alanine may be attributed to hydrazine's inhibitory effect on transaminases.

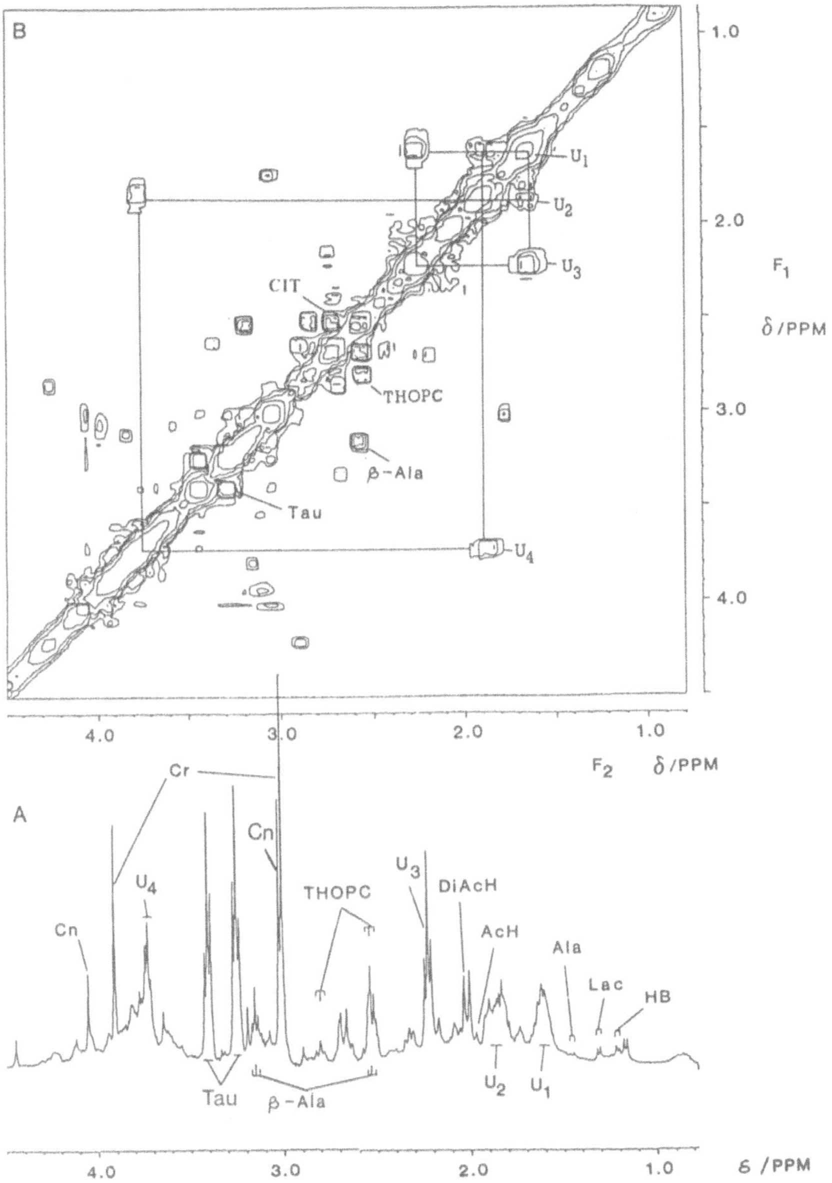

Fig. 2. Aliphatic region of 400 MHz NMR spectra - **A**, *below:* 1-D; **B**: 2-D COSY - of rat urine 6-24 h after a 60 mg/Kg i.p. dose of hydrazine. Note the 2-D simplification of 1-D overlapping resonances. See text for comment on 'cross-peaks' and on the appearance of an unknown hydrazine metabolite. Compounds designated as in Fig. 1; also β-ala, β-alanine; HB, 3-D-hydroxybutyrate; Cr, creatine.

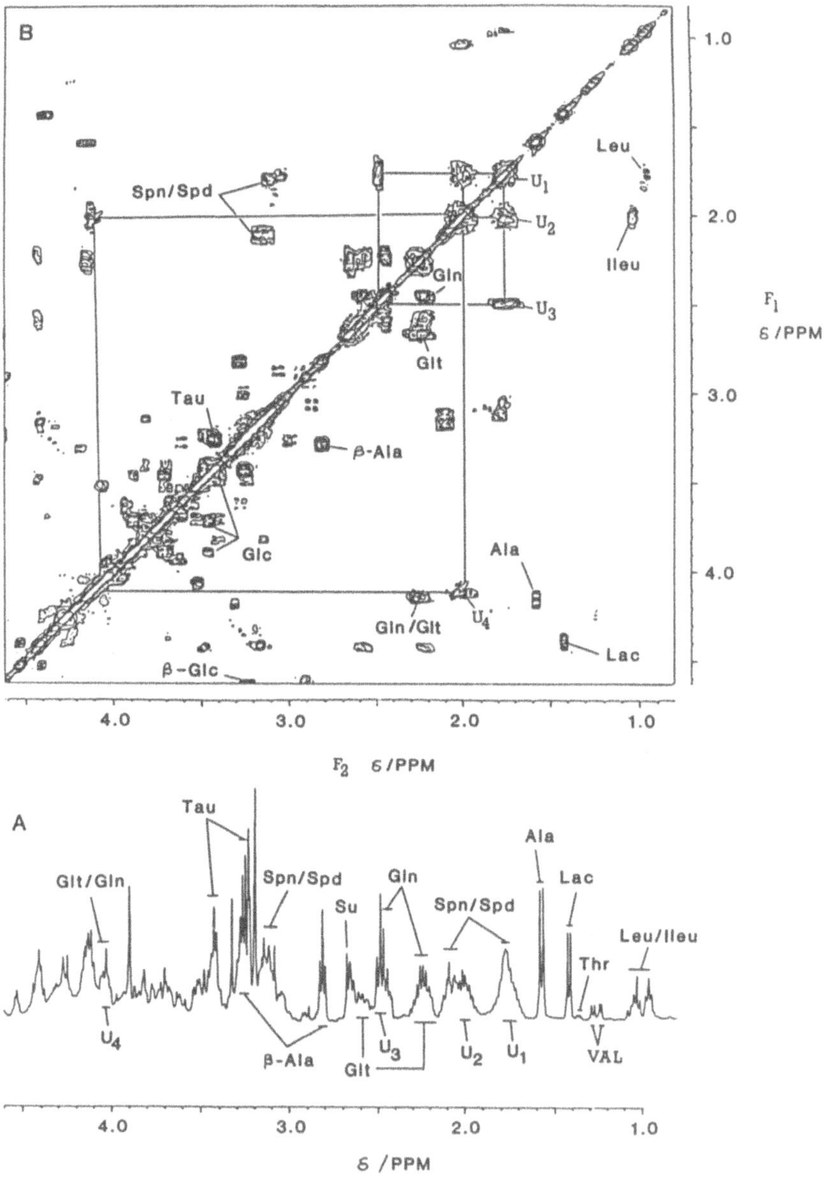

Fig. 3. Aliphatic region of 400 MHz NMR spectra - **A**, *below:* 1-D; **B**, 2-D COSY - for a TCA extract of rat liver 24 h after exposure to hydrazine (60 mg/Kg, i.p.). See text for comments on the unknown metabolite seen in the 2-D 'map' and on the metabolite range shown by 1-D where 2-D allowed definite assignments: Leu, leucine; Ileu, isoleucine; Gln, glutamine; Glt, glutamate; Spn, spermine; Spd, spermidine; β-alanine; taurine; Glc, glucose (β-Glc, β-anomer). Also (& see Figs. 1 & 2 legends): Thr, threonine; Val, valine.

CONCLUDING COMMENTS

In principle, 1- and 2-D ^{1}H NMR methods can be used to elucidate xenobiotic-induced changes in intermediary metabolism in any organ system or biological fluid. NMR techniques are also uniquely suited to the study of the metabolism and toxicity of diverse compounds, and may be of particular value where metabolic profiles and toxic mechanisms are poorly understood.

Acknowledgements

We gratefully acknowledge the support of the SERC, ICI, Wellcome Trust and the National Kidney Research Fund for this and related work.

References

1. Nicholson, J.K., Timbrell, J.A., Bales, J.R. & Sadler, P.J. (1985) *Mol. Pharmacol. 27*, 634-643.
2. Everett, J.R., Jennings, K. & Woodnut, G. (1985) *J. Pharm. Pharmacol. 37*, 869-873.
3. Bales, J.R., Nicholson, J.K. & Sadler, P.J. (1986) *Clin. Chem. 31*, 757-762.
4. Coleman, M.D. & Norton, R.S. (1986) *Xenobiotica 16*, 69-77.
5. Tulip, K., Timbrell, J.A., Wilson, I.D., Troke, J. & Nicholson, J.K. (1986) *Drug Met. Disp. 14*, 746-749.
6. Mclaughlin, A.C., Takeda, H. & Chance, B. (1979) *Proc. Nat. Acad. Sci. 76*, 5445-5449.
7. Jue, T., Arias-Mendoza, F., Gonnella, N.C., Shulman, G.I. & Shulman, R.G. (1985) *Proc. Nat. Acad. Sci. 82*, 5246-5249.
8. Allars, H., Coleman, M.D. & Norton, D.S. (1985) *Eur. J. Drug Met. Pharmacol. 10*, 253-260.
9. Cohen, S.M. (1983) *J. Biol. Chem. 258*, 14294-14308.
10. Cohen, S.M. (1984) *Fed. Proc. 43*, 2657-2662.

#NC(F)-5

A Note on

^{19}F NMR STUDIES OF THE METABOLISM OF TRIFLUOROMETHYLANILINE

K.E. Wade, †J. Troke, †C.M. Macdonald, †I.D. Wilson⊗ and J.K. Nicholson

Department of Chemistry
Birkbeck College
(University of London)
Gordon House
29 Gordon Square
London WC1E 6BT, U.K.

†Department of Drug Metabolism
Hoechst Pharmaceuticals
Research Laboratories
Walton Manor, Walton
Milton Keynes
Bucks. MK7 7AJ, U.K.

There is now much interest in the application of NMR techniques to determine the metabolic fate and disposition of foreign compounds [1-4] (& J.K. Nicholson in Vols. 16 & 17, this series - *Ed.*). For fluorinated compounds ^{19}F NMR can be used with particular advantage due to its large chemical shift range giving good specificity, high sensitivity (83% of ^{1}H) and the absence of endogenous fluorinated compounds in body fluids. In theory, drug recovery studies can be achieved by quantifying the spectra using a known concentration of a standard, thus obviating the need for expensive synthesis of radio-labelled compounds. To illustrate this, we have now investigated the metabolism and urinary excretion of TFMA* and its acetanilide. TLC and GC-MS studies [5] had shown that TFMA is metabolized to its oxanilic acid and a hydroxy metabolite which is probably glucuronidated (Fig. 1).

METHODS AND MATERIALS

Animal dosing and sample collection.- Two groups of 3 adult male Sprague-Dawley rats (wt. range 200-250 g) were dosed orally with TFMA at 16 and 50 mg/kg respectively. Two other rats were dosed with *N*-acetyl-TFMA, ^{13}C-labelled in the acetyl and ^{14}C-labelled in the ring. Over the following 48 h the rats were kept in metabolic cages and urine and faeces collected. The animals were starved for

⊗ now at ICI Pharmaceuticals, Alderley Park, Macclesfield SK10 4TG.

**Abbreviations*.- TFMA, 4-trifluoromethylaniline; Fid, free induction decay; FT, Fourier transformation; MS, mass spectrometry.

Fig. 1. Postulated metabolic pathways of TFMA in the rat.

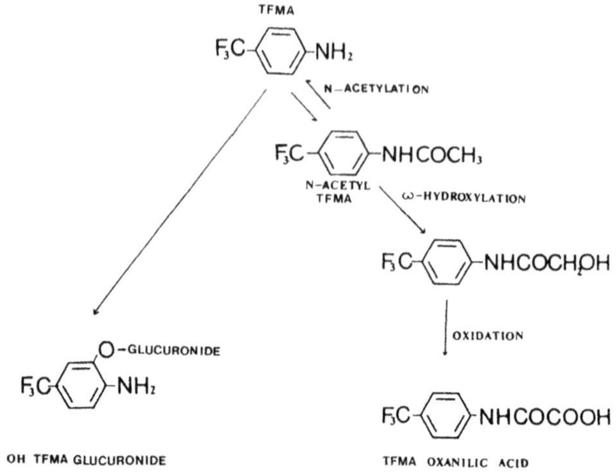

16 h before and for 4 h after dosing, but access to water was unrestricted. Urinary volumes were recorded at 8, 24 and 48 h. The urine was collected in polypropylene tubes over dry ice and stored at -40° till analyzed.

NMR spectroscopy.- Samples (0.5 ml) to which ^2H_2O had been added as an internal field-frequency lock were placed into 5 mm (o.d.) glass NMR tubes. Alternatively, fresh urine samples were freeze-dried and redissolved in ^2H_2O to allow concentration of the samples before acquiring the spectra. NMR spectra of some samples were measured using a wide-bore NMR probe with 15 mm tubes requiring 7 ml of sample. Measurements were made on Varian VXR400 and Bruker WH400 spectrometers operating at 9.4 Tesla field strength.

^{1}H NMR spectroscopy.- ^{1}H spectra were measured at 400 MHz using a 52° flip angle. (Typically, 100 and 100 Fid's were collected into 32 K computer points; acquisition time 3.2 sec). A further delay of 2.5 sec between pulses was used to ensure that the spectra were fully T1-relaxed. The Fid's were multiplied by an exponential function corresponding to 0.5 Hz line-broadening prior to FT. A continuous secondary irradiation field at the water resonance frequency was applied in order to suppress its intense signal. Chemical shifts were referenced internally to sodium 3-(trimethylsilyl)-1-propanesulphonate (δ = 0 ppm).

^{19}F NMR spectroscopy.- Proton-coupled ^{19}F spectra were measured at 376 MHz using a 28° flip angle. Conditions otherwise were as for ^{1}H. Shifts were referenced to α,α,α-trifluoro-*m*-cresol at δ 60.25 ppm with trichlorofluoromethane (δ = 0 ppm). Typically, exponential functions corresponding to the 2 Hz line-broadening were applied prior to FT.

Assignment of NMR signals from metabolites, and MS.- Urine from a TFMA-dosed rat was hydrolyzed with β-glucuronidase (*E. coli*) at pH 6.8 and analyzed by ^{19}F and ^{1}H NMR. The chemical shifts of the metabolites were compared to that of a specimen of the ***N***-oxanilic acid of TFMA. Metabolites were fractionated by solid-phase extraction of urine, pre-acidified to pH 2.2 to retain the postulated acidic metabolites by ion-suppression, and HPLC on a Bond-Elut column with NMR detection (SPEC-NMR). After analysis by ^{19}F and ^{1}H NMR, selected fractions were reapplied to the extraction column to further purify the metabolites. MS examination (Kratos MS80) of these, after freeze-drying, was by fast atom bombardment.

Quantification of the metabolites in the urine was by comparison with a known concentration of trifluoro-*m*-cresol, with cutting-out and weighing of paper traces of NMR peaks. For ***N***-acetyl-TFMA urine these results were compared to liquid scintillation results obtained on a Packard 302A counter using 5 ml of Bifluor scintillant; efficiency was determined by assaying authentic material with varying amounts of chloroform as quenching agent.

RESULTS AND DISCUSSION

For urine following TFMA administration, Fig. 2 shows typical ^{19}F NMR spectra, with 3 metabolite peaks - as found after ***N***-acetyl-TFMA also. The ^{1}H spectrum of neat urine gave no clue to the identity of the metabolites due to the multitude of extensively overlapped signals from endogenous compounds. β-Glucuronidase hydrolysis of post-TFMA urine caused the disappearance of the ^{19}F signal at δ = 59.21 ppm, indicating that it is the ring-hydroxy-TFMA β-D-glucuronide. The chemical shift of the peak at -59.7 ppm was identified as the TFMA-oxanilic acid by comparison with the standard.

Solid-phase extraction of the post-TFMA urine partially separated the metabolites as shown by ^{19}F NMR (Fig. 3). ^{1}H NMR still did not identify the remaining unknown metabolite due to co-eluting contaminants (Fig. 4). However, the *p*-ring-substituted patterns of the oxanilic acid were observed in the 70% and 80% (v/v) methanol fractions, which were essentially free of contaminants. At present the 35 % fraction isolated by solid-phase extraction of the 40% urine fraction is undergoing further analysis by MS.

Good agreement was found between ^{14}C scintillation counting and ^{19}F NMR for the quantification of drug-related material in post-***N***-acetyl-TFMA urine (34% and 40% respectively over 48 h). Therefore quantification of the ***N***-acetyl-TFMA and TFMA in urine can be performed by ^{19}F NMR. Fig. 5 shows excretion profiles for TFMA metabolites. With the two TFMA dosage levels employed, no dose-dependency of metabolism and excretion was evident; TFMA excretion was 36 ±5% after 19 mg/kg and 41 ±10% after 50 mg/kg. ***N***-acetyl-TFMA, via which TFMA is hypothetically metabolized, indeed gave the same 3 metabolites, with no indication of dose-dependency.

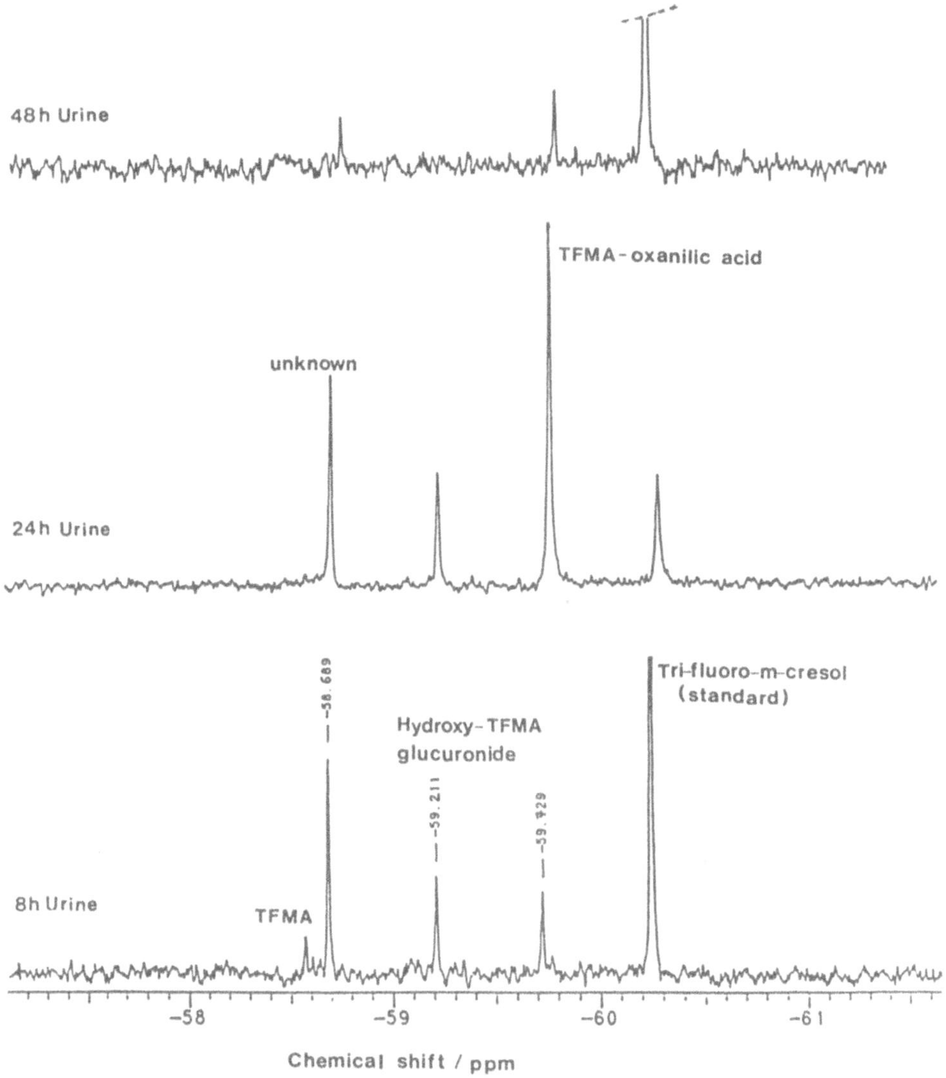

Fig. 2. 376 MHz ^{19}F NMR spectra of urine from a rat given TFMA, 19 mg/kg: collections for 8 h, 8 to 24 h and 24 to 48 h.

Two of the metabolites have been identified as the TFMA-oxanilic acid and the ring-hydroxy TFMA glucuronide. The presence of hydroxy-TFMA after ***N***-acetyl-TFMA dosing indicates that some is deacetylated back to TFMA before hydroxylation and conjugation. ^{19}F NMR provided a quick method of detecting TFMA metabolites in biofluids, superior to the more sensitive proton NMR because of negligible interferences. Excretion profiles and recovery data have been obtained for TFMA and ***N***-acetyl-TFMA by quantification of the ^{19}F NMR spectra. Since many new drugs and novel therapeutic agents are fluorinated, this

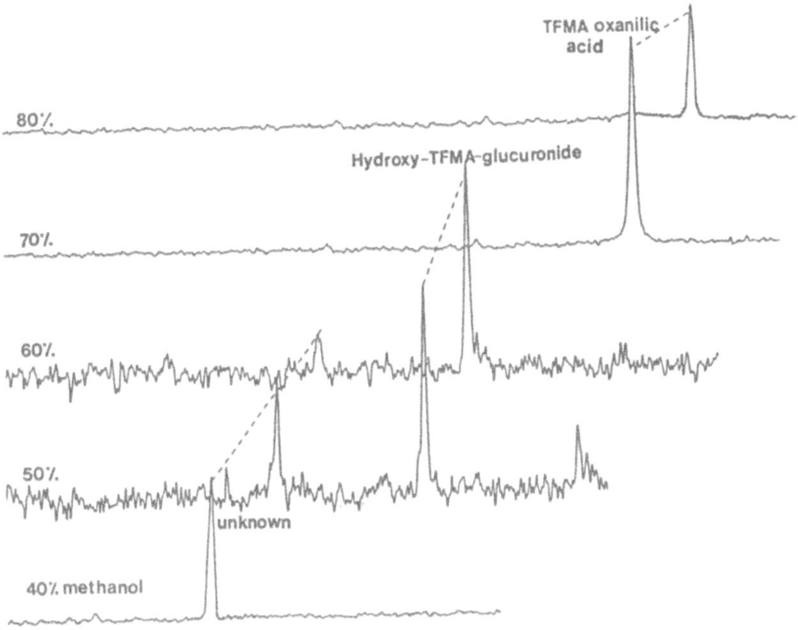

Fig. 3, *above*. As for Fig. 2 but after solid-phase extraction and successive elutions with increasing % methanol.

[**Fig. 4** is *overleaf*.]

Fig. 5, *right*. Urinary excretion of the metabolites of TFMA at dose level 19 mg/kg.

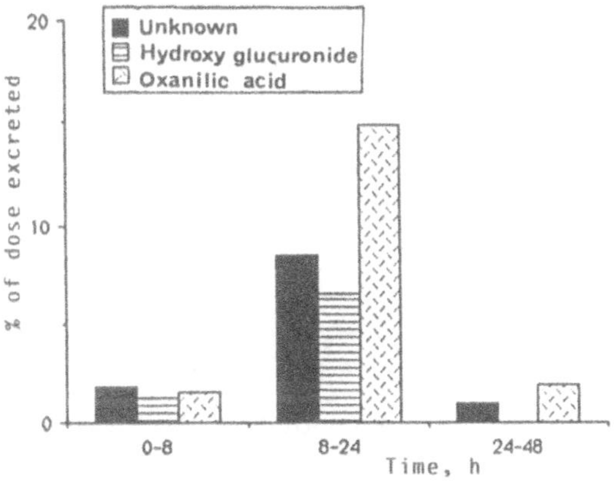

technique could reduce the need for expensive radiolabelled compounds in drug metabolism and excretion studies on fluorinated compounds, and allow metabolic studies to be performed in man at an earlier stage than is currently practicable.

Acknowledgements

The technical expertise of Mr. P. Gilbert and Mr. H.G. Parkes is gratefully acknowledged. We thank the SERC for supporting a CASE Studentship for K.E.W.

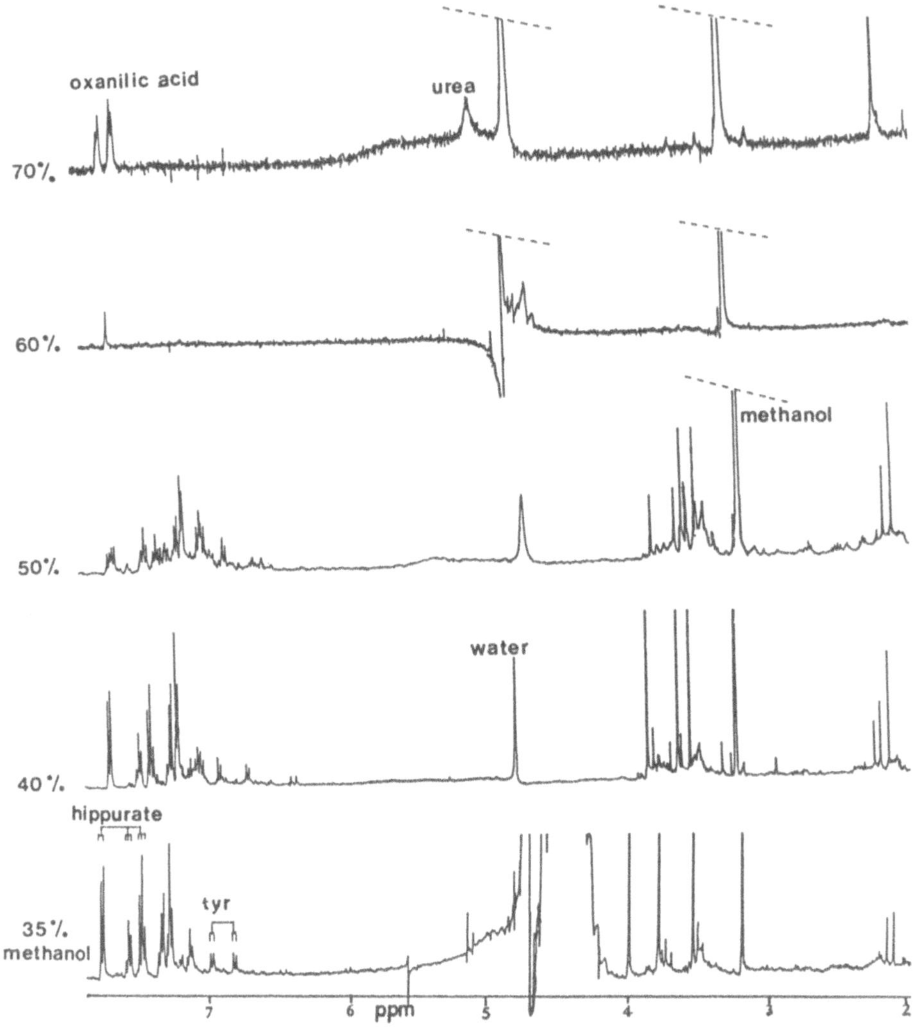

Fig. 4. 400 MHz ^{1}H NMR spectra of eluates as in Fig. 3, post-TFMA.

References

1. Everett, J.R., Jennings, K. & Woodnutt, G. (1985) *J. Pharm. Pharmacol. 37*, 869-873.
2. Tulip, K., Timbrell, J.A., Wilson, I.D., Troke, J. & Nicholson, J.K. (1986) *Drug Metab. Disp. 14*, 746-749.
3. Nicholson, J.K., Timbrell, J.A., Bales, J.R. & Sadler, P.J. (1985) *Molec. Pharmacol. 27*, 634-644.
4. Coleman, M.D. & Norton, R.S. (1986) *Xenobiotica 16*, 69-77.
5. Wilson, I.D., Macdonald, C.M., Fromson, J.M., Troke, J.A. & Hillbeck, D. (1985) *Biochem. Pharmacol. 34*, 2025-2028.

#NC(F)-6

A Note on

NMR AS AN AID IN STUDYING *N*-ALKYLFORMAMIDES AND METABOLITES†

Michael D. Threadgill* and Andreas Gescher

Cancer Research Campaign Experimental Chemotherapy Group
Pharmaceutical Sciences Institute, Aston University
Aston Triangle, Birmingham B4 7ET, U.K.

The common industrial and laboratory solvent DMF≠ and its close analogue NMF share the ability to cause hepatotoxicity in man. They differ, however, in their biological properties towards rodents. NMF is a hepatotoxin towards rats and mice and shows significant activity against tumours of murine [1] and human [2] origin grown in mice. DMF shows neither strong therapeutic nor hepatotoxic activity in mice, whereas NEF in mice shows hepatoxicity only [1, 3]. The explanation for these comparisons and contrasts may be connected with the nature and quantity of the metabolites formed from different *N*-alkylformamides in the various species. The characterization of the urinary metabolites by conventional techniques such as MS and UV spectrometry is hampered by their lability and lack of a suitable chromophore.

NMR spectroscopy was used to overcome these problems in two ways. Firstly, ^{1}H-NMR spectra were obtained using a Bruker WH400 spectrometer at 400 MHz on 0.5 ml samples of urine from mice to which DMF, NMF or other analogues had been administered at ~7.0 mmol/Kg. No other manipulation such as derivatization was involved, but 0.05 ml D_2O was added so as to provide a lock signal for the spectrometer. Sodium 3-(trimethylsilyl)propan-1-sulphonate was also added, as an internal chemical shift standard. Suppression of the H_2O signal (δ 4.80) was achieved by selective presaturation (1.5 sec) and was followed by collection of the free induction decay after 4 dummy

†See also J.K. Nicholson et al., art. #D-1 in Vol. 16, this series.- *Ed.*

*now at MRC Radiobiology Unit, Chilton, Didcot, Oxon. OX11 0RD

≠*Abbreviations.*- Formamide derivs.: ***N*,*N***-dimethyl, DMF; ***N***-methyl, NMF; ***N***-ethyl, NEF; *see over for* HMMF. ***N***-acetyl-*S*-(***N***-methylcarbamoyl)cysteine, AMCC; if Et, not Me: AECC.

scans. This technique permitted satisfactory ^{1}H spectra to be recorded for the δ ranges 0-4 and 6-11 and so resonances arising from most of the protons of the various metabolites of the formamides could be resolved, with the exception of those of $N\underline{CH_2}OH$ groups which have chemical shifts in the δ range 4.3-5 [4].

By comparison of the chemical shifts of the CHO protons(δ 8.03 for Z rotamer, 7.93 for E) and NCH_3 protons (δ 3.05 Z and 2.91 E) in synthetic ***N***-(hydroxymethyl)-***N***-methylformamide (HMMF) with those of signals in the urine of mice which had been treated with DMF, it was possible to show [4] that the major urinary metabolite of DMF in rodents is HMMF. This product is degraded to NMF under GC and MS conditions, leading to previous reports [5, 6] of NMF as the major metabolite of DMF; this demonstrates the value of the non-destructive nature of direct NMR analysis.

Similarly, ***N***-(hydroxymethyl)acetamide was shown to be the almost exclusive urinary metabolite of ***N***-methylacetamide in mice [3]. The corresponding direct ^{1}H-NMR spectra of urine of mice which had received NMF and NEF indicated the presence of the parent compound and the corresponding amine (methylamine and ethylamine respectively). Additionally [cf. observations by J.K. Nicholson et al. (p. 331 in Vol. 16)- *Ed.*], signals corresponding to the SCONHalkyl and $NCOCH_3$ protons of the corresponding mercapturic acids (AMCC and AECC) were observed in thse direct urine spectra, as shown in Fig. 1.

The two principal drawbacks encountered by us with the above straightforward direct method were insufficient sensitivity and overlapping of signals of metabolites with those of endogenous compounds. The latter problem led to difficulties in full characterization of the above mercapturic acids and, in some cases, in relative quantitation. These could be obviated in some respects by the second application of NMR spectroscopy. In this, the metabolites were isolated from urine by TLC or HPLC and the resulting pure or partially pure metabolite was analyzed by ^{1}H-NMR at 400 MHz in an appropriate solvent [usually $CDCl_3$, $(CD_3)_2SO$ or D_2O]. Thus, metabolites present in lower concentration could be characterized, although >300 μg of material was still required. The mercapturic acids AMCC and AECC were fully identified by this less direct approach by comparison of the NMR spectra of the isolated metabolites with those of synthetic material [3, 7].

^{1}H-NMR clearly has an important role in the identification of abundant labile metabolites of non-chromophoric compounds such as formamides or acetamides, but is limited by the relative insensitivity of the technique and could not be applied to urinary metabolites arising from lower doses of the formamides in rodents or in human studies.

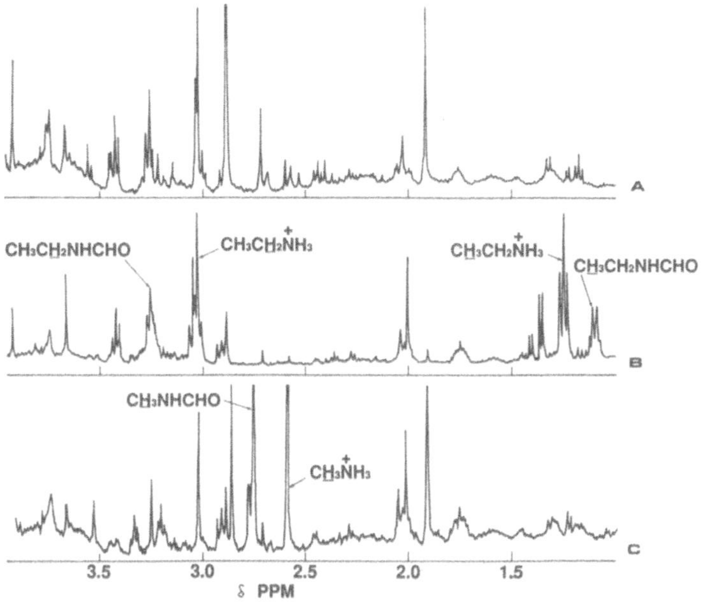

Fig. 1. Partial ^{1}H-NMR spectrum (400 MHz) of 24 h collections of urine from control mice (**A**), mice receiving 400 mg/kg NEF (**B**), and mice receiving 400 mg/kg NMF (**C**).

Acknowledgements

Generous financial support for this work was provided by the Cancer Research Campaign. NMR spectra were obtained at the University of Warwick (Dr. O.W. Howarth), supported by the SERC.

References

1. Gate, E.N., Threadgill, M.D., Stevens, M.F.G., Chubb,D., Vickers, L.M., Langdon, S.P., Hickman, J.A. & Gescher, A. (1986) *J. Med. Chem. 29*, 1046-1052.
2. Lomax, N.R. & Narayanan, V.L. (1981) in *Chemical Structures of Interest to the Division of Cancer Treatment*, Drug Synthesis & Chemistry Branch, Developmental Therapeutics Program, Nat. Cancer Inst.: Bethesda, MD.
3. Kestell, P., Threadgill, M.D., Gescher, A., Gledhill, A.P., Shaw, A.J. & Farmer, P.B. (1987) *J. Pharmacol. Exp. Ther. 240*, 265-270.

4. Kestell, P., Gill, M.H., Threadgill, M.D., Gescher, A., Howarth, O.W. & Curzon, E.H. (1986) *Life Sci.* *38*, 719-724.
5. Brindley, C., Gescher, A. & Ross, D. (1983) *Chem.-Biol. Interact.* *45*, 387-392.
6. Scailteur, V., Hoffman, E., Buchet, J.P. & Lauwerys, R. (1984) *Toxicology* *29*, 221-234.
7. Kestell, P., Gledhill, A.P., Threadgill, M.D. & Gescher, A. (1986) *Biochem. Pharmacol.* *35*, 2283-2286.

#NC(F)-7

A Note on

FLEXIBLE *VERSUS* DEDICATED AUTOMATION

J.C. Pearce and R.D. McDowall

Explanatory note by Senior Editor

It was the authors' intention, now not feasible, to publish an article based on a Forum Discussion initiated by J.C. Pearce [now at Glaxo Group Research, Ware, Herts. SG12 0DJ]. This Discussion was backed by the two Forum abstracts reproduced below; the second, linked with a Poster contribution, was incorporated in the introduction to the Discussion, serving as illustrative material.

The material presented was an up-date of 'Robotics in Drug Analysis' as presented by the authors at the 1985 Bioanalytical Forum and published in an earlier vol. ([1] in the concluding list of pertinent refs.); a description was given of an automated system based on flexible robotics for off-line preparation of Advanced Automatic Sample Processor (AASP) casettes [2]. Subsequently, as amplified by R.D. McDowall et al. *earlier in this book (#D-5, p. 201; Fig. 1 is pertinent), a new approach was investigated, with benefit to flexibility and cost: an autosampler (Gilson Model 222) was linked directly to the AASP valve, enabling the sample to be loaded directly onto the solid-phase cartridge* in situ *[3]. This approach was investigated for the routine assay of drugs, and also for screening different bonded silica phases for assay suitability (using radiolabelled analyte), as an alternative to diverting the robotics system for the latter purpose, as also achieved.*

In the concluding list of refs., [4] relates to the dedicated screening unit based on a Gilson autosampler. The feasibility of dedicating a flexible robotics system to liquid-liquid extraction has been demonstrated [5], although the Varian Autosampler can perform similar tasks on-line [6]. Arndt [7] has commented that robotics automates the status quo *and as such does not represent progress; he expresses the view that the major contribution which flexible systems may make is to trigger new developments in dedicated automation.*

In an ***Addendum****, results of intrinsic interest are summarized, viz. trial of different solid-phase extractants.*

FLEXIBLE *VS.* DEDICATED AUTOMATION IN THE ANALYSIS OF DRUGS IN BIOLOGICAL FLUIDS

J.C. Pearce and *R.D. McDowall

Department of Drug Analysis
Smith Kline & French Research Ltd.
The Frythe, Welwyn, Herts. AL6 9AR, U.K.

The benefits of automation to the busy analytical laboratory are well established. The route that the Drug Analyst takes to realize these benefits has become less clear in recent years. In particular, there has been increasing interest in the use of robotics (flexible automation) which has in the opinion of the authors triggered the development of a new generation of analytical instruments (dedicated automation).

This trend has been especially evident in the development of liquid-solid sample preparation where the initial use of robotic work stations has provoked the manufacture of a crop of dedicated instruments for sample clean-up. The use of the same flexible systems for liquid-liquid extraction has also been successful in a number of laboratories and more recently some manufacturers have configured dedicated autosamplers to carry out this operation.

The experience of the authors in this field bears on the future of both flexible and dedicated automation in drug analysis. Specific attention has been given to cardiovascular drugs (see the following Abstract; cf. #D-5, this vol.), and the ideas are extendable to cover a broad spectrum of xenobiotics.

Senior Editor's abridgement of an Abstract now largely covered in art. #D-5 (laboratory address as above):

QUANTIFICATION OF AN INOTROPIC AGENT IN PLASMA: TOWARDS TOTAL AUTOMATION

G.S. Murkitt, J.C. Pearce and *R.D. McDowall

In order to follow the pharmacokinetics of 5-methyl-6-[4-(4-oxo-1,4-dihydropyridin-1-yl)phenyl]-4,5-dihydro-3(2H)-pyridazinone - SK&F 95654 - a sensitive and specific assay was requested. A fully

* addressee for any correspondence

automated method was developed, integrating sample preparation and analysis. It utilizes a Gilson 222 autosampler with an AASP and an HPLC/UV system.

The method consists of a sample manipulation step, a solid phase extraction (using bonded-silica sorbent) and syringeless injection of extract into an HPLC system (for final separation and quantification of drug and internal standard). The Gilson apparatus prepares plasma samples by the addition of internal standard solution (and, where appropriate, standards or water) followed by mixing. Aliquots of various solutions (methanol, water, diluted sample and water) are then drawn up, in reverse order, into a holding loop with air segmentation. This 'train' of solutions is passed through a C-18 AASP cassette via an injection port connected to the AASP switching valve. The methanol and water condition the phase prior to the application of the sample. The analytes are retained on the C-18 sorbent, whilst the undesired endogenous materials are removed by the subsequent water wash. At the appropriate time, the mobile phase is switched through the cartridge allowing elution of the compounds of interest directly onto the RP-HPLC analytical column; the quantification is achieved by UV monitoring of the eluent.

The method analyzes samples sequentially with concomitant sample preparation and HPLC assay, *i.e.* as one sample is running on the HPLC system the next is being prepared. It is capable of analyzing at least 100 samples per day. Assay validation was performed (with dog plasma) over the range 0.025-2.0 µg/ml, as summarized in art. #D-5. The system is easy to use, has proved to be robust and involves minimal human input.

References *pertinent to flexible* vs. *dedicated automation*

1. Pearce, J.C., Allen, M.P. & McDowall, R.D. (1986) in *Bioactive Analytes, including CNS Drugs, Peptides and Enantiomers* [Vol. 16, this series] (Reid, E., Scales, B. & Wilson, I.D., eds.), Plenum, N. York, 293-296.
2. Pearce, J.C., Allen, M.P., O'Connor, S.A. & McDowall, R.D. (1988) *Chemometrics Intell. Lab. Systems*, in press.
3. Murkitt, G.S., Pearce, J.C. & McDowall, R.D. (1988) *Chromatographia 24*, 411-415.
4. Verillion, F., Pichon, B. & Qian, F. (1987) *Int. Lab. 17*, July/August, 50-54.
5. Plummer, G. (1987) in *Advances in Laboratory Automation Robotics* (Strimaitis, J.R. & Hawk, G.L., eds.), Zymark Corpn., Hopkinton, MA, pp. 47-70.
6. Cunico, R., Mayer, A.G., Wehr, C.T. & Sheehan, T.L. (1986) *BioChromatography 1*, 6-14.
7. Arndt, R.W. (1986) *Chem. Brit. 22*, 974.

SOLID-PHASE EXTRACTION BEHAVIOUR OF THE DRUG STUDIED
- Senior Editor's summary of the authors' findings

The drug for which method development was done by robot possessed a primary amino group and two other amine moieties; being radiolabelled, results could be assessed automoatically by liquid scintillation counting. The drug (in 1 ml) was applied after conditioning the cartridges with methanol (1 ml) and 0.1 M ammonium acetate (1 ml). After washing with 0.01 ml ammonium acetate (1 ml), two elutions (each 1 ml) were performed, with triethylamine (TEA) present in the second which was a putative HPLC mobile phase: methanol/0.01 M phosphate buffer containing 0.01% TEA (1:3 by vol.).

The cartridge types compared, listed in order of observed recoveries (which ranged from 70% to near-nil), were: CN, diol; C-18, C-2, unmodified silica; C-8; CH, CBA; phenyl; SCX. The poor recovery with SCX occurred notwithstanding good extraction efficiency. With no type was there significant break-through or wash-out. Only with the TEA-containing eluent did silica give a fair recovery; the recovery in the first elution was maximal (30%) with C-8 and 20% or much lower with the other cartridge types.

#NC(F)-8

A Note on

DESIGN OF AND FIRST EXPERIMENTS WITH A BIOLUMINESCENCE DETECTOR FOR HPLC

H.M. Ruijten, B.E. Timmerman and H. de Bree

Duphar Research Laboratories, P.O. Box 2
1380 AA Weesp, The Netherlands

As discussed earlier in this book [Brinkman, #F-2; de Bree, #NC(E)-4], HPLC still lacks a universal sensitive and selective detector, and for biomedical and environmental trace-level analyses HPLC is still outclassed by GC (especially capillary with the new chemically bonded phases). The efficacy of enzymes in metabolizing drugs should be exploitable to furnish a selective post-column reactor for HPLC, since mostly the selective enzymatic conversions involve production of NAD, ATP or H_2O_2. Several bioluminescence kits based on light emission initiated by one of these products are now commercially available. Amongst the many bioluminescence procedures found in the literature, most are in the food and clinical chemistry areas and none are HPLC-linked. In the approach now described, both the coenzyme-generating enzymes and those that emit light are immobilized, which entails problems that are now considered along with prospects for the detection approach in general. At the outset our model compound was flesinoxan:

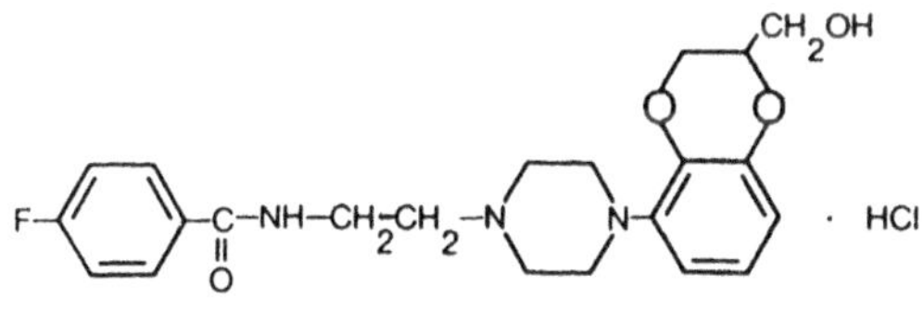

The ultra-sensitive method described by van Berkel et al. in art. #NC(D)-1 involves rather laborious pre-GC steps. In our present 'inject-the-sample' strategy, we aimed to generate NADH by enzymatic attack on the compound's alcohol group, through reactions **I** and **II**:

$$R\text{-}CH_2\text{-}CHR'\text{-}CH_2OH + O_2 \xrightarrow[\textit{oxidase}]{\textit{alcohol}} RC^{H}{=}C^{H}R' + H_2C{=}O + H_2O_2 \qquad \mathbf{(I)}$$

$$H_2C{=}O + NAD^+ + 2\,H_2O \xrightarrow[\textit{formaldehyde dehydrogenase}]{} HCOOH + NADH + H_3O^+ \qquad \mathbf{(II)}$$

With the following reactions light can be produced:

$$\mathrm{NADH + FMN + H_3O^+ \xrightarrow[\textit{oxidoreductase}]{\textit{NADH:FMN}} NAD + FMNH_2 + H_2O} \quad \textbf{(III)}$$

$$\mathrm{FMNH_2 + \mathbf{R}CHO + O_2 \xrightarrow[\textit{luciferase}]{} FMN + \mathbf{R}COOH + H_2O} \quad \textbf{(IV)}$$

where
R = n-alkane
(at least C8)

Fig. 1. Block scheme of the HPLC set-up.

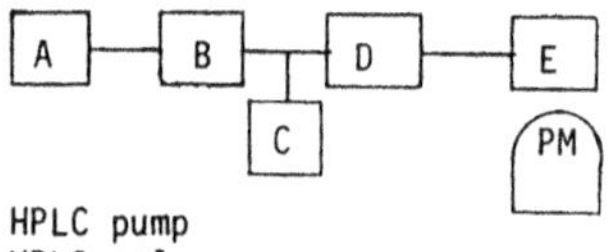

A= HPLC pump
B= HPLC column
C= pump for adding reagents (such as NAD)
D= NADH generator
E= light generator
PM= photomultiplier

The reactions are stoichiometric and thus the light production in reaction **IV** is proportional to the initial amount of analyte. All biochemicals used in our study were from Sigma or Boehringer.

The HPLC set-up is shown diagramatically in Fig. 1. We investigated all elements of the set-up separately, firstly the last part to ascertain whether measurable quantities of light could be obtained with immobilized luciferase and oxidreductase by injecting NADH and proceeding with reactions **III** and **IV**. Having immobilized the enzymes on CNBr-Sepharose separately and put them in a transparent (glass) column, we concluded that the ratio luciferase/oxidreductase must be chosen to favour the luciferase; otherwise autoxidation of the produced $FMNH_2$ may occur. This is one of the so-called 'dark reactions' in the reaction mechanism. Without further optimization we could readily detect 10 pmol of NADH.

Subsequently we checked the NADH generator (**D** in Fig. 1) by 340 nm-absorbance measurement of NADH formed. With the formaldehyde dehydrogenase immobilized on CNBr-Sepharose we obtained 100% conversion of formaldehyde. The alcohol oxidase activity was checked by measuring the amount of formaldehyde produced from methanol. Immobilization using CNBr and epoxy-activated Sepharose deactivated this enzyme almost completely. The remedy was to immobilize it by photosynthesis [1]. We used the 1-butene reaction to produce a photobead from alkylamine coupled to Sepharose, and immobilized alcohol oxidase using 2-acetylbenzoic acid sensitizer (Fig. 2). Thereby immobilized alcohol oxidase retained activity and a conversion of 80% was obtained.

Next we plan further optimization of the light-generator and then to combine the different parts into a system.

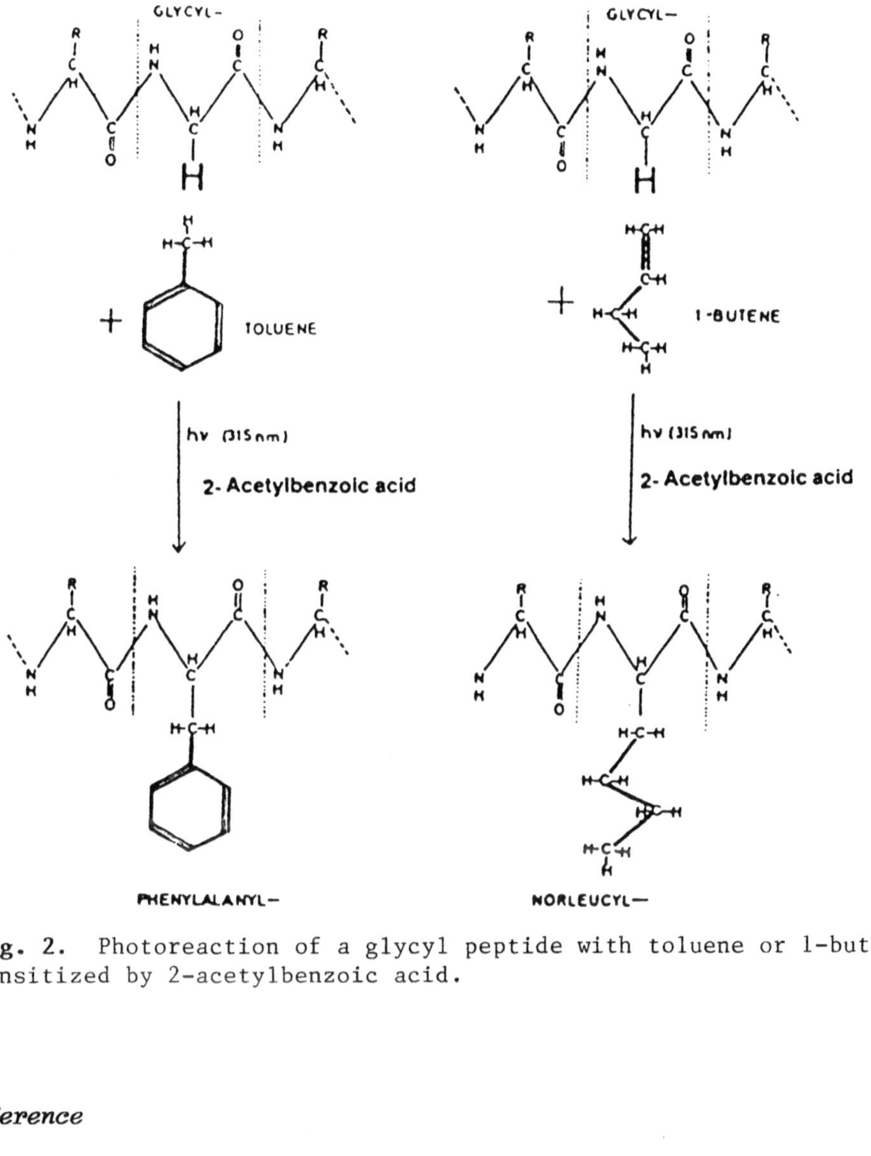

Fig. 2. Photoreaction of a glycyl peptide with toluene or 1-butene sensitized by 2-acetylbenzoic acid.

Reference

1. Krämer, D.M., Lehmann, K., Pennewiss, H. & Plainer, H. (1975) in *23rd Colloquium, Protides of Biological Fluids (Brugge, 1975)*, Pergamon, Oxford, pp. 505-512.

#NC)F)-9

A Note on

A SYSTEM APPROACH TO IMMUNOASSAY USING ENHANCED LUMINESCENCE

M.R. Summers, K. Mashiter, J.C. Edwards, J.K. Martin, G.P. Davidson & J. Holian

Amersham International plc
Amersham, Bucks. HP7 9LL, U.K.

There is increasing awareness of the advantages of non-radioactive labels for heterogeneous, high sensitivity immunoassays. One of the most common systems is the enzyme immunoassay with colorimetric or fluorometric determination. However, because of the steady increase in signal as the reaction progresses, the signal development time (usually 15-30 min) must be accurately controlled, and the reaction stopped by reagent addition at a specific time. Chemiluminescence reactions provide an alternative end-point for such assays.

Components of chemiluminescence reactions have been used in diverse ways to label molecules in immunoassays. Luminescent compounds such as luminol and its closely related analogues can be linked directly to proteins and haptens, although this usually results in loss of light emission from the luminol oxidation reaction. The luminescent detection of compounds such as the acridinium esters is chemically simple and does not suffer from loss of efficiency when the active group is coupled to other molecules. However, the duration of light output from oxidation of luminol or acridinium esters is barely a few seconds, so the test sample must be positioned in a light-tight enclosure in front of the detection system while an initiating agent is added. This adds complexity to the instrumentation due to the requirement for a reagent injection mechanism. Rapid injection and efficient mixing of the reagents are also necessary to obtain acceptable precision.

As a different approach, a catalytic component of the chemiluminescent reaction, such as horseradish peroxidase, can serve as the tracer molecule. Light production from the enzyme-catalyzed oxidation of luminol generally occurs at low efficiency, and in the past a continuous output of light from the enzyme turnover has been achievable only at very low, often undetectable, signal levels. However, the discovery by T.P. Whitehead's group [1] of a series of compounds which enhance the light output from the oxidation of luminol has allowed optimization of this reaction to give a prolonged output of light of high intensity rather than a flash. These compounds may act by accelerating the cycling of activated forms of peroxidase back to the ground state by promoting a more efficient transfer of energy from peroxidase 'Compound II' to the luminol molecules [2]. Because this 'enhanced luminescence' is specific to horseradish peroxidase, the amplification achievable by using an enzyme as the tracer molecule can be exploited without interference from high background signals. Enhanced luminescence offers greater sensitivity than many conventional signal detection systems and can be measured over a wide operating range (several orders of magnitude).

The Amerlite™ system (developed by Amersham International plc) combines the advantages of the high sensitivity of enhanced luminescence and the specificity of monoclonal antibodies with solid-phase (coated well) technology to provide a convenient non-radioactive immunoassay system for measuring haptens and proteins. To date, immunometric and competition assays for a variety of analytes covering thyroid function, reproductive endocrinology, drug monitoring and oncology have been developed. The assays have liquid reagents that store well. For the critical steps such as pipetting, well-washing and signal measurement and processing, special instrumental modules are available to allow automation.

The signal produced by addition of substrate to the wells is long-lived and measurable repeatedly between 2 and 20 min. The analyzer processes the signals from 96 wells in <2 min, assessing assay quality from the curve-fit match of samples to standards. The precision is <1.5% for signal measurement and typically <4% within assays and <8% between assays. The curve-fit programme enables the number of standards to be minimized.

References

1. Whitehead, T.P., Thorpe, G.H.G., Carter, T.J.N., Groucutt, C. & Kricka, L.J. (1983) *Nature 305*, 158-159.

2. Thorpe, G.H.G. & Kricka, L.J. (1986) in *Bioluminescence and Chemiluminescence - New Perspectives* [Proc. 4th Int. Biolum. Chemilum. Symp.] (Scholmerich, J., Andreesen, R., Kapp, A., Ernst, M. & Woods, W.G., eds.), Wiley, Chichester, pp. 199-208.

Comments on material in #F

Comments on **#F-1,** I.D. Wilson - INSTRUMENTAL TLC
& **#F-2,** U.A.Th. Brinkman - FLUORESCENCE etc. IN HPLC

Comments to I.D. Wilson by U.A.Th. Brinkman.- One should realize that OPLC has been with us far longer than AMD (automated multiple development), and that even OPLC practitioners are not really beyond the silica stage and, indeed, rather complicate their nice technique by using stacks of thin layers rather than show interesting routine applications for silica and RP-type places. Moreover, one should realize that when wanting to use OPLC with on-line off-the-plate detection, as tried a few years ago by Prof. Guiochon and colleagues in France, one meets with major technological problems and, besides, needs as many detectors - diode array! - as there are spots applied to the OPLC plate. **Remark to Brinkman by J. Chamberlain.**- The problem of needing 16 detectors to monitor 16 channels in OPTLC (OPLC) can be overcome by using circular TLC with a single detector mounted over the edge of the circular plate which is rotated so that the single detector can rapidly sample the separate channels successively and repetitively as the chromatography proceeds - the chromatograms being recorded and stored on a microcomputer.

Comments on **#F-3,** A.P. Bruins - LC-MS AND LC-MS-MS FOR BIOANLAYSIS
& **#F-4,** T.J.A. Blake - THERMOSPRAY LC-MS AND LC-MS-MS

T. Parton asked A. Bruins which type of interface he would choose for use with a magnetic sector mass spectrometer. **Reply.**- Both the moving belt and the thermospray can be used now. The problems with the thermospray source due to fragmentation in the accelerating field have been overcome by a re-design of the ion optics. **Replies by T.J.A. Blake: (to J. Frölich)** LC-MS can detect down to 1 μg, and we have not yet tried quantification by LC-MS-MS; **(to T. Parton)** we have not yet seen quenching of ion intensity by background material, and when compounds have co-eluted we have observed thermospray ions characteristic of all the components present.

Comments on **#NC(F)-5,** K.E. Wade - ^{19}F-NMR METABOLIC STUDY
#NC(F)-7, J.C. Pearce - AUTOMATION APPROACHES
& **#NC(F)-8,** H.M. Ruijten - HPLC BIOLUMINESCENCE DETECTION

K.E. Wade replied to M.V. Doig who asked about quantitative accuracy: the values obtained by NMR were within 5% of those from balance studies with radiolabel. **J.C. Pearce answered C.W. Vose,** who had asked whether robotics or dedicated systems would be preferable if, notionally, purchase money were available to buy one or the other.-

I would probably look very hard at dedicated systems before opting for robotics, taking account of the time and effort needed to set up a robotics system. **U.A.Th. Brinkman suggested to H.M. Ruijten** that he try (1) post-column ion-pair formation with a highly fluorescent or electroactive counter-ion, and (2) direct derivatization of the analytes with PPD or a similar reagent.

Citations contributed by Senior Editor

In an outline of the '**constantly operating TLC**' (CO/TLC) approach - a term which the author advocates as preferable to 'automated TLC' [cf. art. **F-1**]; the whole TLC process is integrated - the benefits (particularly to resolution and capacity) of a constantly moving flow of samples that are processed in sequence are argued, with little attention to actual instrumentation [1].

Thermospray LC-MS techniques [cf. art. **F**-4] have been investigated, with trial of a number of sulphonamides [2]. **^{15}N-NMR** [cf. arts. **NC(F)-2** to **-6**] was successfully tried, along with FAB-MS, for structural assignment of a urinary ***N***-glucuronide formed by metabolism of 1,2,3,4-tetrahydroisoquinoline-7-sulfonamide (SK&F 29661; inhibitor of a transferase, PNMT) [3]; ^{15}N label was pre-introduced.

The following citations concern **HPLC detection**. The aryloxalate fluorigenic method [cf. art. **F-2**] has been surveyed [4]. (Pertinent to the 'LIF' approach illustrated in art. **F-2**, Fig. 6, is a description [5] of capillary-GC analysis of urine and haemofiltrates for metabolites of **neutral steroids** - androstens, etiocholanolone, corticoid metabolites and degradation products.) Using OTLC (RP column), on-line electrochemical detection has been investigated, and a cleaning procedure devised to obviate buffer-associated problems encountered in catechol oxidation studies; for **acetaminophen** oxidation a comparison was made between d.c. amperometric and differential pulse modes [6]. A polarimeter has been linked to HPLC [7], latterly with a laser **optical rotation detector** (D.M. Goodhall & D.K. Lloyd, reported in [8]. One application has been identifying **enantiomeric pyrethroids** [9].

1. Rogers, D. (1987) *Int. Analyst 1, June issue*, 54 & 56-57.
2. McFadden, W.H. & Lambert, S.A. (1987) *J. Chromatog. 385*, 201-211.
3. Kuo, G.Y., Hwang, B.Y-H. & Staiger, D.B. (1986) *Biochem. Pharmacol. 35*, 1613-1615.
4. Imai, K. & Weinberger, R. (1985) *Trends Anal. Chem. 4*, 70-75.
5. Ludwig-Koehn, H. & Henning, H.V. (1986) *J. High Resol. Chromatog. Chromaatog. Comm. 9*, 35-38.
6. St Claire, R.L. & Jorgenson, J.W. (1985) *J. Chromatog. Sci.23*, 186-191.
7. Yeung, E.S. (1985) *Adv. Instrum. 40*, 319-326.
8. ANON (1987) *Int. Analyst 1, May issue*, 9.
9. Meinard, C., Bruneau, P. & Perronnet, J. (1985) *J. Chromatog. 349*, 109-116.

Analyte Index

Key overleaf to the 10-category **chemical classification** (collation based on some analytically relevant features). Use of a compound as a internal standard is **not** indexed.

Hyphen '-' as in '17-' connotes *et seq.* i.e. treatment in depth.

Prefixes to some page entries, *besides* ch = *chiral distinction:-*

Superscript, e.g. ^{1}, signifies that the study included **metabolite(s)** of the listed compound (**see over**).
Subscript $_r$ signifies that **'real' samples** (animal or human) were assayed, usually including blood or plasma. This may also apply to some entries lacking this prefix; but mostly these concern pure compounds, or non-animal samples (cf. 'Environmental' entry in General Index).
Prefix *p* denotes a study comprehending a **precursor** or **prodrug.**

Any **index search of earlier vols.,** as listed opposite title p. (same 10 categories), can be confined to Vols. 12 (index **cumulative**), 14 and (same prefixes as now used) 16.

....

CATEGORY **I** (no amino group, nor cyclic N except maybe imide)

#**Ia**: acid *other than conjugate*, or ester *(criterion: SEE OVERLEAF)*

ASSIGNMENT 'CATECHISM' *(See previous p. for other guidance)*

Metabolites are not separately listed; the parent molecule's entry is preceded by a superscript: *Phase I* metabolite(s) or, if including *N*-desalkyl or -desacyl,[1] (i.e. **bold**); *Phase II (conjugates)*, [2].

Parent molecules as indexed bear generic names, where applicable, as listed (with formulae) in the *Merck Index*. For certain analytes a group title may also be relevant: e.g.: **Steroids; Amino acids.**

Assignment as 'acidic' (to Ia, IIa or IIIa) applies where the pKa is <6; this excludes phenols, and **conjugates are excluded** since only the parent molecule is listed (prefixed [2]; see above). Also 'acidic' (notional) are **esters** yielding an acidic group in the main moiety if hydrolyzed (as may happen *in vivo*).

Cyclic *N* is never treated as 'amino'; it is 'imide' (possible category: Ic) if -CO-**N**-CO-, but otherwise *may* be basic.

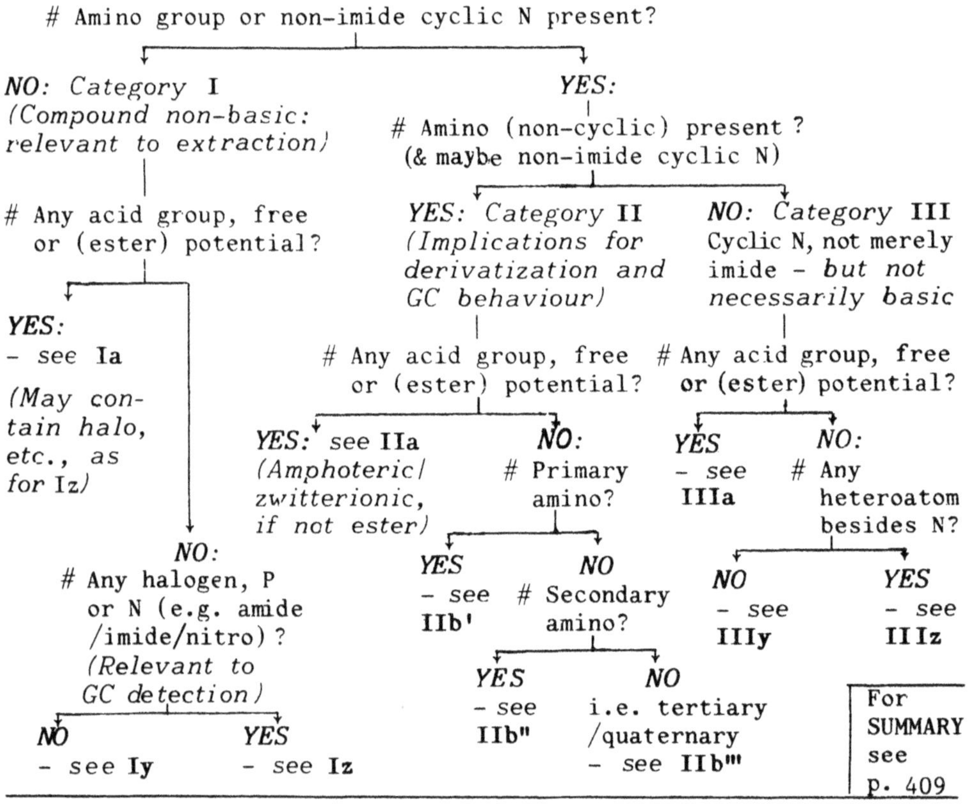

For SUMMARY see p. 409

#II (amines), *continued*

#IIb': primary amino; no acid (unless conjugate) or ester

#IIb": secondary amino, not in a ring; otherwise as for #IIb'

#IIb''': tert. amino or quaternary ammonium, not in a ring; otherwise as for #IIb'

#IIb''', *continued*

CATEGORY **III** (cyclic N, not merely imide; *no* amino)

#IIIa: acid *other than conjugate*, or ester *(main moiety = acid)*

#IIIy: N-hetero (not merely imide) but no other hetero atoms; *no* amino; main moiety *not* an acid

SUMMARY OF CATEGORIES

	I	II	III
Amino?	no	✓	no
Non-imide hetero-N?	no	maybe	✓
Acid or potential acid (not conjugate)?	✓ = **Ia**	✓= **IIa**	✓= **IIIa**
- no! (and not an ester)	Halo, P or N? - no: **Iy** - ✓ = **Iz**	Primary amino? ✓ = **IIb'** If no: 2^y= **IIb"** 3^y or 4^y =**IIb'''**	Hetero atom besides N? - no = **IIIy**, ✓= **IIIz**

Only **parent compound** listed; prefix [1] if Phase I metabolites studied [or **1** (**bold**) if dealkylated amino], & [2] if Phase II (conjugate); prefix *r* = 'real' (biological) sample. ***Full Key:*** *p. 406.* Chiral distinction denoted ch . Prefix *p* denotes prodrug/precursor.

General Index

This Index deals mainly with features studied and with approaches and points of technique, indexed similarly to previous 'A' vols. (listed opposite title p.) so as to facilitate back-searching. The preceding Analyte Index deals with compounds investigated; exceptionally, a few types are also listed below, according to their nature, e.g. 'Prostaglandins'.

In a page entry such as '17-', the '-' means '*et seq.*', i.e. coverage in depth.

GPSR Compliance

The European Union's (EU) General Product Safety Regulation (GPSR) is a set of rules that requires consumer products to be safe and our obligations to ensure this.

If you have any concerns about our products, you can contact us on ProductSafety@springernature.com

In case Publisher is established outside the EU, the EU authorized representative is:

Springer Nature Customer Service Center GmbH
Europaplatz 3
69115 Heidelberg, Germany

Batch number: 10762209

Printed by Printforce, the Netherlands